Dedication

I give thanks first to the Lord, Our God, for blessing me with the opportunity, knowledge and ability to share in this work. Thanks to my Mom, Dolores, for her patience and encouragement; and also my granddaughter, Paige and my wife Marci.

Table of Contents

vii

Acknowledgements

Many thanks are due to all: my Publisher, Industrial Press, Inc., President, Alex Luchars; Editorial Director, John Carleo; Production Manager, Janet Romano; and Product Manager, Suzanne Remore; Richard Jones of Technical Training Systems/Lab Technologies for AutoCad software; Mike Sorich of Technical Training Services and Marc Sullivan of Remote Machine for Mastercam X2 software; Steve Sevitter and David Boucher of Pathtrace, for EdgeCAM 11 software; Martin J. Aguilar of SolutionWare Corporation for MazaCAM Editor software and screen shots; Sandvik Coromant for tooling drawings and other technical data; CarrLane Manufacturing for technical data charts in the appendix; Mazak Corporation for photos used in Conversational Programming. CNC Software, Inc., "Mastercam", for screen shots of the programming process in CAD/CAM; and, GE Fanuc for operation panel, screen shots of the set-up operation and program displays.

Foreword to the Second Edition

It has been said that learning is a lifetime process. In the rapidly evolving computer age, this has never been more true. Manufacturing in general, and machining in particular, has not been immune from the growth of new technologies. CNC programming and CNC machining have not remained untouched, as new materials, new tools, new machine and control features are introduced to the industry. Good learning material to unravel the new approaches and techniques is hard to find. This edition of "Programming of Computer Numerically Controlled Machines" has successfully attempted to fill many voids. As a complete rewrite by Ken Evans of the popular book of the same name by Stanley Gabrel and John Polywka, the book approaches the subject of CNC with 21st century manufacturing in mind. This is a book that has it all.

The best features of the book are its contents and style. The book is very easy to understand – the author shows his skill as a professional communicator on every page. His extensive experience in both industry and educational fields give him a high level of credibility. He tries to be original and, without a doubt, succeeds very well.

The book presents the subject of CNC programming in a practical and well-organized way. Numerous examples, study questions, charts and mathematical formulae complement the extensive text. Illustrations throughout the book lead the reader to the subject of interest.

Written for machinists with little or no CNC experience, this book is a valuable resource for learning CNC programming. The "Operation" section in the early part of the book is designed to ease an experienced machinist into the world of CNC programming. Programming examples are practical, well documented and selected as being typical in machine shops. At the end of the book, the Glossary, the Appendix, and the Index can be easily accessed for instant reference.

As a major update of a popular book, this edition of "Programming of Computer Numerically Controlled Machines" will undoubtedly find its way as a CNC resource for the thousands of machinists, programmers and managers.

Peter Smid

Author of "CNC Programming Handbook"
"FANUC CNC Custom Macros" and
"CNC Programming Techniques"

About the Author

Ken Evans has held a diverse array of machining and related jobs throughout his career and is currently a Machine Tool Technology instructor at Davis Applied Technology College in Kaysville, Utah, where he has been on the faculty for sixteen years. He is responsible for marketing and delivering customized training for industry partners, both onsite and at the college and also teaches foundations through advanced-level courses in the machining curriculum, including Mastercam CAD/CAM classes for students and educators. He is Project Lead the Way certified for Computer Integrated Manufacturing (CIM).

Ken has been a Business Development Manager for a manufacturing solutions provider for the last two years.

He was Training and Applications Specialist and one of the nation's first certified Mazatrol Programming instructor to work with a local distributor teaching Mazatrol Conversational programming classes to their customers. In addition, Ken has trained other educators from around the region in the set-up, programming and operation of CNC machines.

Ken began his teaching career in 1984 at the T.H. Pickens Technical Center in Colorado, at the same time working full-time as a CNC machinist and Quality Control Inspector for a local shop. Ken learned the machinist trade in 1976 at Cessna Aircraft in Wichita, Kansas.

Ken loves the outdoors, including golfing, mountain biking, and farming.

Preface

The author of this book is a full-time Machine Tool Technology instructor at a local College with over 30 years of CNC operation, setup and programming experience. A strong interest in the practical application of CNC is at the heart of this text; therefore, all theoretical explanations are kept to a minimum so that they do not distort an understanding of the programming. Because of the wide range of information available about the selection of tools, cutting speeds, and the technology of machining, we want this book to reach a wide range of readers. Included among these are: Pre-Engineering students, those already involved in programming or maintaining CNC machines, operators of conventional machines who may want to expand their knowledge beyond conventional machining, and, managers or other interested persons who may wish to purchase such machines in the near future. Finally, I hope anyone with an interest in learning about modern CNC machining methods will find the book to be beneficial, as well.

In this third edition, you will notice many changes and enhancements that will improve your reading experience. Chapter objectives are listed at the beginning of each chapter, specific terminology is presented and study questions are added at the end of each chapter to confirm understanding. Throughout the text, figure captions are added to aid clarity. In the first chapter, the foundation is laid with CNC Basics that set the tone for successful programming. The second chapter on CNC Machine Operation gives the reader perspective about CNC Operation and setup procedures, since the first exposure a machinist has to CNC is usually as an operator. Operators will not be concerned right away with programming, but after some time, practice and the confidence of the owner, operators are given greater responsibility, i.e.: changing wear offsets, performing setups and minor program editing. The first and second chapters emphasize the development of machine setup and program editing skills. Students, machinists, supervisors and design and manufacturing engineers will benefit from these chapters by learning foundational skills associated with setup and operation of CNC machine tools, prior to programming.

Chapters three and four focus on the components and development of program code for CNC Turning and Machining Centers, with over 50 programming examples.

Because of the common use of Computer Aided Design and Computer Aided Manufacturing (CAD/CAM) today and the increasing popularity of Conversational Programming at the machine controller, two new chapters were added to the second edition. In this third edition, the CAD/CAM chapter has been updated to include the current version Mastercam X^2 software. Additionally, a new chapter has been added for "Computer Aided Manufacturing from Solid Models" where the latest version of EdgeCAM by

Pathtrace is featured. Because of the effectiveness of CAD/CAM, it is now the conventional method for programming used today.

The Mazatrol Conversational Programming chapter has been expanded to include programming examples and study questions and an example program is created using MazaCAM off-line programming software by SolutionWare. Many new machine tools come standard with some form of Conversational Programming.

The appendix contains many useful charts, techniques and math formulas used for line-by-line programming for user reference and the glossary of terms has been expanded to include more definitions.

The purpose of this book is to expand the reader's current knowledge of CNC programming by providing full descriptions of all program functions and their practical applications. The book contains information on how to program turning and milling machines, which is applicable to almost all control systems. In order to provide clear explanations about one unified system, the controller model referenced here is one of the most widely accepted, popular numerical control systems used worldwide.

Third Edition by Ken Evans

PART 1

CNC Basics

OBJECTIVES:

1. Recognize the importance of Safety when working with CNC Machines.
2. Become familiar with Tool and Work holding methods for CNC Machining.
3. Learn how to calculate proper Feeds and Speeds for CNC Machining.
4. Learn how to plan for CNC programming by using Process Planning Documents.
5. Become familiar with Coordinate Systems and their use in CNC Programming
6. Learn terminology and acronyms associated with the CNC Basics.
7. To learn the ABC's of CNC program format.

SAFETY

As you begin to learn about CNC Programming, it is important to first become aware of and learn how to practice safe working habits. You should not operate any machine without first understanding the basic safety procedures necessary to protect yourself and others from injury, and the equipment from damage. Most CNC machines are provided with a number of safety devices (door interlocks, etc.), to protect personnel and equipment from injury or damage. However, operators should not rely solely on these safety devices, but should operate the machine only after reading and fully understanding the Safety Precautions and Basic Operating Procedures outlined in the maintenance and operation manuals provided with the equipment. The following are some Do's and Don'ts that should be practiced when working with CNC Machines.

Safety Rules for NC and CNC Machines

Do's:
- Wear safety glasses and safety shoes at all times.
- Know how to stop the machine under emergency conditions.
- Keep the surrounding area well lighted, dry and free from obstructions.
- Keep hands out of the path of moving parts during machining operations.
- All setup procedures and loading or unloading of workpieces must be performed with the spindle stopped.
- Follow recommended safety policies and procedures when operating machinery, handling parts or tooling, and when lifting.
- Machine guards should be in position during operation.
- Wrenches, tools and parts should be kept away from the machine's moving parts.
- Make sure fixtures and workpieces are securely clamped before starting the machine.
- Cutting tools should be inspected for wear or damage prior to use.

Don'ts:
- Never operate a machine until properly instructed in its use.
- Never wear neckties, loose fitting long sleeves, wristwatches, rings, gloves or

3

unrestrained long hair, when operating any machine.

- Never attempt to remove metal chips with hands or fingers.

- Never direct compressed air at yourself or others.

- Never operate an NC/CNC machine without first consulting the specific operator manual for the machine.

- Never place hands near a revolving spindle.

- Electrical cabinet doors are to be opened only by qualified personnel for maintenance purposes.

MAINTENANCE

A large investment has been made to purchase CNC equipment. It is very important to recognize the need for proper maintenance and a general upkeep of these machines. At the beginning of each opportunity to work on any Turning or Machining Center, verify that all lubrication reservoirs are properly filled with the correct oils. The recommended oils are listed in the operation or maintenance manuals typically provided with the equipment. Sometimes there is a placard (plate) with a diagram of the machine and numbered locations for lubrication and the oil type is found on the machine. Most modern CNC machines have sensors that will not allow operation of the machine when the Way or Spindle oil levels are too low. Pneumatic (air) pressures need to be at a specified level and regulated properly. If the pressure is too low, some machine functions will not operate until the pressure is restored to normal. The standard air pressure setting is listed in Pounds per Square Inch (PSI) and a pressure regulator is commonly located at the rear of the machine. Refer to the operator or maintenance manuals for recommended maintenance activities.

Coolant Reservoir

The Coolant tank level should be checked and adjusted as needed prior to use. A site glass is normally mounted on the tank for easy viewing. Use an acceptable water-soluble coolant mix, synthetic coolant or cutting oil. Periodically the coolant tank should be cleaned and refilled. And last but not the least important is the cleanliness of the worktable, tools and area. Be sure to clean off any metal chips and remove any nicks or burrs on the clamping or mating surfaces. Always clean the machine after use.

Daily Maintenance Activities

Do's:

- Verify that all lubrication reservoirs are filled.

- Verify air pressure level by examining the regulator on the machine.

- Check the chip pan and coolant level and clean or fill, as needed.

- Make sure that automatic chip removal equipment is operational when the machine is cutting metal.

- Be sure that the worktable and all mating surfaces are clean and free from nicks or burrs.

- Check to see that the Chuck pressure setting is adequate for clamping the work to be machined.

- Clean up the machine at the end of use with a wet/dry vacuum or wash machine guards with coolant to remove chips from the working envelope.

Most new CNC machines are equipped with guards that envelope the worktable. The guards protect the Ways and sensitive micro-switches installed as limit switches for table movement. Guards also help keep the surrounding floor space clean but there is still the task of chip disposal. Some larger production machines incorporate a chip conveyor, which carries the chips to a drum on the floor on either side of the machine for easy removal. Even with these features, there is still a need for chip cleanup inside the working envelope at least once a day. If chips are allowed to gather within the guards, they will eventually find their way around the guards that protect the machine Ways. Over time, some of the chips might become embedded into the Ways and cause irreparable damage. Another problem that may occur as the chips collect, they bunch up and are pushed into contact with the micro-switches. This contact stops the machine from working since the switches send a signal to the control that indicates table travel limit has been exceeded. This message prevents the machine from operating until the chips are removed. If chips get within the guards around the micro-switches, it is necessary to remove those guards and clean. If this extent of cleaning becomes necessary the Machine should be turned off and a Lock-Out/Tag-Out should be incorporated to prevent injury. Remember, it is essential to replace the guards after cleanup.

It is very important to do a thorough machine cleanup when many chips are present. The exterior of the machine usually will only need wiping down with a clean rag. You can cleanup the Ways and the working envelope without damaging the machine by using coolant to wash the machine table and the guards free of chips. Another effective cleaning method is to use a wet/dry vacuum to pick up the chips. Along with the chip conveyor system, these two methods have proven hard to beat.

One cleanup method that is not recommended is to use compressed air to blow away the chips. It is appropriate to use compressed air to remove chips and coolant from the workpiece itself or work holding fixtures such as a vise. The problem with using compressed air to clean up around the Ways is that when chips are blasted away from the table, many are forced behind the guards, further worsening the micro-switch problem described above.

TOOL CLAMPING METHODS

Proper selection of cutting tools and work holding methods are paramount to the success of any machining operation. The scope of this text is not intended to teach all of the necessary information regarding tooling. You must consult the appropriate tooling manuals for selection of tool holders and cutting tools that are relevant to the required operation. Least expensive is not necessarily best.

Sound machining principals require that the most rigid setup possible be used that does not allow large overhangs of tools or workpieces. Ignoring these basic principals can cause tool and workpiece deflection and vibration that will contribute to poor surface finish and, eventually, tool damage which also makes it difficult to maintain dimensional accuracy.

Just as with the rest of the machine tool, there are components used with the actual cutting tool that make it what it is. Obviously, the tool cutting edge is where the metal removal takes place. Without proper tool clamping, the cutting action may not produce the desired results. Therefore, it is very important to carefully select the most effective tool clamping method.

In the case of a simple operation of milling a contour on a part, we may select a collet or a positive locking (posi-lock) end mill holder for the end mill. The correct choice would depend on the actual features of the part to be machined. If the amount of metal to be removed is minimal, then a collet would probably suffice. But if a considerable amount of metal is to be removed (more than two thirds of the tool diameter on a single depth of cut pass), then the posi-lock end mill holder selection is important. The reason for selecting the posi-lock holder is that under heavy cuts, a collet may not be able to grip the tool tightly enough. This situation could allow the tool to spin within the collet while in cut, with the result of ruining the collet and possibly damaging the part being machined. There is a tendency for the tool to dive into the workpiece when the tool spins within the collet and damage to the part may occur. *Note: Most High Speed Steel (HSS) end mills have a flat ground on them to facilitate the use of the posi-lock holder. This flat area allows for a set-screw to lock into it, creating a rigid and stable tool clamping method.* The clamping method for drills could be either a collet or a drill chuck. A keyed drill chuck usually is used for heavier metal removal or larger holes, whereas the keyless- type drill chuck is suitable for small holes. Generally, in the case of larger drills, a collet will be necessary to hold the tool. When holes are to be drilled, remember to center-drill or spot-drill first, so that the tool does not have a tendency to wander off location. The center-drill may be held in the same manner as a drill.

In turning, the selection of the type of tool holder is determined by the finished part geometry and the part material. There are a variety of tool holder styles as well as insert shapes available to accomplish the desired part shape and size.

For more information on the proper selection of inserts and tool holders, refer to the Machinery's Handbook section titled "Indexable Inserts".

Another valuable resource for technical data regarding the selection of inserts and tool holding are the ordering catalogs from the tool and insert manufacturers.

CUTTING TOOL SELECTION

Cutting tools are a very important aspect of machining. If the improper tool and/or tool clamping method is used the result will most likely be a poorly machined part. Always research and use the best tool and clamping method for a given operation. With the high speed and high performance of CNC machines, the proper selection process becomes increasingly important. The entire CNC machining process can be compromised by a lack of good tool planning and improper use.

There are many different types of machining operations performed on either Turning or Machining Centers. The tool is where the action is, so if improper selection takes place here, the whole machining sequence will be affected. Years of study have been dedicated to this subject and are documented within reference manuals and buyer's guides. Using these references will be helpful to correctly choose a tool for a given operation.

Remember that in your selection process, you are searching for the optimum metal-cutting conditions. The best way to understand how to choose the proper conditions is by studying the available data such as: the machine capabilities; the specific type of operation; the proper cutting tool(s) and tool clamping method(s); the geometry of the part to be made; the workpiece and cutter material; and, the method of clamping the part.

It is important to utilize the most technologically advanced methods of metal removal available. Do not hesitate to research this new technology. For example, in recent years, there have been numerous cutting tool innovations that include indexable insert coatings such as: Titanium Nitride (TiN); Titanium Carbon Nitride (TiCN) applied through Chemical Vapor Deposition (CVD), or Physical Vapor Deposition (PVD), and insert materials such as: Ceramic, Cubic Boron Nitride (CBN) and Polycrystalline Diamond (PCD), to name a few. These advances have enabled increased cutting speeds and decreased tool wear providing for higher production throughput. Another tool clamping innovation is modular tooling. This is a standardization of tool holders to facilitate the quick change of tools decreasing setup time. *Refer to the tool and insert ordering catalogs from the tool and insert manufacturers for more information on modular tooling.*

TOOL COMPENSATION FACTORS

Important information about the tool must be given to the machine control unit (MCU), for the machine to be able to use the tool efficiently. In other words, the MCU needs the tool identification number, the tool length offset (TLO), and the specific diameter of each tool. A TLO is a measurement given to the control unit to compensate for the tool length when movements are commanded. The cutter diameter compensation (CDC) offset is used by the control to compensate for the diameter of the tool, for end mill style tools, during commanded movements.

The tool number identifies where the tool is located within the storage magazine or turret and often is the order sequence in which it is used. Each is assigned a tool length offset number. This number correlates with the pocket or turret position number and is where the measured offset distance from the cutting tip to the spindle face, in the case of a milling machine, is stored. For example, Tool No. 1 will have TLO No. 1. Finally, when end milling is necessary the diameter of the tool is compensated for. In most cases, the programmer has taken the diameter of the tool into account. In other words, the programmed tool path is written with a specific tool size in mind. However, more commonly the part geometry is programmed in order to facilitate the use of different tool diameters for a specified operation. When using the part geometry rather than the toolpath centerline for a specific tool diameter, an additional offset is called from within the program called cutter diameter compensation (CDC).

TOOL CHANGING

CNC equipment enables more efficient machining by allowing the combination of several operations into a single setup. This combination of operations requires the use of multiple cutting tools. Automatic Tool Changers (ATC) are a standard feature on most CNC Machining Centers, while many CNC Knee-Mills still require manual installation of the tool. The illustrations in Figures 1 and 2 are two types of tool holders used on

CNC machines; they have some distinct physical differences. Both of the holders are tapered. The one to the left has a single ring at the large end of the taper while the other has two rings. The tool holder that has only one ring is designed for machines that require manual tool changes. The tool with two rings is designed for machines that have Automatic Tool Changers. These rings act as a gripping surface for a tool changer.

The tapered portion of the holder is the actual surface that is in contact with the mating taper of the spindle. These tapers are standardized by the industry and are numbered according to size (No. 30, No. 40, No. 50).

One benefit of these tapers over the standard R-8 Bridgeport style of tool holder is the increased surface area in contact with the mating taper of the spindle. The increased surface area makes the tool setup more rigid and stable.

Another feature on the tool holder is the notch or cutout on centerline of the tool (there is an identical cutout on the opposite side). This enables axial orientation within the spindle and tool changer. As the holder is inserted into the spindle, the cutouts enable it to be locked into place in exactly the same orientation each and every time it is used. This orientation makes a real difference when trying to perform very precise operations such as boring a diameter. These notches also aid the spindle driving mechanism.

On CNC machines with a manual tool change, the holder is inserted into the machine and rotated until the holder pops into place (axial orientation is done by hand) and then, the draw bar is tightened to clamp the tool holder in place. Finally, another component of the CNC Automatic Tool Changing system is the Retention Knob or Pull-Stud. Machining Centers need the retention knob/pull-stud to pull the tool into the spindle and clamp the holder. This knob is threaded into the small end of the taper as shown. *Note: There are several styles of knobs available. The operator should consult the appropriate manufacturer manual for specifications required in their situation.*

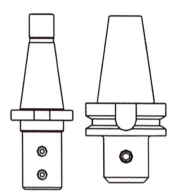

Figure 1&2 Common Millling Tool Holders

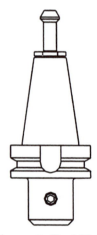

Figure 3 Tool Holder with Retention Knob

Figure 4 Retention Knob

METAL CUTTING FACTORS

Many tool and work holding methods used on manual machines are also used on CNC machines. The machines themselves differ in their method of control but otherwise they are very similar. The major objective of CNC is to increase productivity and improve quality by consistently controlling the machining operation. Knowledge of the exact capabilities of the machine and its components as well as the tooling involved is imperative when working with CNC. It is necessary for the CNC programmer to have a thorough knowledge of the CNC machines they are responsible for programming. This may involve an ongoing process of research and update training with the ultimate goal of obtaining a near optimum metal-cutting process. From this research and training comes a decrease in the cycle time necessary to produce each part lowering per piece cost to the consumer. Fine tuning of the machining process for high-speed production gives more control over the quality of the product on a consistent basis. The following are some of the most important factors that affect the metal cutting process.

The Machine Tool

The machine used must have the physical ability to perform the machining. If the planned machining cut requires 10 horsepower from the spindle motor, a machine with only 5 horsepower will not be an efficient one to use. It is important to work within the capabilities of the machine tool. The stability, rigidity and repeatability of the machine are of paramount importance as well. Always take these things into consideration when planning for machining.

The Cutting Fluid or Coolant

The metal cutting process is one that creates friction between the cutting tool and the workpiece. A cutting fluid or coolant is necessary to lubricate and remove heat and chips from the tool and workpiece during cutting. Water alone is not sufficient because it only cools and does not lubricate, and it will also cause rust to develop on the machine Ways and table. Also, because of the heat produced, water vaporizes and thus compromises the cooling effect. A mixture of lard-based soluble oil and water creates a good coolant for most light metal-cutting operations. Harder materials, like stainless steel and high alloy composition steels, require the use of a cutting-oil for the optimum results. Advancements have been made with synthetic coolants, as well. Finally, the flow of coolant should be as strong as possible and be directed at the cutting edge to accomplish its purpose. Programmers and machine operators should research available resources like the *Machinery's Handbook* and coolant manufacturer data, for information about the proper selection and use of cutting fluids for specific types of materials.

The Workpiece & the Work Holding Method

The material to be machined has a definite effect on decisions about what tools will be used, the type of coolant necessary, and the selection of proper speeds and feeds for the metal-cutting operation.

The shape or geometry of the workpiece affects the metal-cutting operation and determines the type of work holding method that will be used. This clamping method is important for CNC work because of the high performance expected. It must hold the workpiece securely, be rigid, and should minimize the possibility of any flex or movement of the part.

The Cutting Speed

Cutting speed is the rate at which the circumference of the tool moves past the workpiece in surface feet (sf/min) or meters per minute (m/min), to obtain satisfactory metal removal.

The cutting speed factor is most closely related to the tool life. Many years of research have been dedicated to this aspect of metal-cutting operations. The workpiece and the cutting tool material determine the recommended cutting speed. The *Machinery's Handbook* is an excellent source for information pertaining to determining proper cutting speed. If incorrect cutting speeds, spindle speed or feedrates are used, the results will be poor tool life, poor surface finishes and even the possibility of damage to the tool and/or part.

The Spindle Speed

When referring to a milling or a turning operation, the spindle speed of the cutting tool or chuck must be accurately calculated relating to the conditions present. This speed is measured in revolutions per minute, r/min (formerly known as RPM) and is dependant upon the type and condition of material being machined. This factor, coupled with a depth of cut, gives the information necessary to find the required horsepower necessary to perform a given operation. In order to create a highly productive machining operation all these factors should be given careful consideration. Refer to the formulae below needed to calculate r/min.

For Inch Units:

$$r/\min = \frac{12 \times CS}{\pi \times D}$$

For Metric Units:

$$r/\min = \frac{1000 \times CS}{\pi \times D}$$

where

CS = Cutting Speed from the charts in *Machinery's Handbook*

π = 3.1417

D = Diameter of the workpiece or the cutter.

Many modern machine controllers have a feature that allows automatic calculation of feeds and speeds that is based on appropriate operator input of the cutting conditions.

The Feedrate

Feedrate is defined as the distance the tool travels along a given axis in a set amount of time, generally measured in inches per minute in/min (formerly known as IPM) for milling or inches per revolution in/rev (formerly known as IPR) for turning. This factor is dependent upon the selected tool type, the calculated spindle speed and the depth of cut. Refer to the *Machinery's Handbook* for the chip load recommendations and review the formula below that is necessary to calculate this aspect of the metal-cutting operation.

where
$$F = R \times N \times f$$

F = Feed in in/min or mm/min

R = r/min calculated from the preceding formula

N = the number of cutting edges

f = the chip load per tooth recommended from the *Machinery's Handbook*

The Depth of Cut

The depth of cut is determined by the amount of material to be removed from the workpiece, cutting tool flute length or insert size and the power available from the machine spindle. Always use the largest depth of cut possible to ensure the least affect on the tool life.

Cutting Speed, Spindle Speed, Feedrate and Depth of cut are all important factors in the metal-cutting process. When properly calculated, the optimum metal-cutting conditions will result. *Refer to the Machinery's Handbook and the tool and insert ordering catalogs from the tool and insert manufacturers for more information on recommended depths of cut for particular tooling.*

PROCESS PLANNING FOR CNC

Certain steps must be followed in order to produce a machined part that meets specifications given on a blueprint. These steps need to be organized in a logical sequence to produce the finished part in the most efficient manner. Before machining begins, it is essential to go through the procedure called Process Planning. The following are the steps in the process:

1. Study the working drawing or blueprint.
2. Select the proper raw material or rough stock as described on the blueprint.
3. Study the blueprint and determine the best sequence of individual operations needed to machine the required geometry.
4. Transfer the information onto planning charts.
5. Use in-process inspection to check dimensional values as they are completed while the part is still mounted on the machine.
6. Make necessary corrections and deburr.
7. Perform a 100% dimensional inspection when the part is finished and log the results of the first article inspection on the quality control check sheet.
8. Take corrective action if any problems are identified.
9. Begin production.

Planning Documents

A blueprint may be thought of as a map that defines the destination. This destination is the end product. The roads available to get to this destination may be numerous. We do not start the trip without first determining what the destination is and how we are going to get there.

Planning sheets resemble the required path to the destination. They are written descriptions of how to get there (to the end product). The following are descriptions of sample planning documents.

The Blueprint

The information given on the blueprint will include the material, overall shape and the dimensions for part features. The geometry determines the type of machine (mill or lathe) to be used to produce the part. By studying the blueprint, material and opera-

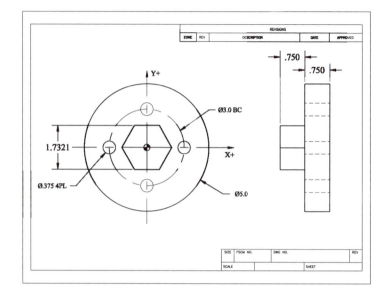

Figure 5
Machined Part Blueprint

Chart 1
Process Planning Operation Sheet

Date:			Prepared By:	
Part Name:			Part Number:	
Quantity:			Sheet ___ of ___	
Material:				
Raw Stock Size:				
Operation Number	Machine Used	Description of Operation		Time

tions (drilling, milling, boring, etc.) can be identified. The tools and work holding method can also be determined. Occasionally, the geometry will require multiple machines to manufacture the part, and thus additional operations will be necessary.

Operation Sheet

The purpose of this planning document is to identify the correct order for operations to be performed and the machine to be used. For example, suppose you are required to produce the part shown in Figure 5 above. Saw cutting the rough stock into blanks would come first. Then the part must be turned on a lathe to create the five-inch diameter and rough turn the diameter for the hexagon. Next a milling machine is used to cut the hexagon and drill the bolthole circle. Before any inspection can be carried out, the part must be deburred. Finally the part can be inspected for accuracy.

The operation sheet is particularly useful when many identical parts are machined (production run). The operation sheet is similar to directions or a how-to approach. The process needed to manufacture the finished part has been decided in advance and is documented for future use.

When small batches of parts are to be made, there may not be an operation sheet. It is the machinist's responsibility to study the blueprint and decide the necessary steps to machine the part. The operation sheet can aid in this decision making process.

With CNC machining, multiple part geometry features can be performed in one setup. In some cases, when using a CNC Mill Turn Center, a part might be machined to its completed status without ever using another machine. This is very efficient and another advantage of the use of CNC equipment.

To complete an operation sheet, study the blueprint; then decide on the steps necessary to machine the part. Document the machining process and refine any problems the process has, then list the operations in the correct sequence in which they will be performed.

The top section of the operation sheet is for reference information and includes:

> The date the document is prepared or revised

> The name of the person preparing it

> The part name and the part number (from the blueprint)

> The quantity of parts to be manufactured

Since some parts may require a large number of operations, it is possible that more than one operation sheet will be needed to document the whole process. The top section also includes a sheet numbering system (Sheet _____ of _____). This information must be included. Other information included on the operation sheet header is the material, the raw stock size for the part, followed by the operations list.

CNC Setup Sheet

The CNC Setup Sheet is the document that tells the machinist what tools are to be used and any specific information related to tools. For instance, it may be necessary to have a certain amount of extension for a drill to be able to completely machine through a part. This document is where the Operator/CNC Machinist finds this information. In Part 5 or 6 of this text you will be introduced to Computer Aided Design and Computer Aided Manufacturing (CAD/CAM) and how you can develop CNC Setup Sheets within the CAD/CAM programs. Many companies today are going to electronic "paperless" factories. The CNC Setup Sheet has two sections. The top section is for reference information and includes:

Part 1 CNC Basics

The date the document is prepared or revised

The name of the person preparing it

The part name and part number (from the blueprint)

The machine being used

Note: If more than one machine is to be used to manufacture a single part, separate setup sheets are completed for each machine.

The CNC part program used in the manufacturing process

Workpiece Zero reference points for the part (program zero)

Work holding devices

Note: If more than one device is needed, the operation number(s) and process are also included.

The lower half of this form lists the tool(s) by number, description and offset. There is a column for comments, remarks or explanations, if needed. Specific tool requirements, like minimum tool length extension, can be entered in the comments section

Quality Control Check Sheet

This planning document is used for the final inspection stage of the machining process. Once the part is completed, it is necessary to check all of the dimensions listed

Chart 2
Process Planning CNC Setup Sheet

Date:	Prepared By:
Part Name:	Part Number:
Machine:	Program Number:

Workpiece Zero: X _____ Y _____ Z _____

Setup Description:

Tool #	Tool Description	Offset #	Comments

on the blueprint to verify they are within the specified tolerance. The Quality Control Check Sheet is an excellent method to document the results of this inspection and a valuable tracking tool.

Reference information is similar to the other planning documents. Included are:

      The date the document is prepared or revised

      The name of the person checking the part

      The part name and part number (from the blueprint).

On the check sheet, 100 % of the blueprint dimensions and their tolerances are written down in list form. Using this method, sequentially go through each of the print dimensions and log the results. This assures that the machined part meets the specifications given on the blueprint. As the part is checked and verified, some dimensions may not meet specifications. It is important to make sure that these incorrect values are noticed. A good method for relating this is to write those dimensions that are out of print in red ink or use a highlighter pen to emphasize them. You could also include details in the comments section of the QC Check Sheet. If dimensions are found that do not meet specifications, corrective action must be taken.

Chart 3
Process Planning
Quality Control Check Sheet

Date:				Checked By:
Part Name:			Part Number:	
Blueprint Dimension	Tolerance	Actual Dimension	Comments	

TYPES OF NUMERICALLY CONTROLLED MACHINES

There are two basic groups of numerically controlled machines, Numerical Control (NC) and Computer Numerical Control (CNC).

In an NC system, the program is run from a punched tape, it is impossible to store such a program in memory. For a punched tape to be used again to machine another part, it must be rewound and read from the beginning. This routine is repeated every time the program is executed. If there are errors in the program and changes are necessary, the tape will need to be discarded and a new one punched. The process is costly and error prone and while this type is still in use, it is becoming obsolete.

Machines with a CNC system are equipped with a computer, consisting of one or more microprocessors and memory storage facilities. Some CNC Machines have hard drives and are network configurable. Program data is entered through Manual Data Input (MDI) at the control panel keyboard, via an RS232 communications interface port or via Ethernet from a remote source like a Personal Computer (PC) network. The control panel enables the operator to make corrections (edits) to the program stored in memory, thereby eliminating the need for new punched tape.

Types of CNC Machines have expanded vastly over the last decade. Turning and Machining Centers are the focus of this book but there are many other types of machines using Computerized Numerical Control. For example there are: Electrical Discharge Machines (EDM), Grinders, Lasers, Turret Punches, and many more. Also, there are many different designs of Machining and Turning Centers. Some of the Machining Centers have rotary axes and some Turning Centers have live tooling and secondary spindles. For this text, the focus will be limited to Vertical Machining Centers with three axes and Turning Centers with two axes. These types of machines are considered the foundation of all CNC learning. All operations on these machines can be carried out automatically. Human involvement is limited to setting up, loading and unloading the workpiece and entering the amounts of dimensional offsets.

WHAT IS CNC PROGRAMMING?

CNC programming is a method of defining machine tool movements through the application of numbers and corresponding coded letter symbols. As shown in the list below, all phases of production are considered in programming, beginning with the technical part drawing and ending with the final product:

Technical Part Drawing

Work Holding Considerations

Tool Selection

Preparation of the Part Program

Part Program Tool Path Verification

Measuring of Tool and Work Offsets

Program Test by Dry Run

Automatic Operation or CNC Machining

All programming begins by a close evaluation of the technical drawing and emphasizes assigned tolerances for particular operations, tool selection, and the choice of a machine. The next step is the selection of the machining process. The machining process refers to the selection of fixtures and determination of the operation sequence. Following that, is a selection of the appropriate tools and determination of the sequence of their application. Before writing a program, spindle speeds and feed rates must be calculated.

When program writing begins, special attention is given to the specific tool movements necessary to complete the finished part geometry, including non-cutting movements. Individual tools are identified and noted in the program manuscript. Miscellaneous functions are noted for each tool such as; flood coolant, spindle direction, r/min and feedrates (these items will be covered in greater detail in the following chapters). Then, once the program is written, it must be transferred to the machine through an input medium like one of the following: punched tape, floppy disk, USB, by RS-232 interface or by Ethernet.

Machining is initiated by preparing the machine for use, commonly called setup (for example, input of Workpiece Zero and Tool Length Offset into CNC memory registers). Many modern controllers have a function for graphical simulation of the programmed tool path on the Cathode Ray Tube (CRT). This enables the machinist or set-up person to verify that the program has no errors, and to visually inspect the tool path movements. If all looks well, the first part can be machined with increased confidence. After completion, a thorough dimensional inspection will compare dimensions of the final product to those on the part drawing. Any differences between the actual dimensions and the dimensions on the drawing are corrected by values inserted into the offset register of the machine. In this manner, the correct dimensions of consecutively machined parts can be obtained.

INTRODUCTION TO THE COORDINATE SYSTEM

All machines are equipped with the basic traveling components, which move in relation to one another as well as in perpendicular directions. CNC Turning Centers are equipped with a tool carrier, which travels along two axes (see Figures 6 and 7).

Note that in the following drawings of lathes, the cutting tool and turret is located on the positive side of spindle centerline. This is a common design of modern CNC Turning Centers. For visualization purposes in this book the cutting tool will be shown upright. In reality it is mounted with the insert facing down and the spindle is rotated clockwise for cutting.

Note: the direction of spindle rotation, clockwise (CW) or Counterclockwise (CCW), in turning, is determined by looking from the headstock towards the tailstock and tool orientation.

Machining Centers are milling machines equipped with a traversing worktable, which travels along two axes, and a spindle with a driven tool that travels along a third axis.

All axes of machines are oriented in an orthogonal (each axis is perpendicular to the other) coordinate system, for example, the Cartesian coordinate system (right-hand rule system). (See Figure 9)

Figure 6 Turning Center Axes

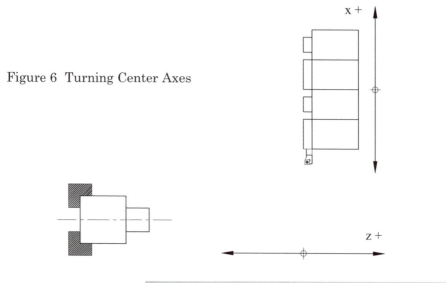

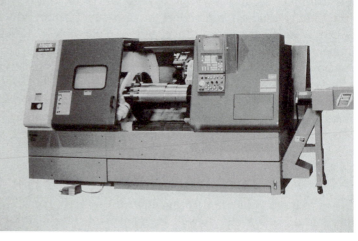

Figure 7
Two Axis Turning Center
Courtesy MAZAK Corporation

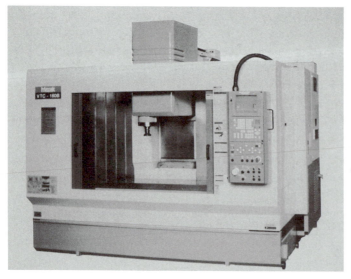

Figure 8
Three Axis Machining Center
Courtesy MAZAK Corporation

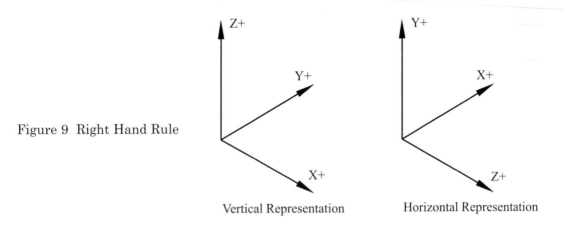

Figure 9 Right Hand Rule

Vertical Representation   Horizontal Representation

The Right-Hand Rule System

In discussing the X, Y, and Z axes, the right hand rule establishes the orientation and the description of tool motions along a positive or negative direction for each axis. This rule is recognized worldwide and is the standard for which axis identification was established.

Use Figure 9 above to help you visualize this concept. For the vertical representation, the palm of your right hand is laid out flat in front, face up the thumb will point in the positive X direction. The forefinger will be pointing the positive Y direction. Now fold over the little finger and the ring finger and allow the middle finger to point up. This forms the third axis, Z, and points in the Z positive direction. The point where all three of these axes intersect is called the origin or zero point. When looking at any vertical milling machine, you can apply this rule. For the Horizontal mill the same steps described above could be applied if you were lying on your back.

COORDINATE SYSTEMS

Visualize a grid on a sheet of paper (graph paper) with each segment of the grid having a specific value. Now place two solid lines through the exact center of the grid and perpendicular to each other. By doing this, you have constructed a simple, two-dimensional coordinate system. Carry the thought a little further and add a third imaginary line. This line passes through the same center point as the first two lines but as vertical; that is, it rises above and below the sheet on which the grid is placed. This additional line would represent the third axis in the three-dimensional coordinate system which is called the Z axis.

Two-Dimensional Coordinate System

A two-dimensional coordinate system, such as the one used on a lathe, uses the X and Z axes for measurement. The X runs perpendicular to the workpiece and the Z axis is parallel with the spindle centerline. When working on the lathe, we are working with a workpiece that has only two dimensions, the diameter and the length. On blueprints, the front view generally shows the features that define the finished shape of the part for turning. In order to see how to apply this type of coordinate system, study the

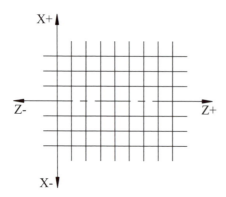

Figure 10
Two-Dimensional Coordinate System

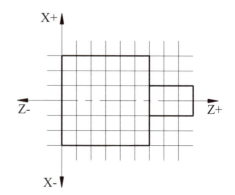

Figure 11
Part Drawing with a 2D Coordinate
System Overlay

Figure 12
Two-Dimensional Turned
Part Drawing

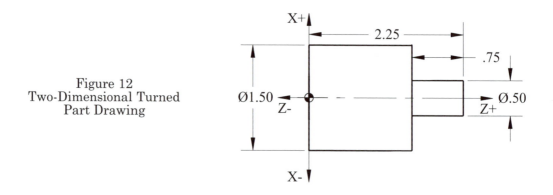

following diagrams. (See Figures 10, 11 and 12)

Think of the cylindrical work piece as if it were flat or as shown in the front view of the part blueprint. Next, visualize the coordinate system superimposed over the blueprint of the workpiece, aligning the X axis with the centerline of the diameter shown. Then align the Z axis with the end of the part, which will be used as an origin or zeropoint. In most cases, the finished part surface nearest the spindle face will represent this Z axis datum and the centerline will represent the X axis. Where the two axes intersect is the origin or zero point. By laying out this "grid," we now can apply the coordinate system and define where the points are located to enable programmed creation of the geometry from the blueprint. Another point to consider: on a lathe, is that the cutting takes place on only one side of the part or the radius, because the part rotates and it is symmetrical about the centerline. In order to apply the coordinate system in this case, all we need is the basic contour features of one-half of the part (on one side of the diameter). The other half is a mirror image; when given this program coordinate information, the lathe will automatically produce the mirror image.

Three-Dimensional Coordinate System

Although the mill uses a three-dimensional coordinate system, the same concept

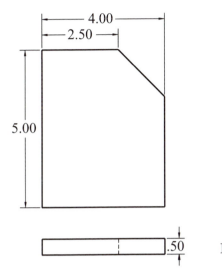

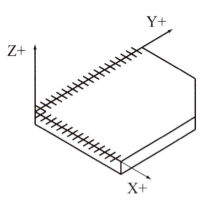

Figure 13 Three-Dimensional Coordinate System

(using the top view of the blueprint) can be used with rectangular workpieces. As with the lathe, the Z axis is related to the spindle. However, in the case of the three dimensional rectangular workpiece, the origin or zero-point must be defined differently. In the example shown in Figure13, the upper left hand corner of the workpiece is chosen as the zero-point for defining movements using the coordinate system. The thickness of the part is the third dimension or Z axis. When selecting a zero-point for the Z axis of a particular part, it is common to use the top surface.

The Polar Coordinate System

If a circle is drawn on a piece of graph paper so that the center of the circle is at the intersection of two lines and the edges of the circle are tangent to any line on the paper this will help in visualizing the following statements. Let's consider the circle center as the origin or zero-point of the coordinate system. This means that some of the points defined within this grid will be negative numbers. Now draw a horizontal line through the center and passing through each side of the circle. Then draw a vertical line through the center also passing through each side of the circle. Basically, we've made a pie with four pieces. Each of the four pieces or segments of the circle is known as a quadrant. The quadrants are numbered and progress counter-clockwise. In Quadrant No. 1, both the X and Y axis point values are positive. In Quadrant No. 2, the X axis point values are negative and the Y axis point values are positive. In Quadrant No. 3, both the X and Y axis point values are negative. Finally, in Quadrant No. 4, the X axis point values are positive while the Y axis values are negative. This quadrant system is true regardless of the axis that rotation is about. The following drawings illustrate the values (negative or positive) of the coordinates, depending on the quarter circle (quadrant) in which they appear.

Although the rectangular coordinate system can be used to define points on the circle, a method using angular values may also be specified. We still use the same origin or zero-point for the X and Y axes. However, the two values that are being considered are an angular value for the position of a point on the circle and the length of the radius joining that point with the center of the circle. To understand the polar coordinate system, imagine that the radius is a line circling around the center origin or zero-point.

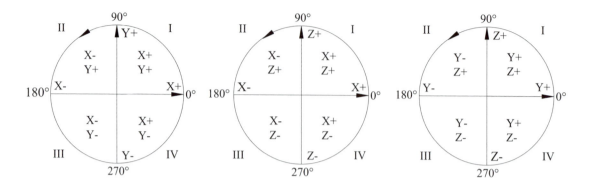

Figure 14 Polar Coordinate System Quadrants

Thinking in terms of hand movements on a clock, the three-o'clock position has an angular value of 0° counted as the "starting point" for the radius line. The twelve-o'clock position is referred to as the 90° position, nine-o'clock is 180°, and the six-o'clock position is 270°. When the radius line lies on the X axis in the three-o'clock position, we have at least two possible angular measurements. If the radius line has not moved from its starting point, the angular measurement is known as 0°. On the other hand, if the radius line has circled once around the zero point, the angular measurement is known as 360°. Therefore, the movement of the radius determines the angular measurement. If the direction in which the radius rotates is counter-clockwise, angular values will be positive. A negative angular value (such as -90°) indicates that the radius has rotated in a clockwise direction. *Note: A 90° angle (clockwise rotation) places the radius at the same position on the grid as a +270° (counter-clockwise) rotation.*

Sometimes the blueprint will not specify a rectangular coordinate but will give a polar system in the form of an angle for the location of a feature. With some basic trigonometric calculations, this information can be converted to the rectangular coordinate system.

The same polar coordinates system applies regardless of the axis of rotation as is shown once again in Figure 14. When rotation is around the X axis, the rotational axis is designated as A, the Y axis, the rotational axis is designated as B, and the Z axis, the rotational axis is designated as C. These are considered an additional axes and are known as the fourth axis.

All operations of CNC machines are based on three axes: X, Y, and Z.

1. *(X0, Y0, Z0)*
2. *(X0, Y0, Z+)*
3. *(X0, Y-, Z+)*
4. *(X0, Y-, Z0)*
5. *(X-, Y-, Z0)*
6. *(X-, Y0, Z0)*
7. *(X-, Y0, Z+)*
8. *(X-, Y- Z+)*

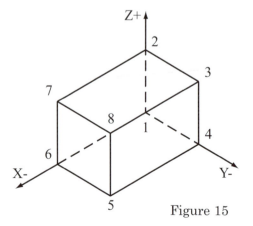

Figure 15

22

Figure 15 illustrates a box-like object in which one vertex (point 1) is located at the origin of the coordinate system. At the side of the drawing, the coordinate signs are given for each the numbered locations. Note the position of the coordinate system on the following machines.

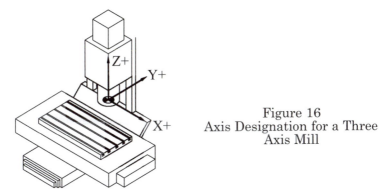

Figure 16
Axis Designation for a Three
Axis Mill

On vertical milling machines the spindle axis is perpendicular to the surface of the worktable.

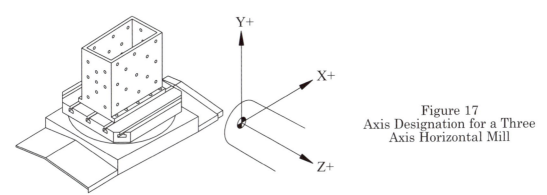

Figure 17
Axis Designation for a Three
Axis Horizontal Mill

On lathes, the spindle axis is also the workpiece axis.

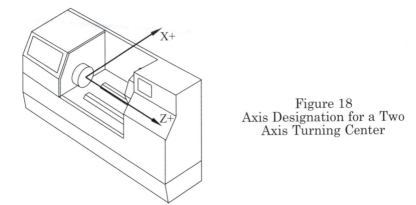

Figure 18
Axis Designation for a Two
Axis Turning Center

On horizontal milling machines the spindle axis is parallel to the surface of the worktable.

POINTS OF REFERENCE

When using CNC machines, any tool location is controlled within the coordinate system. The accuracy of this positional information is established by specific Zero Points (reference points). The first is Machine Zero, a fixed point established by the manufacturer that is the basis for all coordinate system measurements. On a typical lathe, this is usually the spindle centerline in the X axis and the face of the spindle nose for the Z axis. For a milling machine, this position is often at the furthest end of travel in all three axes in the positive direction. Occasionally, this X axis position is at the center of the table travel.

This Machine Zero Point establishes the coordinate system for operation of the machine and is commonly called Machine Home (Home Position). Upon startup of the machine, all axes need to be moved to this position to establish the coordinate system origin (commonly called Homing the machine or Zero Return). The Machine Zero Point identifies to the machine controller where the origin for each axis is located.

The Operator's manual supplied with the machine should be consulted to identify where this location is and how to properly Home the machine.

The second zero point can be located anywhere within the machine work envelope and is called Workpiece Zero and is used as the basis for programmed coordinate values used to produce the workpiece. It is established within the part program by a special code and the coordinates are taken from the distance from the Machine Zero point. The code number in the program identifies the location of offset values to the machine control where the exact coordinate distance of the X, Y and Z axes of Workpiece Zero is in relationship to the Machine Zero. All dimensional data on the part will be established by accurately setting the Workpiece Zero. A way of looking at the Workpiece Zero is like another coordinate system within the machine coordinate system, established by the Home Position.

Tool offsets are also considered to be Zero Points as well and are compensated for with Tool Length and Diameter Offsets. The tool setting point for a lathe has two dimensions; the distance on diameter from the tool tip to the centerline of the tool turret, and the distance from the tool turret face to the tool tip. The tool setting point for the mill is the distance from the spindle face to the tool tip, and the distance from the tool tip to the spindle centerline.

Blueprint Relationship to CNC

The standard called ASME Y14.5-1994 establishes a method for communicating part dimensional values, in a uniform way, on the engineering drawing or blueprint. The drawing information will be translated to the coordinate system in order for dimensional values and part features to be manufactured.

On the blueprint, Datum features are identified as Primary (A), Secondary (B) and Tertiary (C). Dimensions for the workpiece are derived from these datum features. On the drawing, the point where these three datum features meet is called the origin or zero point for the part. When possible, this same point should be used for Workpiece Zero. This allows the use of actual blueprint dimensions within the part program and often results in fewer calculations. Most drawings are developed using an absolute dimensioning system based on Datum dimensions derived from the same fixed point (origin or zero point). Occasionally, some features may be dimensioned from the location of another feature. An example of this might be a row of holes exactly one half of an inch apart. This type of dimensioning is called relative or incremental.

Note: A thorough knowledge of blueprint reading is imperative for successful results using manual or CNC equipment.

Machine Zero

Each CNC machine is assigned a fixed point, which is referred to as Machine Zero (or Machine Home). For most machines, Machine Zero is defined as the extreme travel end position of main machine components that are oriented in a given coordinate system. From Machine Zero, we can determine the values of the coordinates that, in turn, determine the position of the points commanded in a CNC program. Electromechanical sensors called micro-switches (limit switches) are located in the extreme end positions of traveling machine components. These sensors send a signal to the controller when they are activated and thus setting the Home position. In the case of milling machines, Machine Zero on the table is set with respect to the X and Y axes. Machine Zero on the spindle is set with respect to the Z axis, whereas Machine Zero of the tool carrier on lathes is set with respect to the X and Z axes. Positioning the traveling components at zero can be performed manually, as well as with the use of the control panel or directly from within the program by employing a Reference Point Return function (G28). At the initial start up of any CNC Machine, it is required that the machine be "Homed" or sent to Machine Zero before proceeding any further. From that point on, all machine components will always automatically return to the same exact position when commanded to do so in the program.

Machine Zero is frequently the position in which tool changes take place. Therefore, if you intend to change the tool before a given operation, then the machine must be positioned at Machine Zero for the Z axis on vertical machines and the Y axis on horizontal machines.

Workpiece Zero

So far, for all main traveling components of CNC machines, we have assigned an oriented axis within the coordinate system. Any movement of machine components must be described by points, which actually determine the traveling path of the tool. Changes in the position tool are determined with respect to the stationary reference point of Machine Zero.

In order to better understand this concept, this situation can be illustrated with a rectangular plate in which all coordinates are described at their four corners (P_1, P_2, P_3, P_4).

$$P_1 = X\text{-}15.0, Y\text{-}10.0$$

$$P_2 = X\text{-}15.0, Y\text{-}12.0$$

$$P_3 = X\text{-}20.0, Y\text{-}12.0$$

$$P_4 = X\text{-}20.0, Y\text{-}10.0$$

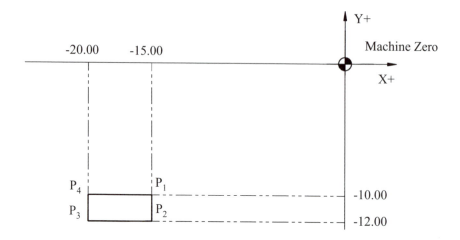

Figure 19 Machine Zero to Workpiece Zero

Determine the coordinates of these points. The rectangle has been placed in such a manner that each side is parallel to one axis of the coordinate system. If the distance from Machine Zero is measured to any point on the workpiece, the coordinates of the remaining points can be determined from the dimensions given on the drawing.

All programmed point coordinates, whose values are determined with respect to Machine Zero, must be calculated with respect to Machine Zero every time which is time consuming. It may also cause errors due to the fact that all the given dimensions determining the points do not always refer to those on the drawing. As previously mentioned, in order to determine the coordinates for the four corners of the rectangular part illustrated, it is necessary to find the distance between Machine Zero and a specific point of reference on the part. Then, all the remaining dimensional data to be used are taken from the blueprint.

For all CNC machines, we follow certain principles to define the method of selecting Workpiece Zero from within the part program. At the beginning of the program, we input the value of the distance between Machine Zero and the selected Workpiece Zero, by employing function G92 or G54 through G59 for Machining Centers and function G50 or G54 for Turning Centers. These measured values are either input directly into the program, as in the case of G92 for Mills and G50 for Lathes, or in offset registers in the control for G54 through 59. Let us review the same situation as above and note the changes of the point coordinates when applying Workpiece Zero.

G92X15.0Y10.0   or   G54-59X0.0Y0.0

P_1 = X0, Y0

P_2 = X0, Y-2.0

P_3 = X-5.0, Y-2.0

P_4 = X-5.0, Y0

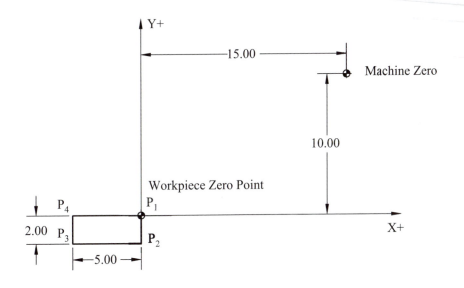

Figure 20 Workpiece Zero Point

The values X15.0 and Y10.0 for G92 or G54 through 59 are valid until they are recalled by the same function, but with different coordinates, for *X* and *Y*. When programming Machining Centers, we place function G92 or G54 through 59 only at the beginning of the program, whereas the values assigned to function G50 for Turning Centers will need to be added to the program with respect to each tools position. Once this activation is read by the control, all coordinates will be measured from the new Workpiece Zero allowing the use of part dimensions for programmed moves.

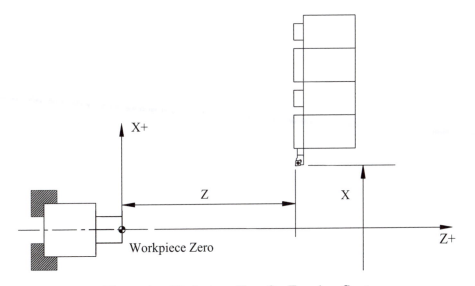

Figure 21 Workpiece Zero for Turning Centers

Part 1 CNC Basics

With Turning Centers, Workpiece Zero in the direction of the Z axis is most often on the face surface of the workpiece, and the centerline axis of the spindle is Workpiece Zero in the direction of the X axis.

On Machining Centers, Workpiece Zero is frequently located on the corner of the workpiece or in alignment with the Datum of the workpiece.

The application of Workpiece Zero is quite advantageous to the programmer because the input values of X, Y, and Z in the program can be taken directly from the drawing. If the program is used another time, the values of coordinates X and Y (assigned to functions G50 and G92 or G54 through G59) will have to be inserted again, prior to automatic operation.

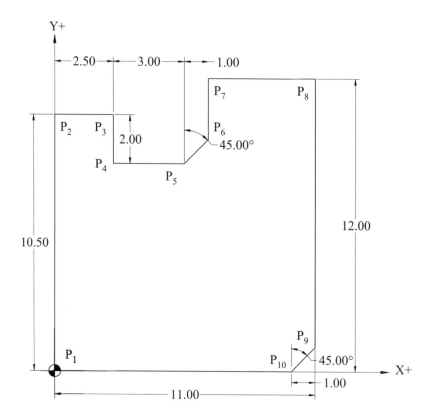

Figure 22 Absolute and Incremental Coordinate System Points

Absolute and Incremental Coordinate Systems

When programming in an absolute coordinate system, the positions of all the coordinates are based upon a fixed point or origin of the coordinate system. The tool path from point P_1 to P_{10}, for example, is illustrated on the next page:

	X	Y
P_1	0.0	0.0
P_2	0.0	10.5
P_3	2.5	10.5
P_4	2.5	8.5
P_5	5.5	8.5
P_6	6.5	9.5
P_7	6.5	12.0
P_8	11.0	12.0
P_9	11.0	1.0
P_{10}	10.0	0.0

Programming with an incremental coordinate system is based upon the determination of the tool path from its current position to its next consecutive position and in the direction of all the axes. Sign determines the direction of motion. Based on the drawing from the previous example, we can illustrate the tool path in an incremental coordinate system, starting and ending at P_1.

	X	Y
P_2	0.0	10.5
P_3	2.5	0.0
P_4	0.0	-2.0
P_5	3.0	0.0
P_6	1.0	1.0
P_7	0.0	2.5
P_8	4.5	0.0
P_9	0.0	-11.0
P_{10}	-1.0	-1.0
P_1	-10.0	0.0

Coordinate Input Format

CNC machines allow input values of inches specified by the command G20, or millimeters specified by the command G21, and degrees with a decimal point and significant zeros in front of (leading) or at the end (trailing) of the values. When using inch programming the two ways distances can be specified:

Programming with a decimal point

1 inch = 1. or 1.0

1 1/4 inch = 1.250 or 1.25

1/16 inch = 0.0625 or .0625

Programming with significant trailing zeros

In this case, the zero furthest to the right corresponds with the ten thousandths of an inch.

$$1 \text{ inch} = 10000$$

$$1 \, ^1/_8 \text{ inch} = 11250$$

$$1 \, ^1/_{32} \text{ inch} = 10313$$

These two coordinate input formats are the standard on all CNC machines

With modern controllers neither leading nor trailing zeros are required, the decimal placement is the significant factor. In this case the input is as follows:

$$1 \text{ inch} = 1. \text{ or } 1.0$$

$$1 \, ^1/_4 \text{ inch} = 1.25$$

$$^1/_{16} \text{ inch} = .0625 \text{ or } 0.0625$$

PROGRAM FORMAT

The language described in this book is used for controlling machine tools is known informally as "G-Code". This language is used worldwide and is reasonably consistent. The standard it is governed by was established by the Electronics Industries Association and the International Standards Organization called EIA/ISO for short. Because of this standardization a program created for a particular part on one machine may be used on other similar machines with minimal changes required.

Each program is a set of instructions that controls the tool path. The program is made up from blocks of information separated by the semicolon symbol (;). This symbol (;) is defined as the end of the block (EOB) character. Each block contains one or more program words. For example:

Word	Word	Word	Word	Word
N02	G01	X3.5	Y4.728	F8.0

Each word contains an address, followed by specific data. For example:

Address	Data	Address	Data	Address	Data
N	02	G	01	X	3.5

The following chart is a list for all of the letter addresses that are applicable in programming, along with brief explanations for each:

Address Characters

CHARACTER	MEANING
A	Additional rotary axis parallel and around the X axis
B	Additional rotary axis parallel and around the Y axis
C	Additional rotary axis parallel and around the Z axis
D	Tool radius offset number
	(Turning) Depth of cut for multiple repetitive cycles
E	User macro character
F	Feed rate
	(Turning) Precise designation of thread lead
G	Preparatory function
H	Tool length offset number
I	Incremental X coordinate of circle center or parameter of fixed cycle
J	Incremental Y coordinate of circle center
K	Incremental Z coordinate of circle center
	(Turning) parameter of fixed cycle
L	Number of repetitions (subprogram, hole pattern)
	Fixed offset group number
M	Miscellaneous function
N	Sequence or block number
O	Program number
P	Dwell time, program number, and sequence number designation in sub-program(Turning) Sequence number start for multiple repetitive cycles
Q	Depth of cut, shift of canned cycles
	(Turning) Sequence number end for multiple repetitive cycles
R	Point R for canned cycles, as a reference return value
	Radius designation of a circle arc
	Angular displacement value for coordinate system rotation
S	Spindle-speed function
T	Tool function
U	Additional linear axis parallel to X axis
V	Additional linear axis parallel to Y axis
W	Additional linear axis parallel to Z axis
X	X coordinate
Y	Y coordinate
Z	Z coordinate

Common Symbols Used in Programs

SYMBOL	MEANING
-	Minus Sign, Used for Negative Values
/	Slash, Used for Block Skip Function
%	Percent Sign, Necessary at program beginning and end for communications only
( )	Parentheses, Used for comments within programs
:	Colon, Designation of Program Number
.	Decimal Point, Designation of fractional portion of a number

Part 1 Study Questions

1. Programming is a method of defining tool movements through the application of numbers and corresponding coded letter symbols.

    T or F

2. A lathe has the following axes:

    a. X, Y & Z

    b. X & Y only

    c. X & Z only

    d. Y & Z only

3. Program coordinates that are based on a fixed origin are called:

    a. Incremental

    b. Absolute

    c. Relative

    d. Polar

4. On a two axis turning center the diameter controlling axis is:

    a. B

    b. A

    c. X

    d. Z

5. The letter addresses used to identify axes of rotation are:

    a. U, V & W

    b. X, Y & Z

    c. A, Z & X

    d. A, B & C

Part 1 CNC Basics

6. The acronym TLO stands for:

        a. Tool Length Offsets

        b. Total Length Offset

        c. Taper Length Offset

        d. Time Length Offset

7. When referring to the polar coordinate system, the clockwise rotation direction has a positive value

        T or F

8. In Figure 15 of Part 1, which quadrant is the part placed in?

9. A program block is a single line of code followed by an end-of-block character

        T or F

10. Each block contains one or more program words

        T or F

11. Using Figure 13 in Part 1, list the X and Y absolute coordinates for the part profile where Workpiece Zero is at the upper left corner. (The corner cutoff is at a 45° angle).

12. Using Figure 13 in Part 1, list the X and Y incremental coordinates for the part profile where Workpiece Zero is at the upper left corner.

13. How often should the machine lubrication levels be checked?

PART 2

CNC MACHINE OPERATION

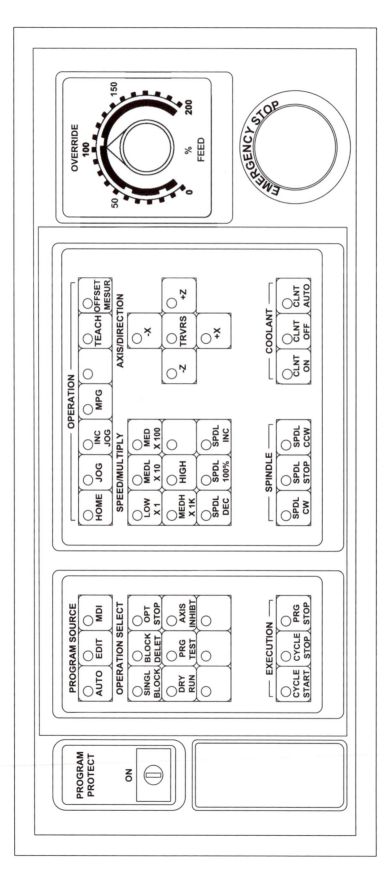

Figure 1 Common Operators Panel

Part 2 CNC Machine Operation

OBJECTIVES:

1. To become familiar with common CNC Machine Operation Panel functions.
2. To become familiar with common Machine Control Panel functions.
3. To learn common operations performed at the Machine Control.
4. Learn how to use the Controls to input setup data including Tool and Work Offsets.
5. Learn how to use the Control to edit programs.
6. Examine some common cases of problem situations and learn how to solve them.

Every CNC Machine Tool has an Operation Panel and a Control Panel that are the interface between the Operator and or Programmer and the Machine Tool, sometimes referred to as the Human Interface (HMI). The Operator Panel is the method with which we physically manipulate the working components of the machine to do what we need and the Control Panel is where the program data are entered and stored. A thorough understanding of each is necessary for successful CNC Machine use. First, we will study and example of an Operation Panel.

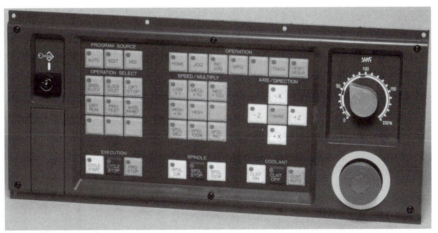

Figure 2 Operation Panel   Courtesy GE FANUC

OPERATION PANEL DESCRIPTIONS

The following descriptions for the above diagram represent the configuration for a common Operators Panel. Some differences do exist for each manufacturer Operation Panel, but they generally contain the same features. The illustration shows a panel for a two-axis lathe. The panel used for a mill would be essentially identical, except for the added keys for the additional axes. The user should consult the applicable manufacturer manual for detailed descriptions that match their needs. Please also note that the Handle (Pulse Generator) and the Rapid Traverse Override buttons are not shown in this view of the controller, although it is described in the text. Another common item not shown here are the switches used to change the chucking direction from external to internal and are specifically for lathes.

OPERATOR PANEL FEATURES
FEEDRATE OVERRIDE

The Feedrate override dial allows control of the feedrate when the operator adjusts the position. At the 12:00 o'clock position the feed, during auto-cycle, will occur at 100% of the programmed value. This allows the control of the work feeds defined by the F-word in the program. The percentage of the value entered in the program can be increased or decreased. This feature offers the operator the control needed to fine tune feeds. It can also be used to control feedrates while using the manual jog mode function.

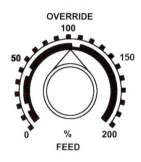

Figure 3 Feed Override

EMERGENCY STOP

The EMERGENCY STOP is the large, red, mushroom-shaped button used to stop machine function when an emergency situation occurs. Some example situations are: overloading of the machine, a clamping of the machined part has come loose, or incorrect data in the program or work or tool offsets have caused a collision (crash) between the tool and the workpiece. When this button is pressed, all program commanded feedrates and spindle revolutions are halted immediately. Recovery from an "E-Stop" condition requires resetting the program controller and Homing of the machine axes. To reset the EMERGENCY STOP button, turn it clockwise. It should "pop-out" of the depressed condition. Check the monitor for any alarm signals and take note of the Alarm # and description, and then eliminate the cause that forced the use of the EMERGENCY STOP button. Press the Reset button to clear all pending commands and Home the machine axes when no interference conditions are present.

Figure 4 Emergency Stop

PROGRAM PROTECT

When this key-switch is in the ON condition (vertical), it prohibits any program changes to be made. The condition does not affect work or tool offset adjustments. Some shops set this condition to ON, remove this key and allow only the programmer or set-up person access to the key. This is especially true in larger shops with multiple shifts and many workers. Some Quality programs like ISO9000 require that CNC program integrity is insured.

Figure 5 Program Protect Keyswithch

PROGRAM SOURCE

On some operator panels, a rotary switch referred to as Mode Select is used instead of buttons that are shown here. This switch includes both automatic (AUTO) and man-

PROGRAM SOURCE		
AUTO	EDIT	MDI

Figure 6
Program Source

ual operation functions. The position of this switch determines whether the machine utilizes the automatic or the manual control. This switch can also be positioned to allow the entry of data into the control manually (Manual Data Input or MDI) or to make changes to the program through the EDIT mode. For this purpose, operator panel buttons are used to specify the control or operational mode.

Note: When the buttons are pressed, they are active and remain so until another mode button is pressed. In some conditions, multiple Light Emitting Diodes (LED's) may be lit simultaneously. The LED in the upper left corner of the button is lit when the mode is ON and active.

Auto

By pressing this button, the control mode enables the CNC commands stored in the memory to be executed for automatic operation. When the Cycle Start button is pressed, and this mode is active, automatic operation will occur.

Edit

By pressing this button, the program edit mode is selected. The EDIT mode enables the user to enter the part program to control memory, enter any changes to the program and transfer data from the program via RS232 interface to an offline storage device or check the program file memory and storage capacity.

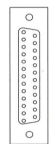

Note: The RS232 interface is a 25-pin serial cable connector (in Figure 7) is located behind the flip-up door just below the Program Protect key switch. This connection is also used for DNC (Direct Numerical Control) program operations when the program is too large for the controller memory and is executed directly from a remote Personal Computer (PC).

Figure 7
RS232
Communications
Interface Port

MDI-Manual Data Input

By pressing this button, the MANUAL DATA INPUT mode is selected. The MDI mode enables the automatic control of the machine, using information entered in the form of program blocks without interfering with the basic part program. This mode is often used during the machining of workpiece holding equipment such as soft-jaws and during setup. It corresponds to single moves (milling surfaces, drilling holes), descriptions of which need not be entered to memory storage. MDI mode can also be used during the execution of the program. For example, suppose the program is missing the command M03 S350; needed to turn on the spindle, clockwise, at 350 r/min and the End-Of-Block character (;). In order to correct this omission, press the SINGLE BLOCK button and then the MDI button. Using MDI, the user can enter functions M03 and S350 from the control panel keypad followed by the EOB character. Then, enter this command by pressing INPUT on the control panel. Press AUTO to reenter the program auto-cycle mode and then press CYCLE START to continue execution of the program from memory.

OPERATION SELECT

The following buttons are related to the automatic operation of the machine. Activating one of these buttons has an effect on the operation and is described on the next page:

Part 2 CNC Machine Operation

Single Block

The execution of a SINGLE BLOCK (SINGL BLOCK) of information is initiated by pressing this button to turn it ON. Each time the CYCLE START button is pressed, only one block of information will be executed. This switch can also be used if you intend to check the initial performance of a new program on the machine or when the momentary interruption of a machine's work is necessary.

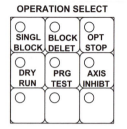

OPERATION SELECT

Figure 8
Operation Selection Buttons

Block Delete

BLOCK DELETE (BLOCK DELET) is sometimes called Block Skip. When this button is pressed and is active simultaneously with the auto-cycle mode, the controller skips execution of the program blocks that are preceded by the slash (/) symbol and that end with the end of block (;) character. For instance, if a section of the part program or a particular block of the program is not presently needed, but you would like to keep this information for future use, then place the block skip symbol (/) at the beginning of each such block. The BLOCK DELET button is located on the control panel. If it is activated, then information contained in the blocks that are preceded by the symbol (/) will not be executed.

Example:

N100G01X2.810Y3.256

/N105X3.253Y2.864

/N110X3.800

(Blocks N105 and N110 will be skipped)

Notes: The symbol (/) should be placed at the beginning of the block. If it is not, then all the information contained in the block preceding the symbol (/) will be executed, while the information following this symbol will be omitted.

If the BLOCK DELET is in the OFF condition, all blocks, regardless of the symbol (/), will be executed.

When transferring the program to punched tape or external computer, all program information, regardless of symbol (/), is transferred.

Opt Stop-Optional Stop

When this button is pressed, the OPTIONAL STOP mode is active. The OPTIONAL STOP function interrupts the automatic cycle of the machine if the program word M01 appears in the program. Quite often, function M01 is placed in the program after the work of a particular tool is completed or before a tool change. This enables the operator to perform routine measurements directly on the machine and, if necessary, make adjustments and then rerun the same tool to correct inaccuracies.

Dry Run

By pressing the DRY RUN button during automatic cycle, all of the rapid and work feeds are changed to the rapid traverse feed set in the parameters instead of the programmed feed. Consult the manufacturer manual for specific directions on the use of this function.

DRY RUN is also used to check a new program on the machine without any work actually being performed by the tool. This is particularly useful on programs with long cycle times so the operator can progress through the program more quickly.

Caution: When using this function, it is NOT intended for metal cutting.

Prg Test-Program Test

This function is also known as MACHINE LOCK. Activating this mode inhibits axis movement on all of the axes. This button is used to check a new program on the machine through the controller. All movements of the tool are locked, while a program check is run on the computer and displayed on the screen. The operator can observe the position display on the screen and if any program errors are encountered, an alarm will be displayed. This function is especially useful for checking very large programs requiring a long cycle time to complete. This test is normally the first in a series (Program Test, Dry Run and Single Block) of preliminary actions to be executed before full auto cycle mode is attempted. For any program test, all offsets should be set first.

Axis Inhbt-Axis Inhibit

This function is identical to MACHINE LOCK for all axes. Activating this mode inhibits axis movement on all of the axes. A common situation would be to inhibit the axes to allow for internal checking of the program. Some controllers have additional buttons or switches that enable inhibiting of only one axis at a time. This function is especially useful when inhibiting the Z axis so that all X, Y movements can still be observed.

EXECUTION

These three buttons are related to automatic operation of the machine. The first button starts, the second temporarily stops automatic operation and the last key merely indicates when a program stop is encountered. Their specific functions are described below:

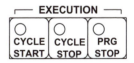

Figure 9
Execution Buttons

Cycle Start

The CYCLE START button is used to start automatic operation. Use this button in order to begin the execution of a program from memory. When the CYCLE START button is pressed, the LED located above this button goes on and the active program will be executed to the end.

Cycle Stop (Feed Hold)

Pressing the CYCLE STOP button during automatic operation will halt all feed movements of the machine. It will not stop the spindle r/min or affect the execution of tool changes on some machines. When the CYCLE STOP button is pressed, the LED located on the button goes on, and the LED located on the CYCLE START button goes off. When the CYCLE STOP button is pressed, all feeds are temporarily stopped; however, the spindle rotation is not affected. This button is used when minor problems are encountered, such as coolant flow direction or when checking DISTANCE-TO-GO during setup. When the problem is remedied, press CYCLE START again to resume automatic cycle operation. It is not recommended using this button to interrupt a cut because the spindle does not stop and damage to the tool or part may occur. When pressed during the execution of the tapping or threading cycle, CYCLE STOP will take effect after the thread pass or the tap is withdrawn. If the tap breaks during the tapping cycle, the only way to stop the machine is by pressing the RESET button on the controller or the EMERGENCY STOP button.

Prg Stop-Program Stop

When a Program Stop is commanded in the program by the program word M00, automatic operation is stopped and the LED on this button is turned on. This button does not have an ON/OFF function that affects the program stop condition. It is merely an LED indicator lamp to indicate when a program stop condition is active.

OPERATION

The keys in the Operation section of the control are used for manual operation of the machine during setup and initial startup. Their specific functions are described below:

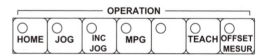

Figure 10 Operation Buttons

Home

Pressing this button on and then pressing the X or Z (X, Y or Z for Machining Centers), buttons causes the machine to return to the Machine Zero position for each axis in relation to the machine coordinate system.

Jog

Pressing the JOG button activates a manual feed mode that allows the selection of manual feed movements along single axes X or Z (X, Y, or Z for machining centers). With the button activated, use the Axis/Direction buttons and the Speed/Multiply buttons to move the desired axis at the chosen feed rate (in/min). On some controls Speed/Multiply is a rotary type switch that activates this function.

INC Jog

Press this button (Incremental JOG) to activate the JOG mode in Incremental steps at feed as per selection using the Speed/Multiply buttons.

MPG-Manual Pulse Generator

Pressing this button activates the manual handle feed mode for the selected axis. This handle is known as the Manual Pulse Generator. Pressing the MPG button places the machine in the HANDLE mode. This mode enables manual control of axis movements (for *X, Y* or *Z,* or for rotational axes *A, B* or *C)* by use of the handle after activating their respective axis buttons. For instance, press MPG then press *X* and then use the handle to move to the desired position along the *X* axis. By turning the handle clockwise, you can move the tool in a positive direction with respect to the position of the coordinate system. By turning the handle counterclockwise, the tool is moved in a negative direction with respect to

Figure 11
Manual Pulse
Generator (MPG) Handle

the position of the coordinate system. The handle contains 100 notches, each of which corresponds to an increment (distance to be moved). Turning the handle, you can feel the displacement from one notch to the next.

To set the magnitude for the distance to be moved, press one of the Speed/Multiply buttons as described in more detail below.

Caution: If the handle is rotated quickly while the magnitude is set at X100 or X1K the tool will move at a rapid feed rate and a crash could occur!

Teach

This button activates the Teach-in Jog or Teach-in Handle mode. When this mode is used, the movements of the axes are recorded while in either Jog or Handle mode. Machine positions along the *X, Y,* and *Z* axes obtained by this manual operation are stored in memory as a program position and are used to create a program. These movements can then be executed just as with any program. Consult the manufacturer operator's manual for detailed descriptions on its proper use. Not all controls have this feature.

Offset Mesur-Offset Measurment

When this button is pressed, the OFFSET MEASURE mode is selected and the position of the tool in relation to the coordinate system is written into the tool or work offset for the active axis.

SPEED/MULTIPLY

When INC JOG is selected from the Operation Mode buttons, the incremental step selected by the Speed/Multiply buttons determines the magnitude of the displacement along the chosen axis in the selected direction. When one of these buttons is pressed and released, the movement will be as follows:

X1 = a movement of .0001 inch or .0025 millimeters (mm)

X10 = a movement of .001 inch or .0254 mm

X100 = a movement of .010 inch or .254 mm

X1K = a movement of .100 inch or 2.54 mm

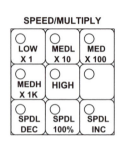

Figure 12
Speed/Multiply Buttons

The buttons used to select the axis and the direction of movement are located on the mid to upper right part of the operator panel as shown in Figure 1 for this controller.

For example, to displace the tool along the *X* axis in a positive direction by the value of .010 inch, follow these steps:

- Press the INC JOG button.
- Press the X100 button.
- Press the button X once.

Each time the button is pressed, a displacement of the selected value results.

If the JOG mode is selected when one of these buttons are activated and the selected axis button is pressed and held in, the movement will occur at feed as indicated by LOW, MEDL, MED, MEDH or HIGH. The operator can also use the % Traverse Feed override dial to further control this feed rate.

When the MPG Operation mode is selected, the incremental step selected by the Speed/Multiply buttons determines the magnitude of the movement along the chosen axis in the selected direction.

Each button setting corresponds with the scale of the Handle. One full revolution of the Handle (360°) corresponds to 100 units on the scale. On the button X1 the X means "times" the minimum increment.

When the button is pressed for:

X1, turning the handle by one unit corresponds to the movement of .0001 inch or .0025 mm.

X10, turning the handle by one unit corresponds to the movement of .001 inch or .0254 mm.

X100, turning the wheel by one unit corresponds to the movement of .01 inch or .254 mm.

X 1K, turning the wheel by one unit corresponds to the movement of .100 inch or 2.54 mm.

Usually, X1 is used when you are precisely dialing-in the zero of the workpiece and when you are determining the tool length offset.

In manual control, you must also use the AXIS/DIRECTION buttons to determine the axis of displacement.

For example: if you need to move the machine table, with respect to the tool by 1.00" along the *X* axis in a positive direction, follow these steps:

- Press MPG Operation mode button.
- Press the AXIS/DIRECTION button X+
- Press the SPEED/MULTIPLY button X10.

Turn the handle one full revolution (100 units) and then check the value of the displacement on the screen. It should indicate a movement of 1.00".

LOW
X1

The button, LOW indicates feed rate at a LOW speed while in the JOG Operation mode. The button, labeled as LOW X1 indicates that turning the handle by one unit corresponds to the displacement of .0001 inch or .0025 mm while in the MPG Operation mode.

MEDL
X10

The button, MEDL indicates feed rate at a MEDIUM LOW speed while in the JOG Operation mode. The button, labeled as MEDL X10 indicates that turning the handle by one unit corresponds to the displacement of .001 inch or .0254 mm while in the MPG Operation mode.

MED
X100

The button, MED indicates feed rate at a MEDIUM speed while in the JOG Operation mode. The button, labeled as MED X100 indicates that turning the handle by one unit corresponds to the displacement of .01 inch or .254 mm while in the MPG Operation mode.

MEDH
X1K

The button, MEDH indicates feed rate at a MEDIUM HIGH speed while in the JOG Operation mode. The button, labeled as MEDH X 1K indicates that turning the handle by one unit corresponds to the displacement of .100 inch or 2.54 mm while in the MPG Operation mode.

HIGH

This button is used to indicate the feed rate at a HIGH speed while in the JOG Operation mode.

SPDL DEC-Spindle Speed Decrease

Pressing this button causes the spindle speed to decrease.

SPDL 100%

Spindle override 100%: Sets an override of 100% for the spindle motor speed.

SPDL INC- Spindle Speed Increase

Pressing this button causes the spindle speed to increase.

SPINDLE

These buttons are used exclusively during the manual operation of the machine for setup functions. The descriptions below explain their specific function:

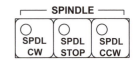

Figure 13 Spindle Buttons

SPDL CW-Spindle Rotation CW

By pressing this button while in one of the Operation modes; HOME, JOG, INC JOG or MPG the spindle will start rotation in the clockwise (CW) direction. The spindle r/min is adjusted by using the SPDL DEC, SPDL 100%, or SPDL INC buttons. The last r/min commanded in the program or used while in this mode is retained and will restart upon pressing the SPDL 100% button. When the machine is first started, there has been no value established for the r/min, so if one of these buttons is pressed while in the modes described above an alarm will result. An r/min must be input via MDI or by activating the program. From that point on, as long as the machine is not turned off, the r/min will be activated at the last commanded value when one of these buttons is pressed.

SPDL STOP-Spindle Stop

Pressing this button stops spindle motor rotation while in one of the Operation modes listed above. Pressing this button will NOT stop the spindle while in any of the Execution modes.

SPDL CCW-Spindle Rotation CCW

By pressing this button while in one of the Operation modes HOME, JOG, INC JOG or MPG the spindle will start rotation in the counterclockwise (CCW) direction. The spindle r/min is set by using the SPDL DEC, SPDL 100%, or SPDL INC buttons. The last r/min commanded in the program or used while in this mode is retained and will restart upon pressing the SPDL 100% button.

AXIS/DIRECTION

These buttons are used to select the Manual feed axis direction. Pressing these buttons executes movement along the selected axis in the selected direction by jog feed (or step feed) when the corresponding button is set to on in the jog feed mode (or step feed mode). The same is true for each of the axes where buttons are present.

AXIS/DIRECTION

Figure 14 Axis Direction Buttons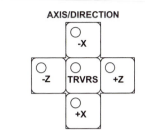

–X

When pressed, this button executes movement along the X axis in the negative direction with respect to the coordinate system.

+Z

When pressed, this button executes movement along the Z axis in the positive direction with respect to the coordinate system.

–Z

When pressed, this button executes movement along the Z axis in the minus direction with respect to the coordinate system.

+X

When pressed, this button executes movement along the X axis in the positive direction with respect to the coordinate system.

Note: The Y axis buttons for milling machines are not depicted in the operator panel in Figure 1. The following are their descriptions.

–Y

When pressed, this button executes movement along the Y axis in the negative direction with respect to the coordinate system.

+Y

When pressed, this button executes movement along the Y axis in the positive direction with respect to the coordinate system.

TRVRS-Rapid Traverse

Caution: When using this rapid traverse function be sure that all machine movements will not cause interference during motion!

Press this button to activate jog feed at a rapid traverse rate. Use this button to perform rapid movements along a previously chosen axis. For example, press the button TRVRS then the X positive button, the X axis will move at rapid traverse until the button is released.

Rapid Traverse Override (%)

(Not shown on this operator panel model)

Some operator panels include a % override dial or buttons that can be used to control the rapid feed rate. This switch or these buttons are used to reduce the rapid feed

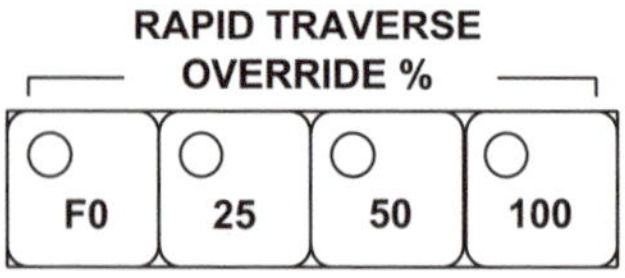

Figure 14b
Rapid Traverse
Override Buttons

RAPID TRAVERSE
OVERRIDE %

| F0 | 25 | 50 | 100 |

rate (G00). If it is positioned at 100, this corresponds to 100% of the rapid feed rate that the machine can generate. Buttons are commonly incremented in steps of 10, 25, 50 and 100%.

In the above illustration, F0 corresponds with 10% (as determined by parameter).

COOLANT

During manual or automatic operation, these buttons may be used to activate or stop the flow of coolant.

CLNT ON-Coolant ON

When this button is pressed the supply of coolant is started.

CLNT OFF-Coolant OFF

When this button is pressed the supply of coolant is stopped.

CLNT AUTO-Coolant Auto Mode

This button is pressed to activate the automatic start and stoppage of coolant flow during program execution when called by program words M08 and M09 respectively.

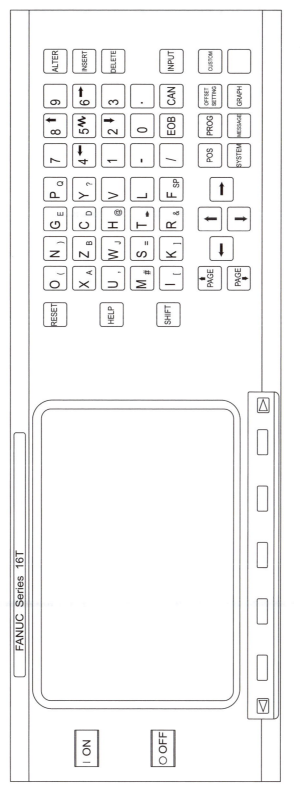

Figure 16 Common Control Panel

CONTROL PANEL DESCRIPTIONS

CONTROL PANEL

The control panel described here is quite typical of the control panels used on CNC machines. The control panel switches and buttons may be distributed differently on the panel for each individual machine; however, the purpose and function of each switch and button remains the same. Some control panels are equipped with additional buttons or switches not shown here. Definitions and applications of these buttons or switches can be found in the manufacturer instruction manuals for the machines.

The control panel is located at the front of the machine and is equipped with a CRT and with various buttons and switches, as illustrated in Figure 16.

Two items not shown on this controller that are common on many modern controls are a 3.5 floppy disk and PCMCI (Portable Computer Memory Card Interface) slot. These are both used as a file storage medium and a method for file transfer. The floppies typically hold a maximum of 1.44 mega bytes (MB) of data while PCMCI cards range from 40MB to 200MB. The 1.44 MB floppy can store the equivalent of 3600 meters of paper tape. Both are widely used to store part programs offset data and NC parameter information. Because of emerging technologies in computer industry, the storage and data transfer medium described here are changing and improving rapidly and some controllers are now being equipped with a Universal Serial Bus (USB) drive interface.

A detailed description for the use of each button and the purpose of the particular sections of the control panel are presented in the following sections:

POWER-ON & POWER-OFF

These buttons are used to activate/deactivate the power to the control. The ON button is typically green in color and when it is ON, the key is lit. The OFF button is typically red in color and when the power is turned OFF to the control, the key is lit. At startup of the main power to the machine, the OFF button is lit.

Figure 17
On Button

Press theses buttons to turn CNC power ON and OFF.

Notes: The control is always turned ON after the MAIN POWER switch is ON. The switch is located on the door of the control system and is typically at the back or side of the machine. The control is always turned OFF before the MAIN POWER switch is turned OFF.

Figure 18
Off Button

CRT DISPLAY

This is the television-like screen on which all the program characters and data are shown. Sizes vary from around 9 inches to approximately 15 inches. Displays are color, monochrome or Liquid Crystal Displays (LCD).

Reset Key

Pressing this button resets or cancels an alarm and can be used

Figure 19
Reset
Button

for cancellation of an automatic operation. An alarm can only be cancelled if its cause has been eliminated.

When the reset button is pressed during automatic operation, all program commanded axis feeds and spindle revolutions are cancelled. The program will return to its starting block when this button is pressed.

Help Key

Pressing this key gives the operator help screens on how to operate the machine functions such as MDI key operation, or details related to an alarm that has occurred in the control.

HELP

Figure 20
Help Key

SOFT KEYS

The soft keys have numerous functions, depending on the applications selected with other keys. The specific functions of the soft keys are displayed at the bottom of the CRT screen as shown in Figure 21. The purpose of the soft keys is to minimize the use of dedicated keys on the control panel.

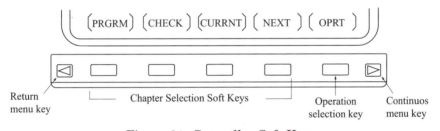

Figure 21 Controller Soft Keys

By use of the soft keys; the machine ORIGIN register can be reset, READ soft keys allow entering the program to memory from a punched tape or other storage medium, and the PUNCH soft key allows program readout from memory to a punched tape or other storage medium.

By pressing a soft key, the function selections that belong to it appear. These selection choices are called chapters. These selection soft keys are the first four rectangular keys under the CRT. By pressing one of the chapter selection soft keys, the screen for the selected chapter appears. If the soft key for a target chapter is not displayed, you must press the continuous menu key located at the right end of the soft keys (sometimes referred to as next-menu). In some cases, there are additional chapters that can be selected from within a chapter. When the desired screen is displayed, press the soft key under operation selection (OPRT) on the display for data to be manipulated. To reverse through the chapter selection soft keys, press the return menu key located at the left end of the soft keys.

The general screen display procedure is explained above; however, the actual display procedure varies from one screen to another. For details, see the description of individual operations.

Note: The operator should consult the manufacturer manual for more specific detailed instructions on the use of the soft keys.

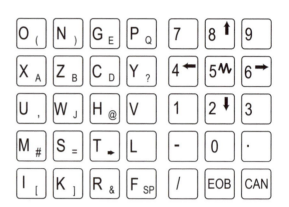

Figure 22
Alpha-Numerical Keypad

ADDRESS AND NUMERIC KEYS (ALPHA-NUMERICAL KEYS)

This keypad of letters, numbers, and symbol characters is used to input data while writing or editing programs at the control. These keys are also used to enter numerical data and offsets into memory. Many of the keys are used in conjunction with other keys.

Shift Key

Because there is not enough space on the control for all keys necessary, some keys have two characters on them. When the letter or symbol indicated in the bottom right corner of the key is needed, the operator first presses the SHIFT key, which switches the key to that character. This sequence must be followed each time an alternate letter is needed. The shift key functions the same way as its equivalent on a computer keyboard.

Figure 23
Shift Key

On the display, a special character will be shown when a character indicated at the bottom right corner on the key can be entered.

Input Key

The INPUT key is used for MDI operation and to change the offsets. After the data are entered via the keypad the INPUT key is pressed. The data are entered into the offset register or the program for execution.

Figure 24
Input Key

Cancel Key

This key is used while inputting data in the MDI mode. It is essentially a destructive backspace key and can be used to correct an erroneous entry. Press this key to delete the last character or symbol input to the key input buffer. For instance, when the key input buffer displays:

Figure 25
Cancel Key

N5X12.00Z

and then the cancel key is pressed, the address Z is erased and N5Xl2.00 is displayed.

EOB Key-End of Block Key

Figure 26
End of Block
Symbol Key

This is the END-OF-BLOCK key. When pressed while in the MDI mode the EOB character (;) is inserted into the program at the cursor location.

Note: The (;) symbol is never part of the program manuscript. The control system will automatically show the EOB character, for every "Enter" key used on a keyboard.

PART PROGRAM EDIT KEYS

These keys are used to enter new program data (Insert), to make program changes (Alter), or to delete program data in memory (Delete). They are used while editing programs.

Figure 27
Program Edit Keys

FUNCTION KEYS

These buttons correspond to particular display modes (active mode).

By pressing any one of these buttons, the display will be switched to the corresponding screen. Then the soft keys may be used to display the needed data.

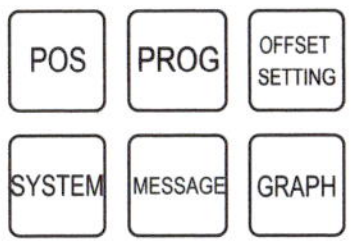

Figure 28
Function Keys

- Press the POS key to display the Position Screen.

- Press the PROG key to display the program list screen.

- Press the OFFSET/SETTING key to display the screen used to set offsets or adjust parameter settings.

- Press the SYSTEM key to display the system screen.

- Press the MESSAGE key to display the message screen.

- Press the GRAPH key to display the graphics screen.

CURSOR

This symbol is in the form of a blinking dash on the display, which is located below the position of a particular address while in one of the Edit modes. On many controls, the cursor highlights the whole word, for example **X7.777**

Cursor Move Keys (Navigation Keys)

In order to navigate through the program, there are four keys used to move the cursor.

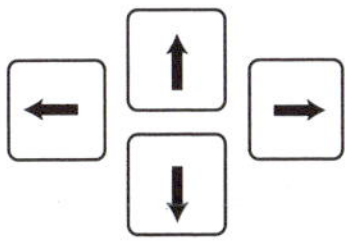

Figure 29
Cursor Move Keys

The right pointing arrow key is used to move the cursor to the right or in the forward direction. When this key is pressed, the cursor moves only one space each press of the button, in the forward direction.

The left pointing arrow key is used to move the cursor to the left or, in the reverse direction. Just as with the prior described case, when this key is pressed, the cursor moves only one space each press of the button, in the reverse direction.

The downward pointing arrow key is used to move the cursor downward through the program in the forward direction. When this key is pressed, the cursor moves one full line downward in the forward direction, each time.

The upward pointing arrow key is used to move the cursor upward through the program in the reverse direction. When this key is pressed, the cursor moves one full line upward in the reverse direction each time.

Page-UP/DOWN Keys

Usually the length of the program exceeds what the height of the screen will display. The CURSOR move keys can be used to scroll through the program. When you press and hold the CURSOR button with the down or up arrow, the cursor will move through the program line-by-line. A more effective method to move a large amount is to use the two PAGE keys. Using these keys will advance in the direction selected by the number of lines the screen can display. The last block of a given page becomes the first block of the next page. Use the PAGE button with the arrow pointing up to change pages in the opposite direction.

Figure 30
Page Keys

Example:

O0001

N̲1G50X7.777 Z7.777 S1000

N2T0100M39

N3G96S600M03

In the above example, with the CURSOR resting below N then pressing the CURSOR button three times with the right-pointing arrow causes the cursor to be located below the letter (address) G.

By pressing the CURSOR button (with the arrow pointing up) repeatedly, the prompt will move to the first word of program O, which corresponds to the upper limit of cursor movement. Another fast way to return to the program head is to press the RESET key.

By pressing the CURSOR button once, with the arrow pointing down, the cursor will move down one line. If the cursor must be moved over a few or many words, you need not press the button repeatedly. Just press and hold this button down; the cursor automatically jumps one word at a time in the given direction. The PAGE keys allow for scrolling through long programs more effectively.

OPERATIONS PERFORMED AT THE CNC CONTROL

The following explanations are for operations considered routine for operators of CNC machine tools and are given in their sequence of use.

Please note that the following procedures are specific to the type controller depicted here (Fanuc 16 or 18 series). The procedures for another type control may be similar. Be sure to consult the manufacturer manuals specific to your machine tool operation and control panel.

The Machine is Turned on and Homed (Machine Zero)

Turn on the main power switch, and then press the ON Power button on the controller. Most modern machine tools will automatically start-up in the HOME mode. This means that before any automatic or manual operation may begin, it will be required to Home the machine first.

If the Operation selection LED, HOME, is not lit, press it now.

Using the Axis/Direction keys, press the direction necessary to HOME the machine. Note that many machine tools will have LED's for each axis that are lit to indicate when an axis is HOMED.

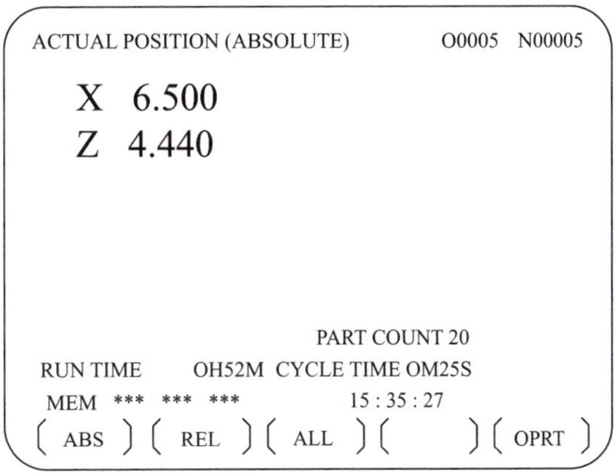

```
ACTUAL POSITION (ABSOLUTE)          O0005  N00005

  X  6.500
  Z  4.440



                        PART COUNT 20
RUN TIME      OH52M  CYCLE TIME OM25S
MEM  ***  ***  ***              15 : 35 : 27
( ABS  ) ( REL  ) ( ALL  ) (         ) ( OPRT )
```

Figure 31
Actual Position (Absolute) Screen

At machine start-up, a common screen displayed is ACTUAL POSITION (ABSOLUTE). If it is not displayed, press the function key labled POS then the soft key ABS. The displayed coordinate values represent the relationship between the Workpiece Zero and the Machine Zero (HOME). When the machine is HOME, press the soft key OPRT, then press ORIGIN and then press ALLEXE to zero each of the coordinate axes.

By pressing the soft keys, other display screens can be activated. For instance, when we press the button ABS (which corresponds to position), the digital counter appears on the screen for the X and Z axes, which is the absolute coordinate system for a given workpiece (for milling machines X, Y, and Z will be displayed). The position (POS) function is assigned four display screens and can be found by pressing the soft keys labeled; ABS, REL, ALL and OPRT. The first screen corresponds to a position change in the ABSOLUTE (ABS) system for X, Z, as illustrated. The second screen RELATIVE (REL) corresponds to position changes in the incremental system, U and W (for milling machines X, Y, and Z). The third, ALL gives representation of all four of the displays simultaneously on one screen as shown in Figure 32.

The values listed in the readout for MACHINE represent the distance from Machine Home position.

The DISTANCE TO GO readout is the most significant part of the third display. The coordinates in this quarter of the display of the screen correspond to the path that will be followed by the tool in order to complete the execution of a given block of information while under automatic operation.

Example:

N20G00Z0.

N22G01Z-12.000F.015

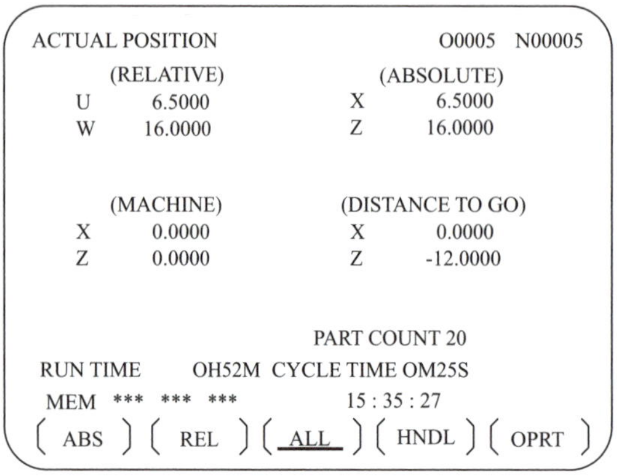

```
ACTUAL POSITION              O0005  N00005
       (RELATIVE)            (ABSOLUTE)
  U      6.5000         X      6.5000
  W     16.0000         Z     16.0000


       (MACHINE)            (DISTANCE TO GO)
  X      0.0000         X      0.0000
  Z      0.0000         Z    -12.0000


                    PART COUNT 20
RUN TIME     OH52M CYCLE TIME OM25S
MEM  ***  ***  ***          15 : 35 : 27
( ABS )  ( REL )  ( ALL )  ( HNDL )  ( OPRT )
```

Figure 32
Position, Actual Position, Screen

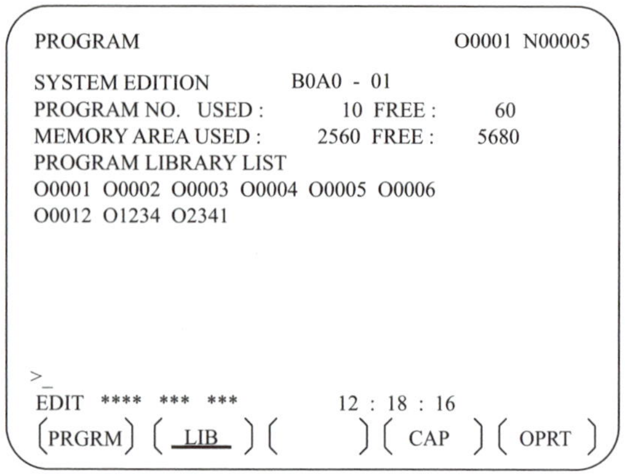

```
PROGRAM                            O0001  N00005
SYSTEM EDITION           B0A0 - 01
PROGRAM NO.  USED :       10  FREE :    60
MEMORY AREA USED :     2560  FREE :   5680
PROGRAM LIBRARY LIST
O0001 O0002 O0003 O0004 O0005 O0006
O0012 O1234 O2341


>_
EDIT  ****  ***  ***          12 : 18 : 16
(PRGRM)  ( LIB )  (        )  ( CAP )  ( OPRT )
```

Figure 33
Program Library Screen

When block N22 is first read by the control, the value Z-12.000 will appear under DISTANCE TO GO readout in the lower right corner of the screen. After moving a distance of 1 inch, the value of coordinate Z changes to Z-11.000, and so on. The other displays, "ABSOLUTE" and "RELATIVE" correspond to the first two display screens, but this time they are smaller so that all four may fit on one screen. All of the displays may be changed to read in millimeters, with respect to Machine Zero, by changing a machine parameter or by using a G-Code in the program.

A Program is Loaded From Memory

The program may be in the program directory but not activated for automatic operation. Follow these steps to activate a program.

1. Press the EDIT button to enter the EDIT mode.

2. Press the PROG function key.

3. Either the program contents or program file directory will be displayed.

4. Press the OPRT soft key.

5. Press the rightmost (continuous-menu key) soft key.

6. Use the keypad to enter the desired program number preceded by the letter address O.

7. Press the FSRH (forward search) and the EXEC soft keys.

The program will now be in the active status and ready to use for automatic operation.

An NC Program is Loaded Into Memory

Follow the steps below to load a program into the controller from an NC tape.

Be sure that the input device is ready for reading (tape entry to tape reader if used).

Part 2 CNC Machine Operation

1. Press the EDIT button on the operator's panel to enter the Edit mode.

2. Press the PROG function key.

3. Either the program contents or program file directory will be displayed.

4. Press the OPRT soft key.

5. Press the rightmost (continuous-menu key) soft key.

6. Use the keypad to enter the desired program number to load preceded by the letter address O.

7. Press the READ soft key and then the EXEC soft key.

    The program will be loaded into the controller's memory.

A Program is Saved to an Offline Location

For example; an NC Tape, Floppy disc, PCMIA card, or PC hard disk connected via RS232.

Follow the steps below to save a program to an NC tape.

Be sure that the output device is ready for output.

If the NC tape output is EIA/ISO, it needs to be specified by using a parameter.

1. Press the EDIT button on the operator's panel to enter the EDIT mode.

2. Press PROG function key.

    Either the program contents or program file directory will be displayed.

3. Press the OPRT soft key.

4. Press the rightmost (continuous-menu key) soft key.

5. Use the keypad to enter the desired program number to save preceded by the letter address O.

6. To save all programs stored in memory Press –9999.

7. To save multiple programs at one time enter their program numbers separated by a coma i.e.: O1234, O1235.

8. Press the PUNCH and then EXEC soft keys.

    The program will be saved to the offline location media.

A Program is Deleted From Memory

To delete a program for the controller memory, follow these steps:

1. Enter the EDIT mode.

2. Press the PRGRM soft key.

    The program directory will be displayed.

3. Press the OPRT soft key.

    The screen with soft keys labeled F SRH, READ, PUNCH, DELETE and OPRT will be displayed.

    The program directory is displayed only while in the EDIT mode.

4. Press the DELETE soft key.

5. Enter the program file number (preceded by the letter address O) you wish to delete.

6. Press the EXEC soft key.

The file is deleted.

To delete all programs from memory use steps one through three, and then the following steps in place of the last three steps as above:

Caution: Be sure the program files you are deleting are backed up prior to executing these steps because they will not be recoverable!

O–9999

DELETE

EXEC

MDI OPERATIONS

The operator may input small programs via the keypad at the control. The size of the program is limited to 10 lines on the control that is described in this book and is determined by the parameter setting from the manufacturer. It is an excellent method of executing simple commands like tool changes, controlling the spindle r/min and its rotation direction, etc. To enter the MDI mode of operations follow these steps:

1. Press the MDI button on the operator panel.

2. Press the PROGRAM function key.

3. Enter the desired program number preceded by the letter address O.

4. Enter the data to be executed by using the methods described later in PROGRAM EDITING FUNCTIONS.

As soon as the program number is entered, you can begin to enter program data. If a program number is not input, the control assumes O0000, and the data may be entered. Each block ends with end of block (EOB) character (;) so that individual blocks of information can be kept separately. For example: N1G50S1000;

5. Press the EOB function key to insert the semicolon at the end of each line.

6. Press the INSERT edit key.

7. Press CYCLE START to execute the program information.

If a typographical error is made while entering a given block, you can eliminate the error by pressing the CAN key to CANCEL the error and then reenter the correct value.

The MDI program may be executed just as with automatic operation and the same control functions apply except that an M30 (tape rewind) command does not return the control to the program head instead M99 is used to perform this function. Please refer to the machine tool manufacturer manual for specific instructions.

Erasure of an entire program created in MDI mode may be accomplished as follows:

Part 2 CNC Machine Operation

1. Use the alphanumeric keypad to enter the address O.

2. Press the DELETE key on the MDI panel.

The same result may be accomplished by pressing the RESET key.

Also the program will be erased when execution of the last block of the program is completed by single–block operation.

To perform an individual MDI operation, use the methods described above. For the control described here use the display screen shown in Figure 34.

Example 1:

1. Turn on the spindle at 500 RPM in the clockwise direction.

2. Key in the following command:

3. S500M03

4. EOB

5. INPUT

6. CYCLE START

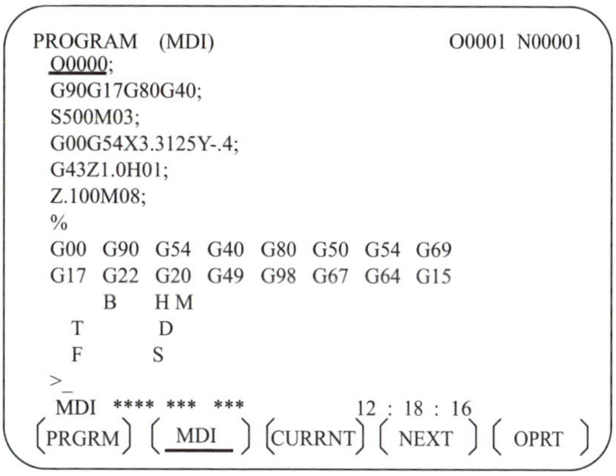

```
PROGRAM   (MDI)                    O0001 N00001
 O0000;
 G90G17G80G40;
 S500M03;
 G00G54X3.3125Y-.4;
 G43Z1.0H01;
 Z.100M08;
 %
 G00  G90  G54  G40  G80  G50  G54  G69
 G17  G22  G20  G49  G98  G67  G64  G15
      B    H M
   T        D
   F        S
 >
 MDI  ****  ***  ***               12 : 18 : 16
 (PRGRM)  ( MDI )  (CURRNT) ( NEXT ) ( OPRT )
```

Figure 34
Program Screen in MDI Mode

Example 2:

Follow these directions to position tool number 5 to the active position on the turret (or to install tool 5 into the spindle on a milling machine).

Key in the following command:

1. T0500 (or T5M06 for a mill)

2. EOB

3. INPUT

4. CYCLE START

The codes listed along the bottom of the display pictured in Figure 34 are G-Codes that are active upon start up of the machine called defaults. They are also rein-stated upon pressing of the RESET key.

MEASURING WORK OFFSETS, TURNING CENTER

It is necessary to establish a relationship between the machine coordinate system and the workpiece coordinate system. The following steps are necessary to input the measured values for the workpiece zero to the controls Work Coordinates offset page.

Measure the Z Axis Work Coordinate

1. Manually position the cutting tool and make a cut on the face of the workpiece.

2. Without moving the Z axis, stop the spindle and move the tool away from the part in X-axis direction.

3. Measure the distance along the Z axis from cut surface to the desired zero point.

4. Press the WORK soft key to display the WORK COORDINATES display screen.

5. Position the cursor to the desired Workpiece offset to be set.

6. Use the letter address key Z to select the axis to be measured

7. Use the value of the measurement taken to input the Z axis Work Coordinate

8. Press the MEASUR soft key.

   The Work Coordinate for the Z axis will be input.

Measure the X Axis Work Coordinate

1. Manually position the cutting tool and make a cut along Z axis to create a diameter on the workpiece.

2. Without moving the X axis, stop the spindle and move the tool away from the part in Z axis direction.

3. Measure the diameter you just cut on the workpiece.

4. Use the value of the measurement taken to input the X axis Work Coordinate (enter the diameter).

5. Follow the same procedure for setting the Z axis Work Coordinate value as stated in steps six and seven above.

The Work Coordinate for the X axis will be input.

MEASURING WORK COORDINATE OFFSETS, MACHINING CENTER

Following is the procedure for setting the Work Offsets for each workpiece coordinate system G54 to G59. When the values are known you can:

1. Press the OFFSET/SETTING function key.
2. Press the WORK soft key.

The WORK COORDINATES setting screen is displayed as shown in the Figure 35. There are two display screens needed to handle the six offsets G54 to G59. To display a desired page, follow either of these two methods:

1. Press the PAGE up or PAGE down keys until the desired offset is shown.
2. Enter the offset number, G54 – G59.
3. Press the NO.SRH soft key.

To change the coordinate values of the offsets, use the following method:

1. Use the arrow keys to position the CURSOR on the appropriate offset number.

```
WORK COORDINATES                    O0001  N00000

     NO.   DATA              NO.    DATA
     00    X    0.000        02     X    0.000
    (EXT)  Y    0.000       (G55)   Y    0.000
           Z    0.000               Z    0.000

     01    X    0.000        03     X    0.000
    (G54)  Y    0.000       (G56)   Y    0.000
           Z    0.000               Z    0.000

     >  _                      S  0 T0000
    MEM **** *** ***              16 : 20 : 44
    (OFFSET) (SETING) ( WORK ) (        ) ( OPRT )
```

Figure 35
Work Coordinates Display Screen

Part 2 CNC Machine Operation

2. Use the alphanumeric keypad to enter the new value for the offset.

3. Press the INPUT soft key.

> *Note: When the INPUT key is used to enter values, the amount entered will replace any amount in the register. When the +INPUT or -INPUT key is used, the existing amount in the offset register will be added or subtracted, whichever applies, by the amount entered into it.*

Once the value is entered here, it is the Workpiece Zero or origin for the workpiece coordinate system.

To change an offset by a specific amount, use the alphanumeric keypad to enter the desired value then press the +INPUT soft key.

Measured Values

Work Offsets can be measured manually by positioning an edge-finding tool to contact with the workpiece zero-surface in both X and Y axes sequentially. This procedure is called Edge Finding and is nearly always the perpendicular edges (secondary and tertiary datum) of the workpiece that is referenced.

Follow these steps for Work Coordinate Offset measuring:

1. Position the machine to HOME.

2. Use the procedure above (steps 1 – 3) to find the WORK COORDINATES setting display screen.

3. Use the arrow keys to position the cursor on the offset you wish to use.

4. Press 0 INPUT for the X value.

5. Press 0 INPUT for the Y value.

6. Install an Edge-Finding tool into the spindle using MDI or manually.

7. Start the spindle RPM clockwise at approximately 1000 by using MDI or manually.

8. Manually position the tool tip edge to contact the workpiece zero-surface along the X or Y axis.

9. Use the alphanumeric keypad and press X or Y and then INPUT, to enter the axis to be measured.

The desired axis should be blinking on the display screen and the soft key options NO. SRH and MEASUR should be present.

10. Press the MEASUR soft key. The absolute position value will be input to the offset.

11. Manually retract the Edge-Finding tool and repeat the same operation for the remaining axis. In most cases, the operator will be required to input the difference between the value input and the Edge-Finder radius (typically 0.100 or 3mm) before automatic operation can be executed.

TURNING CENTER TOOL OFFSETS

On Turning Centers, the tool offsets are measured in two directions: Z and X.

These values represent the difference between the reference position (Machine Home) of the tool turret used in programming and the actual position of a tools tip used as the programmed tool point. The amount of Tool Nose Radius is input on the OFFSET display screen where R is indicated for each tool. An incorrect value here will have an affect on the finished part where tapers and radii are turned. Refer to Part 3, Tool Nose Radius and Tip Orientation for more details.

OFFSET			O0001	N00000
NO.	X	Z	R	T
001	0.000	6.500	0.000	0
002	0.000	0.000	0.000	0
003	2.500	-6.000	0.000	0
004	0.000	0.000	0.000	0
005	0.000	0.000	0.000	0
006	0.000	0.000	0.000	0
007	0.000	0.000	0.000	0
008	0.000	0.000	0.000	0

ACTUAL POSITION (RELATIVE)
U 4.200   W 8.000
> _
MDI **** *** ***   16 : 20 : 44
(OFFSET) (SETING) (WORK) (   ) (OPRT)

Figure 36 Offset Display Screen

Measured Values

If the position register commands (G50 for turning or G92 for milling) are used, the values for each tool that have been measured will be input into the program for each tool with the G50 or G92 command.

The more commonly used method today is to input these values into the OFFSET/GEOMETRY register for each tool (See Figure 37). Follow these steps for input of the measured tool offset value.

OFFSET/GEOMETRY			O0001	N00000
NO.	X	Z	R	T
G 001	2.500	1.500	0.000	0
G 002	2.500	-6.500	0.000	0
G 003	2.500	-6.000	0.000	0
G 004	2.500	0.000	0.000	0
G 005	2.500	-6.500	0.000	0
G 006	2.500	-6.500	0.000	0
G 007	2.500	-6.500	0.000	0
G 008	2.500	-6.500	0.000	0

ACTUAL POSITION (RELATIVE)
U   0.000   W   0.000
V   0.000   H   0.000
>MZ12. _
MDI **** *** ***   16 : 20 : 44
(NO. SRH) (MEASUR) (INP. C.) (+INPUT) (INPUT)

Figure 37 Offset/Geometry Display Screen

Measure the Z Axis Offset

1. Manually position the cutting tool and make a cut on the face of the workpiece.

2. Without moving the Z axis, stop the spindle and move the tool away from the part in X axis direction.

3. Measure the distance along the Z axis from cut surface to the desired zero point.

   Use this value to input the Z axis offset for the desired tool number, with the following procedure:

4. Press the OFFSET/SETTING function key.

5. Press the OFFSET soft key until the required tool offset compensation display screen is found.

6. Use one of the search methods or use the cursor keys to move the cursor to the offset number to be set.

7. Use the alphanumeric keypad to select the letter address Z.

8. Use the alphanumeric keypad to key in value of the measurement taken.

9. Press the MEASURE soft key.

The difference between measured value and the coordinate will be input as the offset value.

Measure the X Axis Offset

1. Manually position the cutting tool and make a cut along Z axis to create a diameter on the workpiece.

2. Without moving the X axis, stop the spindle and move the tool away from the part in Z axis direction.

3. Measure the diameter just cut on the workpiece.

4. Follow the same procedure for setting the X offset value as stated above (steps 5 – 10).

Apply this method for all of the remaining tools used in the program. The offset values are automatically calculated and set.

Tool Sensor Measuring

On some newer machines, a method of tool-offset measurement exists where a tool sensor is used as opposed to machining the diameter and face of the material. In this case, all of the programmed tools are manually or automatically positioned to contact the sensor for each axis and the offset values are automatically input to the control. The operator still must manually enter Tool Nose Radius compensation values in the "R" column of the OFFSET/GEOMETRY register.

ADJUSTING WEAR-OFFSETS FOR TURNING CENTERS

Wear-Offsets are used to correct the dimensions of the workpiece that change because of cutting tool wear.

For a Turning Center, the X direction offset corresponds to the diameter, for example: if the X wear offset for a tool is .01, an incremental change of minus .01 refers to a decrease of the diameter by .01 and an incremental change of plus .01 refers to an increase of the diameter by .01.

To adjust the WEAR-offsets:

Press the OFFSET/SETTING button until the screen display shown in Figure 38 appears.

Examples of Adjusting Wear-Offsets

For the following examples, the operator should display the OFFSET screen for WEAR offsets and the cursor should be positioned to the tool and axis requiring adjustment.

Example 1: The Absolute System

If after machining the workpiece

```
OFFSET                          O0001  N00000
 NO.   GEOM (H)   WEAR (D)   GEOM (H)   WEAR (D)
 001    0.000      0.000      0.000      0.000
 002    0.000      0.000      0.000      0.000
 003    0.000      0.000      0.000      0.000
 004    0.000      0.000      0.000      0.000
 005    0.000      0.000      0.000      0.000
 006    0.000      0.000      0.000      0.000
 007    0.000      0.000      0.000      0.000
 008    0.000      0.000      0.000      0.000
ACTUAL POSIITON (RELATIVE)
     X     0.000        Y   0.000
     Z     0.000
 >  _
MEM **** *** ***              16 : 20 : 44
(OFFSET) (SETING) (      ) (      ) ( OPRT )
```

Figure 38 Offset Display Screen
for Wear-Offsets

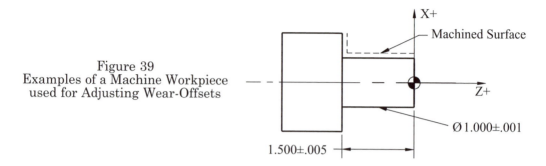

Figure 39
Examples of a Machine Workpiece
used for Adjusting Wear-Offsets

shown in Figure 39, the measured external diameter exceeds the value of tolerance (i.e. 1.003), enter the offset with a negative sign assigned to the value -.003 in the wear offset by following these steps.

1. Press X

2. Key in -.003

3. Press INPUT

Then, after machining several more pieces, the diameter increases due to tool wear. If the measured diameter is 1.002, enter the offset as follows:

1. Press X

2. Key in -.005

3. Press INPUT

Please note that it was necessary to add a value of .002 into the Offset register to the previously entered value of .003. A similar approach is applicable in the direction of the Z axis.

If the measured length is 1.492, then the value of the offset entered is -.008.

1. Press Z

2. Key in -.008

3. Press INPUT

A new measured length of 1.494 gives an entered value of the offset of -.006.

1. Press Z

2. Key in -.014

3. Press INPUT

Example 2: The Incremental System

To gain a better understanding, let us examine identical cases when the incremental coordinate system is used. The measured value is = 1.003.

Offset: U

1. Key in -.003

2. Press INPUT

Following that, the diameter is = 1.002.

1. Press U
2. Key in -.002 (on the screen)
3. Press INPUT (X-.005)

And Z = 1.492

Offset: W
    1. Key in -.008
    2. Press INPUT

After machining a few pieces, Z = 1.94.

Offset: W
    1. Press W
    2. Key in -.006 (on the screen)
    3. Press INPUT (Z-.014)

MACHINING CENTER TOOL OFFSETS

Tool Length Offsets "TLO" are called in the program by the H-word. The values input into the corresponding TLO# register are needed for proper positioning of the tool along the Z axis. Similarly, the Cutter Diameter Compensation "CDC" values are entered on the Offset display register under column "R" and are called in the program by the D-word. These compensations are important for proper radial (X, Y) positioning of the tool. If the values are known, the following sequence can be used to input them into the offset page. When the setup values are known, you may:

1. Press the OFFSET/SETTING function key.

2. Press the OFFSET soft key. It may be necessary to press this key several times until the desired offset display is present.

3. Use the arrow direction or page keys to position the cursor to the tool number to be set.

The search method may also be used by entering the tool number whose compensation is to be changed and the pressing the NO.SRH soft key.

Enter the numerical value of the offset (including sign) and press the INPUT soft key.

To add or subtract from an existing offset value key in the amount (a negative value to reduce the current value), then press the +INPUT soft key.

Diameter compensation values are input, as known, after measuring their actual size. Depending on the parameter setting for the specific machine used, the value is entered as either tool diameter or radius. Consult the appropriate manufacturer operation manual for exact conditions.

Measured Values

Tools length offsets can be measured by manually positioning the tool tip to contact with the Workpiece Zero surface (Z axis). This procedure is called "Touching-Off" and is nearly always the topmost surface, primary datum, of the workpiece. All tools used in the program must have their offsets recorded in the Offset register. If there is not a value in the offset register for a programmed tool, the control will not execute for that tool call, an alarm will occur and the machine will stop. If a value of zero is in the offset register, the control will accept the zero offset and over travel will result. Conversely, if a value in the offset register is incorrect, the control will execute the tool call as if it was correct and the result could be a collision. For this reason, it is a good idea to delete tool offset data from the offset register when the tool for which it was intended is removed. To do this, follow the directions above, in Example 1, to search to the tool and erase the tool data. Input a value of zero for the tool offset register desired.

Following is a description of the steps needed for the tool offset measuring procedure:

1. Manually position the tool tip to contact the workpiece zero-surface (Z axis).

2. Press the POSITION function key.

3. Press this key several times to get to the ACTUAL POSITION (RELATIVE) display screen.

4. Use the alphanumeric keypad and press Z and then INPUT, to enter the axis to be measured. The axis should be blinking on the display screen and the soft key options PRESET and ORIGIN should be displayed.

5. Press the ORIGIN soft key. The value in the RELATIVE position display will be changed to 0.

6. Press the OFFSET/SETTING function key several times until the offset page for tool compensation is displayed.

7. Manually position the tool tip to contact the workpiece zero surface (Z axis).

8. Use the arrow direction keys or search method described above to position the cursor to the desired offset.

9. Use the alphanumeric keypad and press Z.

10. Press the INPUT soft key.

The relative Z value for the tool offset will be input to the offset register. Repeat for each tool used in the program.

ADJUSTING WEAR-OFFSETS FOR MACHINING CENTERS

For Machining Centers, the wear offset is assigned only in the direction of the Z axis for tool length compensation. Variations in the X and Y axes are compensated for by adjusting the Cutter Diameter Compensation CDC values. The method for inputting adjustment data is similar to Adjusting Wear-Offsets for Turning Centers.

TOOL PATH VERIFICATION OF THE PROGRAM

One of the optional features of modern controllers that help the operator verify that the program is ready to use, is the graphic display of the programmed tool path as is shown in Figure 40. This visual representation of the programmed tool path provides yet another check enabling the operator to catch any errors before machining takes place.

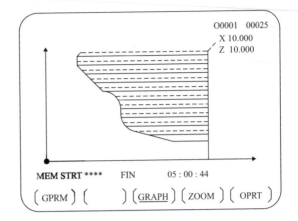

Figure 40
Tool Path Graphic Display Screen

Follow these steps to access this display screen:

1. On the controller, press the GRAPH function key.

2. Press the GPRM soft key.

The graphics parameter screen will be displayed (not shown). Use the cursor to position to each parameter and the INPUT key to insert all of the required data.

3. Press the GRAPH soft key.

Simulation of the programmed tool paths will be displayed on the screen.

The operator has the added ability to adjust the magnification, change views, and display a solid model of the workpiece on the display.

DRY RUN OF PROGRAM

Under the DRY RUN condition the tool is moved at the rapid traverse feedrate regardless of the feedrate in the program. The actual feedrate is determined by a parameter setting and also by a rotary dial or override buttons on the operation panel.

This function is used for checking the programmed movements of tools when the workpiece is not present in the work holding device. The rapid traverse rate can be adjusted by using the rapid traverse feed override or by pressing F0, 25, 50, 100% buttons on the controller (see *Rapid Traverse Override*). Consult the manufacturer manual for specific instructions.

Caution: This function is not intended for metal cutting!

Another form of DRY RUN is to execute the program cycle without a part mounted in the work holding device at the programmed feedrates.

EXECUTION IN AUTOMATIC CYCLE MODE

When all the prior steps have been completed, the program is ready to be executed under automatic cycle. A helpful and informative display screen to use during this is the Program Check. The display shown in Figure 41 is convenient because the operator can see program lines as they are called i.e.: the Absolute Position display, a Distance-To-Go display and all the active commands.

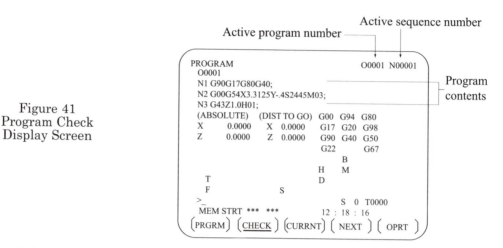

Figure 41
Program Check
Display Screen

To begin execution in the automatic cycle mode follow these steps:

1. Be sure that the desired program is in the control and active, and that all set-up procedures have been completed. Use the steps above to activate if it is not, by following the directions as stated in the section *"A Program is Loaded From Memory"*.

2. Press the RESET key on the controller.

3. Press the PRGM soft key on the controller.

4. Press the CHECK soft key on the controller.

5. Press the AUTO button on the operator panel.

6. Press the CYCLE START button on the operator panel.

The automatic cycle will begin.

DNC OPERATION

Occasionally, it will be necessary to run the program from a remote storage device (Floppy Disc, PCMCI card or a computer hard disk). This is typical for instances when the program is very large and will not fit in the control system memory. Direct Numerical Control (DNC) allows the program to be executed from the offline storage location. When using the computer hard drive method, the offline Personal Computer (PC) is required to have the necessary communications software and RS232 cabling hardware connected or have an Ethernet configuration. Please consult the manufacturer manual for specific instructions. To execute a DNC operation, follow these steps:

1. Call the program number to be executed by one of the search methods.

2. Press the RMT button on the Operator Panel to set REMOTE execution mode.

3. Note that this button is not shown on the controller we depict.

4. Press CYCLE START to begin automatic operation.

PROGRAM EDITING FUNCTIONS

Editing of part programs includes inserting, deleting and altering of program words. Understanding the techniques for program number searching, sequence number searching, word searching and address searching, is required before any editing of the program can begin (described next). The control needs to be in the proper mode and the program called.

The following steps are descriptions of how each is accomplished:

1. Press the RESET key.
2. Press the EDIT key to activate the EDIT mode.
3. Press the PROG function key.
4. If the program to be edited is not active, you must call it now. Follow the directions as stated in the section "*A Program is Loaded from Memory*".

Setting the Program to the Beginning

Method one is accomplished by pressing the RESET key while in the EDIT or MEMORY mode, the active program will be returned to the beginning line of the program (program head).

The second method is accomplished by doing the following steps:

1. Press the letter address O while in either the MEMORY or EDIT modes.
2. Using the alphanumeric keypad input the program number.
3. Press the O SRH soft key.

For the third method:

1. From the EDIT or MEMORY mode
2. Press the PROG function key.
3. Press the OPRT soft key.
4. Press the REWIND soft key.

Cursor Scanning

The program may be scanned to an editing location by using the cursor and the page keys. Follow the directions as stated in this section under arrow direction, page and cursor. If the program is very large, using the cursor scanning method is not the most efficient method to search through the program for edit locations.

```
PROGRAM                                   O0001 N00005

  N1 G90G17G80G40;
  N2 G00G54X3.3125Y-.4S2445M03;
  N3 G43Z1.0H01;
  N4 Z.1M08;
  N5 G01Z-.240F5.0;
  N6 Y1.9F15.0;
  N7 Z-.250;
  N8 Y-.4;
  N10G00Z1.0M09;
  N11X.1675Y1.9;

  >_
  MEM STOP *** ***              12 : 18 : 16
 ( PRGRM ) ( CHECK ) (CURRNT) ( NEXT ) ( OPRT )
```

Figure 42
Program Scanning for Editing

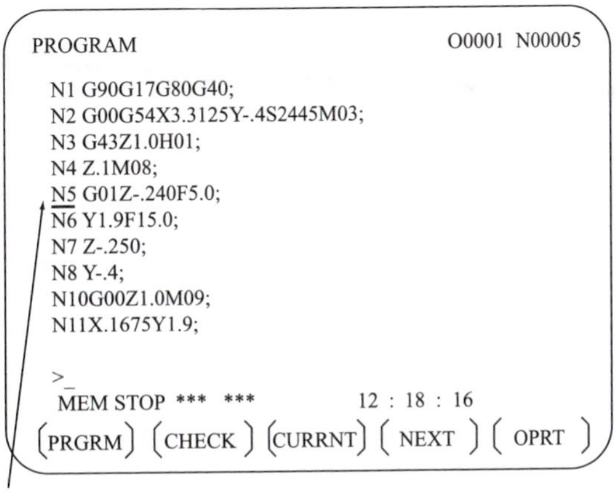

The cursor indicates the currently executed location

69

Part 2 CNC Machine Operation

Sequence Number Searching

If the sequence number in the program that requires editing is known, the operator can search directly to that program location by following these steps:

1. From the EDIT mode, use the alphanumeric keypad to input the sequence number preceded by the letter address N.

2. N____ (sequence number)

3. Press the SRH soft key forward or reverse for the direction needed.

   The cursor will be moved to the identified sequence number.

Word Searching

Much like sequence number searching, the operator can search to a specific word in the program. For instance, to search to a specific word in the program, like T5, follow these steps:

1. From the EDIT mode using the alphanumeric keypad key in the letter address T.

2. Press the number 5.

3. Press the SRH soft key forward or reverse for the direction needed.

   The cursor will be moved to the identified word T5.

Address Searching

Like word or sequence number searching, the operator can search to a specific address in the program. For instance, to search to a specific address in the program, like M06, follow these directions:

1. From the EDIT mode using the alphanumeric keypad key in the letter address M.

2. Press the SRH soft key forward or reverse for the direction needed.

   The cursor will be moved to the identified address of word M06 or the first instance of the M-address that is found.

Inserting a Program Word

From the EDIT mode, use a searching method to scan the program to the word immediately before the word to be inserted. Follow these steps to Insert a program word.

1. Use the alphanumeric keypad to key the address and the data to be inserted.

2. Press the INSERT Edit key

3. The new data are inserted.

   Example: To insert the program word Z.2 on sequence number N4 of the program listed below:

1. Press the EDIT key.

2. Press the PRGRM soft key.

3. Key in the word, X1.2.

4. Press the SRH soft key in the forward direction.

5. Key in the new word, to insert, Z.2.

6. Press INSERT

> **O1234**
>
> **N1G50S1000**
>
> **N2T0100**
>
> **N3G96S600M03**
>
> **N4G00<u>X1.2</u>**

The result will be as follows: **N4G00X1.2Z.2**

Altering Program Words

From the EDIT mode, use a searching method to scan the program to the word to be altered.

1. Use the alphanumeric keypad to key the new address and new data to be inserted.

2. Press the ALTER Edit key

3. The new data are changed.

4. To change the program word, Z.2, in the example to Z.3, follow these steps.

5. Press the EDIT key.

6. Press the PRGRM soft key.

7. Key in the word, Z.2.

8. Press the SRH soft key in the forward direction.

9. Key in the new word, Z.3.

10 Press ALTER.

> The result will be as follows: **N4G00X1.2Z.3**

Deleting a Program Word

From the EDIT mode use a searching method to scan the program to the word that needs to be deleted. Then press the DELETE Edit key.

To delete the program word, Z.3 from the example, follow these steps.

1. Press the EDIT key.

2. Press the PRGRM soft key.

3. Key in the word Z.3.

4. Press the SRH soft key in the forward direction.

5. Press DELETE.

SETTING

The SETTING soft key is accessible by first pressing the OFFSET SETTING function key on the control panel. The setting soft key is the second key from the left. The PAGE key may be used to view multiple display screens. By pressing this key, access is gained to the Parameter setting, sequence number comparison setting, run time and parts count display, etc. By pressing the soft key, this display page allows the operator to adjust settings to enable or disable parameter writing called Parameter Write Enable (PWE), set automatic insertion of program sequence numbers, change from inch to metric units, and set any mirror image data.

From the MDI mode:

1. Press the SETING soft key. The SETTING display will appear on the screen.

2. Set values as necessary for the desired results.

Note that some of the information about the basic parameters of the machine is shown on the screen. To change any of these parameters, perform the following steps:

The mode is set to MDI.

1. Press the function key OFFSET SETTING.

2. Press the soft key SETTING.

3. Use the PAGE keys to display the desired screen.

Use the arrow keys to position the CURSOR under the parameter that you wish to change.

1. Enter the desired new value.

2. Zero (0) or 1 is entered, where 1 indicates the ON condition and 0 the OFF condition.

3. Then press INPUT.

PARAM

This soft key is used to access the display of a set of codes that control certain constants assumed during programming and operation. These codes are numerical values that usually exclude a decimal point. Also, they are sometimes hexadecimal numbers representing an ON/OFF condition for each place in the number having multiple functions. Access to most parameters is not allowed without unprotecting their access. This is done through SETTING, as described above. Consult the manufacturer manual for specific directions to unprotect parameters.

An example of a parameter setting is the distance a drill will retract during the chip breaking process that is assigned to canned cycle G83.

Caution: Parameters should only be changed when the results of such change are understood completely. The changes will affect all programs that are executed!

To access the Parameters for adjustments, follow these steps:

1. Enter the MDI mode.

2. Press the SYSTEM button.

3. Press the soft key PARAM.

4. Move the cursor to the desired parameter screen by using the PAGE keys.

5. Move the cursor to the desired position to change by using the arrow keys.

6. Enter the new value of the desired change.

7. Press the soft key INPUT.

DGNOS

By pressing the DGNOS soft key the diagnostics screen is displayed. It defines a set of coded digits, which allow a quick determination of the cause of any machine damage or required maintenance. Maintenance personnel use this display screen to obtain needed information.

TAPE CODE

The tape code is usually made of 1-in-wide paper that includes eight channels with several combinations of punched holes. Part programs written on process sheets are punched EIA (Electronic Industries Association) or ISO (International Organization of Standardization) codes. On older machines that use this method of program file transfer, the operator must be sure the control is switched to the same code as used on the tape. With the EIA tape coding system an odd number of holes is punched, whereas with the ISO system an even number of holes is punched.

COMMON OPERATION PROCEDURES

In this book, we want to include explanations concerning the situations that may arise during actual machining. We will concentrate on the procedures that should be followed when repetition of particular parts of the program for a specific tool is required. We will also review cases when there is a need to use an EMERGENCY STOP button and recovery from this condition. The following program is used for the case studies:

CNC Turning Center Program
   O1234

     N1G50S2000

     N2T0100

     N3G96S400M03

     N4G00X1.25Z.2T0101M08

     N5 ...   .

     N6 ...   .

     ... ...

     N17M01

     N18G50S1000

     N19T0200

     N20G96S200M03

     N21G00X.75Z.1T0202M08

     N22...   ...   .

     ... ...

     N39M01

     N40G50S2500

     N41T0300

N42G96S600M03

N43G00X2.2Z.05T0303M08

N44...   **...**   **...**   **...**   **...**

......   **...**   **...**   **...**   **...**

N45M30

Using the above program example, let us review a procedure which should be followed if you need to repeat a part of the program for tools T01, T02, and T03.

Turning Center Case No. 1:

Problem:

Execution of the program was interrupted in block N17 and you need to repeat from the beginning all operations performed by tool T01.

Solution:

From the AUTO, MEMORY or EDIT mode:

1. Press the RESET button.

This will cause a cancellation of CNC commands under control and return to the program to the beginning.

2. From the AUTO or MEMORY mode, Press CYCLE START.

Turning Center Case No. 2
Problem:

Execution of the program was interrupted in block N39, and you need to execute a program for tool T02.

Solution:

From the EDIT mode:

1. Press the RESET button.

2. Using alphanumeric keypad on the control panel and the search methods described earlier, search to block N18.

3. From the AUTO or MEMORY mode, Press CYCLE START.

Note: In both cases, if you intend to execute the program to the end without interruption, the OPTIONAL STOP button must be turned OFF. However, if you intend to execute only part of the program corresponding to work of tool T01 or T02, the procedure is as follows:

1. Press the OPTIONAL STOP button to the ON condition.

After work is completed by the desired tool,

2. Press the RESET key while in the AUTO, MEMORY or EDIT mode.

The machine is ready for automatic cycle once again from the program head. The program will stop after reading an M01 code.

Turning Center Case No. 3
Problem:

Execution of the whole program is completed but you need to repeat operations performed by tool T03.

Solution:

From the EDIT mode:

1. Press the RESET button (if the program is completed and at its head the RESET is not required).

2. Using alphanumeric keypad on the control panel and the search methods described earlier, search to block N40.

3. From the AUTO or MEMORY mode, Press CYCLE START.

Turning Center Case No. 4
Problem:

Execution of the program is interrupted by the use of the EMERGENCY STOP button. In this case, follow the same procedure as mentioned above. However, you must HOME the machine to reset the machine coordinate system with respect to the X and Z axes.

CNC Machining Center Program

O2345
N1G40G80G90
N2G54G00X0.Y1.5S1520M03
N3G43Z1.0M08H01
N4 ...   .
... ...
N29G91G28Z0.
N30M01
N31T02
N32M06
N33G90G54G00X.5Y1.3S1500M03
N34G43Z1.0M08H02
N35G81G98Z-.47F6.0R.1
... ...
... ...
N38G91G28Z0.
N39M01
N40T03

N41M06

N42G90G54G00X-4.125Y0.S2000M03

N43G43Z1.M08H03

N44...   ...   .

.. . .

N55G91G28Z0.

N56T01

N57M06

N58G28X0.Y0.

N59M30

Machining Center Case No. 1
Problem:

Execution of the program was interrupted in block N30, and you need to repeat operations performed by tool T01.

Solution:

From the AUTO, MEMORY or EDIT mode:

1. Press the RESET button

2. From the AUTO or MEMORY mode, Press CYCLE START.

Machining Center Case No. 2
Problem:

During the work of tool T02, the tool was damaged, and you should change the tool and repeat all operations performed by this tool.

Solution:

1. Press the CYCLE STOP (Feed Hold) button.

2. Press RESET to stop spindle rotations and coolant flow.

3. Change to one of the OPERATION MODES, JOG or MPG.

4. Move the axes to a clearance point from the part.

5. Press the HOME button.

6. Use the axis jog direction keys to HOME the axes.

7. Change the tool and clear the wear offset tool 2.

8. Press the EDIT key.

9. Using alphanumeric keypad on the control panel and the search methods described earlier, search to block N33

10. From the AUTO or MEMORY mode, Press CYCLE START.

> *Note: In both cases, in order to repeat the work of the remaining tools, the OPTIONAL STOP button should be in the OFF condition.*

If you need to repeat the work of only one tool, follow the steps listed below:

OPTIONAL STOP ON

After machining is completed, for tool T01:

1. Press EDIT.

2. Press the RESET.

3. Zero the machine with respect to X, Y, and Z axes.

4. From the AUTO or MEMORY mode, Press CYCLE START.

After machining is completed, for tool T02 or T03:

1. Press EDIT.

2. Press the RESET.

3. Zero the machine with respect to X, Y, and Z axes.

4. Using alphanumeric keypad on the control panel and the search methods described earlier, search to block N31 for T02 or N40 for T03.

5. From the AUTO or MEMORY mode, Press CYCLE START.

Machining Center Case No. 3
Problem:

Execution of the whole program is completed, but you need to repeat the operations performed by tool T03.

Solution:

1. Press EDIT.

2. Using alphanumeric keypad on the control panel and the search methods described earlier, search to block N40.

3. From the AUTO or MEMORY mode, Press CYCLE START.

Note: Every time that you have used the EMERGENCY STOP button, make sure to HOME the machine axes. Then follow the procedures listed above.

In this section of the book we have covered CNC Machine Operation. Please note that there are hundreds of situations possible and there is not enough space to cover everything here. The intent of this section was to give a basic understanding of the Operation function for Turning and Machining Centers. For complete details on operation features specific to your machine, consult the manufacturer manuals.

Part 2 Study Questions

1. The counterclockwise direction of rotation is always a negative axis movement when referring to the handle (pulse generator).   T or F

2. Which display includes the programmed Distance-To-Go readouts?

3. When the machine is ON and the program check screen is displayed, there is a list group of G-Codes displayed. What does this indicate?

4. Describe the difference between the function of the Input and the +Input soft keys.

5. Which button is used to activate automatic operation of a CNC program?
      a. Emergency Stop
      b. Cycle Stop
      c. Cycle Start
      d. Auto

6. Which display screen lists the CNC program?
      a. Position page
      b. Offset page
      c. Program check
      d. Program page

7. When the machine is turned on for the first time, it must be sent to its home position.   T or F

8. Which operation selection button allows for the execution of a single CNC command?
      a. Dry run
      b. Single block
      c. Block delete
      d. Optional stop

9. Which mode switch/button enables the operator to make changes to the program?
      a. Edit   c. Auto
      b. MDI   d. Jog

10. What does the acronym MDI stand for?

11. Which display screen is used to enter tool information?

12. If the Reset button is pressed during automatic operation, spindle rotations, feed and coolant will stop.   T or F

13. During setup, the mode switch used to allow for manual movement of the machine axes is:

> a. Auto
>
> b. MDI
>
> c. Edit
>
> d. Jog

PART 3

PROGRAMMING
CNC TURNING CENTERS

Chart 1 Preparatory Functions (G-Codes) Specific to CNC Turning Centers

Code Group Function

Code	Group	Function
*G00	01	Rapid Traverse Positioning
G01	01	Linear Interpolation
G02	01	Circular and Helical Interpolation CW (clockwise)
G03	01	Circular and Helical Interpolation CCW (counterclockwise)
G04	00	Dwell
G09	00	Exact Stop
G10	00	Data Setting
G20	06	Input in Inches
G21	06	Input in Millimeters
*G22	09	Stored Stroke Limit ON
G23	09	Stored Stroke Limit OFF
G25	08	Spindle Speed Fluctuation Detection ON
G26	08	Spindle Speed Fluctuation Detection OFF
G27	00	Reference Point Return Check
G28	00	Reference Point Return
G29	00	Return From Reference Point
G30	00	Return to Second, Third, and Fourth Reference Point
G32	01	Thread Cutting
*G40	07	Tool Nose Radius Compensation Cancel
G41	07	Tool Nose Radius Compensation, Left Side
G42	07	Tool Nose Radius Compensation, Right Side
G50	00	Coordinate System Setting/ Maximum Spindle Speed Setting
G52	00	Local Coordinate System Setting
G53	00	Machine Coordinate System Setting
G54	14	Work Coordinate System One Selection
G68	04	Mirror Image for Double Turrets ON
*G69	04	Mirror Image for Double Turrets OFF
G70	00	Finishing Cycle
G71	00	Stock Removal in Turning
G72	00	Stock Removal in Facing
G73	00	Pattern Repeating
G74	00	Peck Drilling Cycle
G75	00	Groove Cutting Cycle
G76	00	Multiple Thread Cutting Cycle
*G80	10	Canned Drilling Cycle Cancellation
G83	10	Face Drilling Cycle
G84	10	Face Tapping Cycle
G86	10	Face Boring Cycle
G90	01	Outer/Inner Diameter Turning Cycle
G92	01	Thread Cutting Cycle
G94	01	Face Cutting Cycle
G96	02	Constant Surface Speed Control
*G97	02	Constant Surface Speed Control Cancellation
G98	05	Feed per Minute
*G99	05	Feed per Revolution

More detailed descriptions and application examples are given later in this section under "Preparatory Functions for CNC Turning Centers (G-Codes)".

Notes:

1. In the table, G-Codes marked with an asterisk (*) are active upon startup of the machine.
2. At machine startup or after pressing reset, the inch (G20) or metric (G21) measuring system last active remains in effect.
3. G-Codes of group 00 represent "one shot" G-Codes, and they are effective only to the designated blocks.
4. Modal G-Codes remain in effect until they are replaced by another command from the same group.
5. If modal G-Codes from the same group are specified in the same block, the last one listed is in effect.
6. Modal G-Codes of different groups can be specified in the same block.
7. If a G-Code from group 01 is specified within a canned drilling cycle block, the cycle will be cancelled just as if a G80 canned cycle cancellation were called.

OBJECTIVES:

1. Learn G-Codes associated with Turning Center Programming.

2. Learn M-Codes associated with Turning Center Programming.

3. Apply the proper use of feeds and speeds within Turning Center programs.

4. Learn how to properly use coordinate systems for programming the Turning Center.

5. Learn program structure.

6. Learn how to use Multiple Repetitive Cycles.

7. Learn the mathematical method for offsetting tool path to compensate for the Tool Nose Radius.

8. Learn how to use Tool Nose Radius Compensation (G40, G41 and G42)

9. Examine several practical examples of Turning Center programs.

In this section, we will focus our attention on programming of the CNC Turning Center. The programming examples given here will be limited to two axis lathes. These machines are the base configuration for CNC Turning and a solid foundation is laid for further learning about more advanced machinery by gaining full understanding of the techniques presented here. We begin by introducing the program codes in the language commonly called "G-Code".

PREPARATORY FUNCTIONS (G-CODES)

Preparatory functions are programmed with the letter address G, normally followed by two digits, to establish the mode of operation in which the tool moves. In the preceding chart and throughout this text, the G-Codes listed and explained refer to the most commonly used Fanuc system, Type A, and as used on the 16T control. There are some variations in the use of the other two control types, B and C, but most of the codes are identical. Consult the programming manual for the specific control prior to selection of the type when system programming.

MISCELLANEOUS FUNCTIONS (M-CODES)

Miscellaneous functions are used to command various operations. Two commonly used M-Codes are M08 for activating coolant flow and M03 for starting spindle rotation in the clockwise direction. The code consists of the letter M typically followed by two digits. Normally, one block will contain only one M-Code function; however, up to three M-Codes may be in a block depending upon parameter settings. Most of the common M-Codes are listed in the following chart Miscellaneous Functions, M-Codes for CNC Turning; however, many machine tool builders assign others for specific purposes relative to their equipment. Always consult the manufacturer manuals specific to the machine in use for pertinent M-Codes.

Chart 2 Miscellaneous Functions, M-Codes for CNC Turning

M-Code	Function
M00	Program Stop
M01	Optional Stop
M02	Program End Without Rewind
M03	Spindle ON Clockwise (CW) Rotation
M04	Spindle ON Counterclockwise (CCW) Rotation
M05	Spindle OFF Rotation Stop
M08	Flood Coolant ON
M09	Coolant OFF
M10	Chuck Close
M11	Chuck Open
M12	Tailstock Quill Advance
M13	Tailstock Quill Retract
M17	Rotation of Tool Turret Forward
M18	Rotation of Tool Turret Backward
M21	Tailstock Direction Forward
M22	Tailstock Direction Backward
M23	Threading Finishing with Chamfering
M24	Threading Finishing with Right-Angle
M30	Program End With Rewind
M41	Spindle LOW Gear Range Command
M42	Spindle HIGH Gear Range Command
M71	Bar Feed ON – Start
M72	Bar Feed OFF – Stop
M73	Parts Catcher Advance
M74	Parts Catcher Retract
M98	Subroutine Call
M99	Return to Main Program From Subroutine

The following descriptions are given for many of the M-Codes seen in the chart. Also observe their use within each of the example programs. Please consult the appropriate operator and programming manual for the machine you are working on for specific information on M-Codes for your application.

PROGRAM STOP (M00)

When this code is encountered spindle r/min, feeds, and coolant flow stop. This function interrupts the automatic work cycle in order to allow the following:

1. In-process inspection and gauging.

2. Visual inspection of tool wear and other components.

3. Removal of chips.

4. Interruption of the cycle, in order to relocate the workpiece when the workpiece is being machined from both sides during one operation.

The control system identification of function M00 is accompanied by the following events:

1. Spindle revolutions are stopped.

2. Flood coolant flow is deactivated.

3. All feed movement is stopped.

4. The CYCLE START light will still be on.

After use of this code, the program block following must restate M03 or M04 and M08 to reactivate these functions. Functions G96, G97, S, and F are NOT canceled by M00.

Optional Program Stop (M01)

This function is the nearly the same as M00, with one significant difference: it is applied only by pressing or switching the OPTIONAL STOP button or toggle switch ON, for example, to stop the machine so that measurements can be taken, or to remove chips at the discretion of the machine operator. If the OPTIONAL STOP button is not active, the machine will ignore this command even if it is present in the program.

Program End (M02)

Function M02 cancels the automatic work cycle, interrupts revolutions, feeds the coolant flow and cancels the control system of the NC. The CYCLE START light goes off in some types of controls and the PROGRAM END light comes on. The cycle is not repeated; repetition of a programmed operation is not possible. For this to occur, the M30 function is used. This command is primarily used on NC tape machines.

Spindle On (Clockwise) (M03)

This function signals the machine to activate the spindle with clockwise revolutions. M03 is cancelled by M04, M05, M00, M01, M02, and M30.

Spindle On (Counterclockwise) (M04)

To activate or cancel this function, follow instructions as for M03.

Spindle Off (M05)

This function is cancelled by M03 and M04.

Flood Coolant On (M08)

M08 activates the flood coolant flow.

Coolant Off (M09)

M09 deactivates the coolant flow.

Chuck Close and Open (M10 and M11)

M10 automatically closes the spindle chuck jaws and M11 automatically opens the spindle chuck jaws. M10 and M11 are used in certain cases when there is a bar feeder, special part puller or gripper used for the insertion and removal of the workpiece from the spindle chuck. Such devices are used in automatic operations and where mass production is the primary focus.

Tailstock Quill Advance and Retract (M12 and M13)

During the process of turning long shafts, a tailstock is often used to support the shaft. A center drilling operation must be programmed first. Then, by applying functions M12 and M13, the workpiece can be supported with a center in the tailstock and the turning of the workpiece can be performed. When the operation is completed, the tailstock is returned to its original position by the M13 command. All operations are performed automatically, without interruption. If the time required for the extension of the tail spindle is noticeably long after function M12, apply a dwell function G04.

Rotation of the Tool Turret Forward and Reverse (M17 and M18)

Miscellaneous function M17 rotates the tool turret in the normal (clockwise) direction; M18 rotates the tool turret in the opposite direction (counterclockwise). These functions may only be used for some types of machines. Function M17 is valid at machine start-up. Function M18 implies change in direction of the rotation opposite to the one set previously. These commands may be useful when special tooling is mounted in the tool turret or chuck where clearance issues are considered during turret rotation.

Tailstock Direction Forward and Reverse (M21 and M22)

Programmable shifting of the tailstock as a whole is also done in some types of machines, especially if the extended length of the tailstock spindle is not sufficient to perform the operation and/or the lathe is large and has a long Z axis stroke. This is a factory option only.

Thread Finishing (M23 and M24)

Miscellaneous function M23 should be applied to the G92 Thread Cutting cycle when the threading tool usually exits at a 45° angle where M24 is necessary, when the ending thread is followed by a greater diameter or a recess groove. M23 is the default state after machine power is turned on. These functions are used in conjunction with the G92 Tread Cutting Cycle for either a 90° or a 45° retract at a thread end respectively.

End of Program (M30)

This function is similar to M02 however, it also activates the rewinding of the punched tape if the program is run from punched tape (NC), or returns to the beginning of the program, if CNC is used.

Tool Function

Tool pockets, in turrets on CNC Turning Centers, are assigned coded numbers. The coding system is fixed and cannot be erased even if the machine is turned off. In the process of programming, tool function is commanded by the four digits that follow the letter address T. The first two digits are related to the tool number and its corresponding geometry offset. The remaining two digits signify the tools wear offset number, as illustrated below, for example T0101:

Tool pocket number	Tool offset number
Geometry Offset	Wear Offset
T01	01

Tool numbers may vary from 1 to the maximum number of pockets in the tool turret; for example: from 1 to 8, etc. If there are two tool turrets, then the numbers assigned to turret No. 2 are sequenced consecutively from the No. 1 turret. For example: if turret No.1 has 8 tools, then the first pocket in turret No. 2 would be 9. The number of the tool and wear offsets available may vary, depending on the specific control, numbers 1 through 32 are common. This means that each tool number may be assigned 1 out of the 32 offset numbers. If tool offset number 01 is assigned to tool 01, this rules out the possibility of assigning the same tool offset number to another tool. Conversely, the same is true for wear offsets.

Example:

Recommended	NOT recommended
T0101	T0501
T0202	T0602

Usually, tool offset 01 is assigned to tool number 01. This convention simplifies operating procedures.

Practical Application of Tool Wear Offset

Tool wear compensation is a procedure aimed at the correction of dimensional variations along the X or Z axes caused by tool wear or deflection.

Please note that in the above and all remaining figures where a tool is depicted in relationship to the workpiece, it will appear above the part, as is common in practice on rear turret, slant-bed style CNC Turning Centers. In reality, the tool will be a right hand style and the insert will be facing downward. The graphical representation here shows the insert facing upward for clarity.

Keep in mind, tool geometry offset numbers refer to certain values of X and Z, and, their relationship to the Workpiece Zero and Machine Home. These geometry offsets should not be used to make adjustments for wear. Wear offsets, on the other hand, are used solely for this purpose. Small, correctional values are input to a particular tool wear offset number for X or Z to compensate for any variations. In Figure 1, the dashed line refers to the cutting tool path, with the X and Z values for tool wear offset as follows:

$$X = 0; Z = 0$$

The solid line refers to the cutting tools path, with X and Z values for tool wear

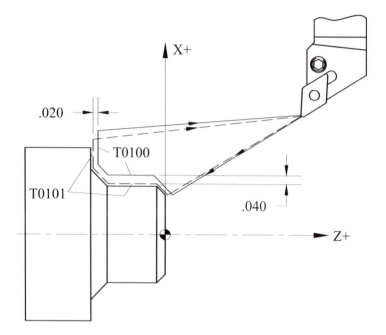

Offset is 0.040 on X axis and 0.020 on Z axis

Figure 1 Tool Wear Compensation by Offset

offset as follows:

$$X = -0.080; \ Z = -0.020$$

Notice the value of the tool offset is twice that of the dimension shown in the preceding figure, because X axis dimensions in CNC programming refer to particular diameters of the workpiece. In most cases, tool function, along with tool offset, appears in the initial phase of the program for each individual tool. The figure demonstrates that if tool function T0101 is used, the programmed tool path is changed by a displacement equal to the value of input to the wear offset.

Command T0100 initiates the cancellation of the tool wear offset. If the T0100 command is omitted, canceling the tool offset, the tool will return to the starting point with a displacement equivalent to the value of the tool wear offset. In this case, the displacement is equal to $X = -0.080$ and $Z = -0.020$ and is indicated on the drawing by the solid line. Such an action will cause the tool path of each subsequent tool to be altered by the offset amount equal to the above-mentioned value. This type of programming error increases the chances for over-cutting of workpieces. Changing the tool wear offset number does not require cancellation of the previous tool offset number.

FEED FUNCTION

Feed function determines the amount of feed rate of the cutting tools in the machining process. Feed is programmed using the letter address "F", followed by up to four digits in the metric system and five digits in the inch system. These digits represent certain values of feed. The following examples are two methods of designating feed rate:

1. Feed rate in inches per revolutions, in/rev (formerly know as IPR) or millimeters per revolution, mm/rev of the spindle (G99). In this case, in order to obtain feed with a certain assigned value of speed, the spindle as well as workpiece must be rotating.

Examples:

F1.1205	in/rev.
F0.05	in/rev.
F0.001	in/rev.

When the spindle speed is changed after the constant surface speed per revolution (G96) has been called, the feed rate will change for a certain period of time. Therefore, feed is directly coupled with spindle speed.

The following notes apply to the feed function, G99 ... F.

a. The values entered into the program for feed remain active until replaced by another feed rate, or cancelled by the G00 rapid traverse call.

b. The input value of speed is equivalent to the actual speed if the feed rate override on the Operation Panel is set to 100%. See Part 2, CNC Machine Operation, for a detailed description on Feed Rate Override.

2. Feed rate per time is measured (programmed), in inches per minute, in/min (formerly know as IPM) or millimeters per minute, mm/min (G98). If feed function in the program contains feed rate per time period, then any change in the spindle speed has no effect on the feed rate because the feed rate and the spindle speed are not coupled.

Examples:

F121.15	in/min
F1.05	in/min
F0.5	in/min

Generally, all CNC lathes are set to a default of feed per revolution of the spindle at machine start-up. In order to establish feed rate per minute (in/min or mm/min), the G98 function must be used, which remains effective until cancelled by function G99, or until the machine is turned off.

The following notes apply to feed function, G98 ... F:

a. The values entered into the program for feed remain active until replaced by another feed rate, or cancelled by the G00 rapid traverse call.

b. The input value of speed is equivalent to the actual speed if the feed rate override on the control panel is set to 100%. See Part 2, CNC Machine Operation, for a detailed description on Feed Rate Override.

c. Feed functions containing feed in inches per minute, are not applicable to threading cycles.

Following are some examples of feed functions:

89

F0.02	in/rev.
F0.004	in/rev.
F0.035	in/rev.
G98	
F2.0	in/min
F0.5	in/min
G99	
F0.012	mm/rev.
F0.008	mm/rev.

SPINDLE FUNCTION

Spindle function is commanded by the letter address S, followed by a number, up to four-digits, as shown below. The spindle rotation direction clockwise (M03) or counterclockwise (M04) is typically in the same block.

S50, S150, S3000

Some machines are assigned two ranges of speeds, low or high, others may have three or more ranges. Depending on the given value of the rotational speed, the machine is adjusted to the appropriate range, by using the following commands:

In practice, most manufacturers overlap the low and high range, for example:

30-1200 (rev/min) low range (M41)

80-4000 (rev/min) high range   (M42)

For Turning Centers, there are two functions applicable to the control of spindle speed. These are:

G96   Constant surface speed control

G97   Constant surface speed control cancellation (sometimes referred to as constant spindle speed).

Both functions appear together with function S, for example:

G97S500 = 500 revolutions per minute (r/min)

G96S400 = 400 Surface Speed (V) in ft/min (or m/min)

Constant spindle speed (G97) is applied in the case of threading cycles and in the machining of a workpiece with the diameter remaining constant. It is also used for all operations on the centerline, like drilling, etc. If the situation calls for several changes of spindle speed in a given program, new values for the S function are assigned.

Example:

G97S1000 is active for diameter one and sets the constant spindle speed.

S800 changes the r/min for diameter two.

S300 changes the r/min for diameter three.

CONSTANT SURFACE SPEED CONTROL (G96)

To further examine this concept, study the following diagram of peripheral speed distribution that appears during a facing cut, using function G97 S1000. The formulas below calculate peripheral surface speed for each diameter of 1.00, 2.00 and 3.00 inches. Because of a decrease in surface speed as the diameter gets smaller, the result is not desirable.

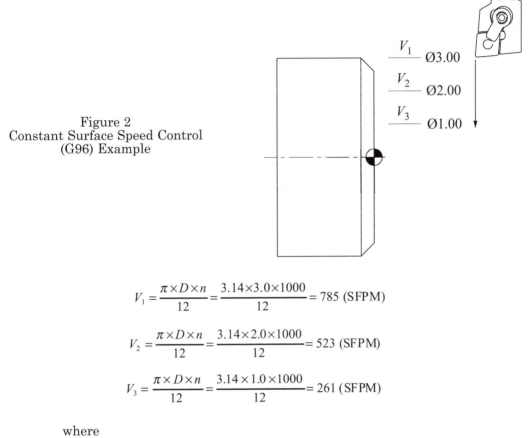

Figure 2
Constant Surface Speed Control
(G96) Example

$$V_1 = \frac{\pi \times D \times n}{12} = \frac{3.14 \times 3.0 \times 1000}{12} = 785 \text{ (SFPM)}$$

$$V_2 = \frac{\pi \times D \times n}{12} = \frac{3.14 \times 2.0 \times 1000}{12} = 523 \text{ (SFPM)}$$

$$V_3 = \frac{\pi \times D \times n}{12} = \frac{3.14 \times 1.0 \times 1000}{12} = 261 \text{ (SFPM)}$$

where

n = RPM or r/min

D = diameter

$\pi = 3.14$

V = Cutting speed (SFPM)

As the diagram shows, surface speed decreases if G97 is used, as a diameter decreases and reaches zero at the centerline of the part. The question is, however, whether this phenomenon is of any advantage to us in the process of facing. Before this question is answered, look at the advice of cutting tool manufacturers who recommend specific cutting speeds for different types of machined materials. In the case of function G97, such a condition will be fulfilled only with respect to one diameter. As mentioned previously, Constant Surface Speed control (G96) is one of the factors that can be included in programs for CNC Turning Centers.

If the surface speed must remain constant, then the spindle speed has to increase with a decreasing diameter. Spindle speed for each consecutive diameter is calculated by the control, according to the formula $n = (12 \text{ x} V)/(\pi \cdot D)$. Thus, a closer look at this formula leads to the conclusion that, theoretically, as the diameter decreases to zero, the spindle speed increases to infinity. In reality, the spindle speed range is limited by the maximum r/min (RPM) capacity of the machine.

In practical terms then, function G96 is very useful during facing and also in all cutting which involves a change in the diameter of the workpiece. As the diameter of the machined workpiece decreases, spindle speed increases, and, conversely, as the diameter increases, the spindle speed decreases.

Maximum Spindle Speed Setting (G50)

G50S ...

When the letter address S, is given within a program block and it is preceded with function G50, it refers to the maximum spindle speed setting that can be applied in the current operation for a given tool. As mentioned above, the spindle speed is calculated according to technological metal-cutting conditions for a given tool, or for any particular material. In some cases, the workpiece holding arrangement may require special equipment, which is mounted onto the conventional holding equipment. Such workpiece holding equipment creates conditions that do not permit utilization of the full range of spindle speeds, especially, maximum spindle speed for a given machine. Because of this fact, a maximum spindle speed for a particular operation can be assigned by using the function G50. This means that, if in the machining process, metal-cutting conditions arise that require a higher spindle speed, an increase of the spindle speed will not take place. This is called "clamping the spindle speed" at a safe maximum r/min.

Note: If G50 is not used in conjunction with the spindle speed command (such as G50S1250), when G96 is commanded, the machine will increase the r/min as the diameter decreases, up to the maximum that the machine is capable of. This condition could result in an excessive r/min and damage could occur.

Coordinate systems for Programming of CNC Turning Centers

Programming in the Absolute Coordinate System

When programming in an absolute coordinate system, input coordinates of programmed points always refer to a fixed zero coordinate point. The actual coordinates of traverse for the tool tip from point 0 to 10 are shown on the following drawing, Figure 3:

In order to simplify programming as well as program readout, values of X are equivalent to each particular diameter of the workpiece.

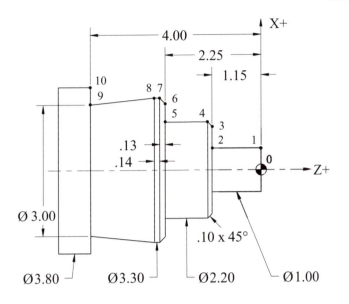

Figure 3
Coordinate Systems for Turning Centers Example

0.	X0.0Z0.0
1.	X1.0Z0.0
2.	X1.0Z-1.15
3.	X2.0Z-1.15
4.	X2.2 Z-1.25
5.	X2.2Z-2.25
6.	X3.04Z-2.25
7.	X3.3Z-2.38
8.	X3.3Z-2.52
9.	X3.0Z-4.0
10.	X3.8Z-4.0

PROGRAMMING IN THE INCREMENTAL COORDINATE SYSTEM

When using an incremental system, the path of the tool from one position to the next is given in the direction of each axis. Using letter address U and W, the point displacements may be input in the direction of the X and Z axes respectively. The sign in front of the value determines the direction of movement. Values of U refer to motion in the X axis direction and refer to changes in the diameter of the workpiece. This next example illustrates the tool movement as programmed using the incremental coordinate system (refer to the same drawing, Figure 3).

0.	U0.0W0.0
1.	U1.0W0.0
2.	U0.0W-1.15
3.	U1.0W0.0
4.	U0.2W-0.1
5.	U0.0W-1.0
6.	U0.84W0.0
7.	U0.26W-0.13
8.	U0.0W-0.14
9.	U-0.3W-1.48
10.	U0.8W0.0

If necessary, combinations of both systems (absolute and incremental) can be used. The CNC control registers the position of the tool, regardless of whichever method of programming is being used.

0.	X0.0Z0.0
1.	X1.0
2.	Z-1.15
3.	X2.0
4.	X2.2W-0.1
5.	Z-2.25
6.	X3.04
7.	X3.3W-0.13
8.	W-0.14
9.	X3.0Z-4.0
10.	X3.8

If the value of one of the coordinates remains the same, then input only the value of the next consecutive changing coordinate.

COORDINATE SYSTEM SETTING (G50)

So far, the meaning of the basic terms Machine Zero and Workpiece Zero have been explained. Setting the absolute zero of the coordinate system in the program is done by activating function G50 and, at the same time, inputting the value of the dimension referenced: the distance between Workpiece Zero and the tools cutting tip in the direction of X and Z axes. It must be done this way because data that refer to the X axis in CNC define the particular diameters. Please refer to Part 2 of this text for a detailed description on the process of setting the G50 coordinates.

G50X . . . Z . . .

Example:

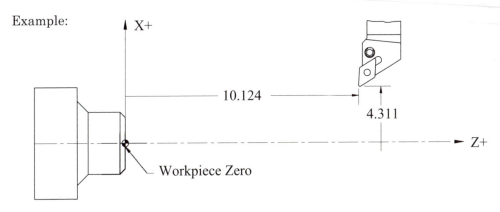

Figure 4 Coordinate System Setting (G50) Example 1

Notes:

1. Values of coordinates assigned to function G50 must refer to the X and Z coordinates from the Workpiece Zero to the tool tip.

2. Assigning smaller values for X and Z will cause an error; therefore, the tool will not approach the workpiece correctly.

3. Assigning larger values for X and Z will cause an error, possibly causing the tool to plunge into the workpiece or the holding equipment, resulting in considerable damage to the part or machine.

4. Remember that in the case of a need for small adjustments to the X or Z axes coordinates, the Wear Offsets should be used to change the values.

5. This method is used on older controls. A more modern method incorporates the use of geometry offsets for each tool, thus, eliminating the need for the position values of each axis in the program.

6. The turret must be at the Machine Home position when function G50 is initiated.

The drawing below (Figure 5) shows the case of smaller values of X and Z than required.

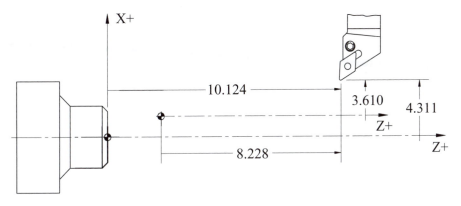

Figure 5
Coordinate System Setting (G50) Example 2

G50X4.311Z10.124   Correct

G50X3.610Z8.228   Incorrect

The following drawing (Figure 6) describes the results of increasing the values of *X* and *Z* in function G50.

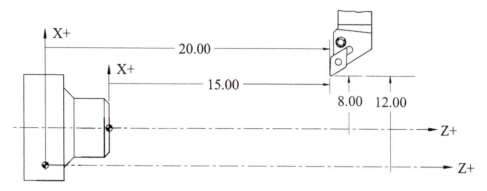

Figure 6 Coordinate System Setting (G50) Example 3

G50X8.0Z15.0   Correct

G50X12.0Z20.0   Incorrect

The following are examples for different shaped tools to illustrate the application of function G50.

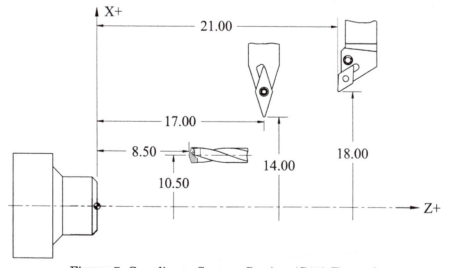
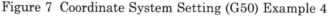

Figure 7 Coordinate System Setting (G50) Example 4

G50X10.5Z8.5 (Drill)

G50X14.0Z17.0 (Profile Tool)

G50X18.0Z21.0 (Finish Turning Tool)

Review of Work Offset (G50) Measurement Technique

The value of the X coordinate assigned to function G50 refers to a diameter on which the tool tip is located with respect to Workpiece Zero. This assumes, of course, that the axis of the spindle is the axis of symmetry, or centerline. In order to correctly obtain this value, apply the following procedures.

1. Set the tool at Machine Home.

2. Reset the readout displaying the value of displacement in the direction of the X axis to zero.

3. Turn on the spindle at the required speed, depending upon material type and tool type, then, manually position the cutting tool to make a cut along Z axis to create a diameter on the workpiece.

Without moving the X axis, measure the diameter of the cylindrical workpiece and add the value of this measurement to the value shown on the position readout. The result is a calculation of the desired dimension. Please refer to Part 2 CNC Machine Operation, Measuring Work Offsets, for the detailed procedure.

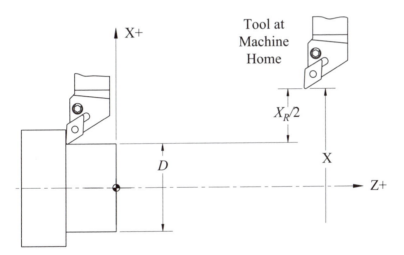

Figure 8 Work Offset (G50) Measurement Technique

$$X_R = X \text{ readout}$$

X_R = value of the displacement registered in the readout

$$X = 2 \times \frac{X_R}{2} + D = X_R + D$$

For $X_R = 5.6263$ and $D = 4.83$, then G50 = X10.0093. If the diameter of the workpiece is greater than the diameter on which the tool is positioned, proceed as follows:

$$X = D - X_R$$

For D = 20.126, and $X_R = 3.822$, so 20.126 −3.822 = 16.304 then G50X16.304 is required. See Figure 9 on next page.

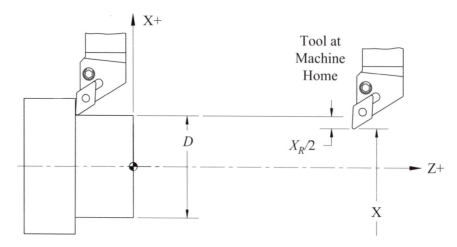

Figure 9
Work Offset (G50) Adjustment Technique

Attention:

The diameter obtained through initial machining used to set the values of the X coordinate assigned to function G50 is arbitrary. Measurement of the value of Z is relatively simple and is obtained from the position readout. The readout provides us with the distance traveled by the tool, along the Z axis, from machine zero to workpiece zero. Workpiece Zero is chosen by the programmer in the direction of the Z axis and is typically the <u>finished</u> front face of the part.

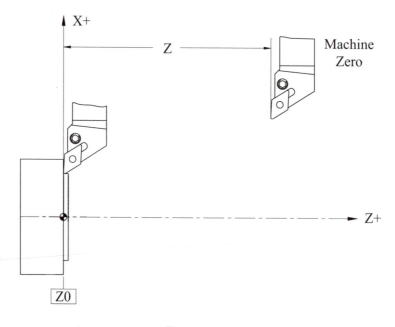

Figure 10
Work Offset (G50) Z Axis Measurement Technique

Values of X and Z assigned in the program to function G50 are valid only for a particular tool. They are replaced with the same function to which different values of X and Z coordinates are assigned for consecutive tools. In other words, there are new coordinate values for G50 for each tool.

The choice of sign (+ or −) for the G50 coordinates depends on the position of the tool tip with respect to the chosen workpiece zero, as shown in the following Figure 11:

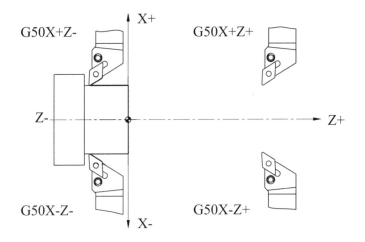

Figure 11 Work Offset (G50) Coordinate Signs

When creating a program, the programmer will not know the values of X and Z that are assigned to function G50 because they are measured and dependent upon the tool used, so random values may be assigned temporarily. These values are variable and are input as described earlier by the operator, or setup person. Therefore, whenever the need to use these values arises in the remainder of this book, logical values will be assigned to the program. Also, note that when G50S, maximum spindle speed setting is used, it is independent, but can be combined into the same block.

G50X4.311Z10.124S1000

Notice the similarities to setting geometry offsets and be sure to refer to Part 2 of this text, "Tool Offsets are Measured" for detailed instructions on measurement techniques.

In reality, this (G50) setting method is seldom used anymore. Because of the potential problems of incorrect entry of dimensional data into the actual program, the method is prone to mistakes. The more modern technique uses the geometry offset register to store the dimensional data for each tool.

Work Coordinate System Setting (G54)

With modern CNC Machines, a much more reliable method of coordinate system setting is used. Many machines today have a tool measurement sensor that is used as a fixed reference position. Each tool is measured to the electronic sensor and the coordinate values are input directly into the offset register specific to that tool. A separate measurement is taken for the X and Z axes. A parameter in the machine control is used to register the location of the tool sensor so that the X axis values are related back to the centerline of the spindle and the Z axis values to a fixed position. This enables the setup person to merely identify and measure the finished face of the part along the Z axis, to establish the Work Coordinate System and input the position in the WORK COORDINATES display for the G54 register. Consult the appropriate Operation manual for the machine you are working on for specific details on setting methods.

PROGRAM STUCTURE FOR TURNING CENTERS

In order to gain a better understanding of the basic structure in CNC Turning Center programming, the following sample program is given.

Sample Program

O1234	**Program Number**
N1G50X4.311Z10.124S1000	
N2T0100M42	
N3G96S500M03	
N4G00X1.2Z.2M08	
. . .	**Program for Tool No. 1**
. . .	
N16G00X20.126Z10.182T0100M09	
N18M01	
N19G50X18.624Z6.146S800	
N20T0200M41	
N21G96S600M03	
N22G00X.8Z.1M08	**Program for Tool No. 2**
. . .	
. . .	
N30G00X18.624Z6.146T0200	
N31M30	

Program Number

The program number typically consists of a four-digit integer following the letter O. This number is used to identify the programming procedures. The range of numbers that can be used as program numbers is 0001 to 9999. Be careful not to mistake the letter O, which precedes the actual program number for a zero. The 9000 series of program numbers are reserved for Macro programs and may not be used.

Example of program numbers:

O0001

O2160

O4004

O1261

Block Composition

A block is basically one line of the part program. It contains the commands used to simultaneously execute operations on CNC machines. The block is composed of program words and always ends with a semicolon, known in CNC programming as the end of block (EOB) character. The semicolon is never part of the written or disk copy of the program. Refer to Part 1, for additional information on the block format.

Examples of program blocks:

N10G50X20.126Z10.182S1000

N20T0200M42

N35M30

Arrangement of the program words in a given block can be random, but the sequence number address N must be first.

Block Number

The block number (also called the sequence number), is defined by the letter N, followed by one to four or five digits, and is limited to these four or five digits. A block number provides easier access to information contained in the program. The arrangement of block numbers in a given program can be random, but typically is sequenced in increments of some amount of other than one. The most common step increments are five or ten.

Examples of block or sequence numbers:

N1000

N10

N15

N20

Block or sequence numbers can be omitted in a block, or in the program itself. This saves storage space in the controller memory. It is important to note, however, that a program searching technique requires a sequence number for the restart of a program from somewhere other than its beginning or head. The logical location then, for sequence numbers, is at tool changes enabling the restart of that tool. It is also possible however, to search for a specific program word, such as T02, as well.

Part Program

The part program is that section of the program that contains essential information needed to control the cutting tool and auxiliary equipment.

Subprogram

The subprogram is a subordinate program. It is also registered in the controller memory with the letter O, followed by a four or five-digit number, just as the main program is. In the main program, a subprogram is called by using the M98P.... function and then M99 (called for in the subprogram), is the function that ends a subprogram and returns to the main program.

Example subprogram:

O1234

N5 ...   .

.. ..

N20   M99

The subprogram is called with function M98 in the main program, while the number of the subprogram is called with the letter address P.

Example of a subprogram call:

M98P1300

In this example the program number for the subprogram is O1300.

In some cases address L may be added to the program block and it indicates how often the subprogram is to be repeated. The amount of repeats for L can = from 1 to 999. If the address L is omitted from the program, the subprogram is executed only once. Examine the following example to obtain a better understanding of the structure.

Example of main and subprogram structure:

Main Program	Subprogram	Subprogram
O1200	O1300	O1400
N5G50X10.0Z6.0	N10G00X2.0	N10G01Z.2
N10	N20	N20 ......
N20	N30	N30 ......
N30M98P1300	N40M98P1400	N40 ....
N40	N50	N50M99
N50	N60M99	
N60		
N70M30		

Consecutive subroutines may be called from a current subprogram by applying the above-mentioned methods. The number of subprograms linked together to form a program may be as high as 9999. In block N30 of the main program example above, execution of the program begins with subprogram O1300 and at N40 for the subprogram O1400. After the execution of each of these subprograms is completed, further execution of the program begins with the next consecutive block in the previous program. This is the block (N40) following the block that called the subprogram.

Program Example

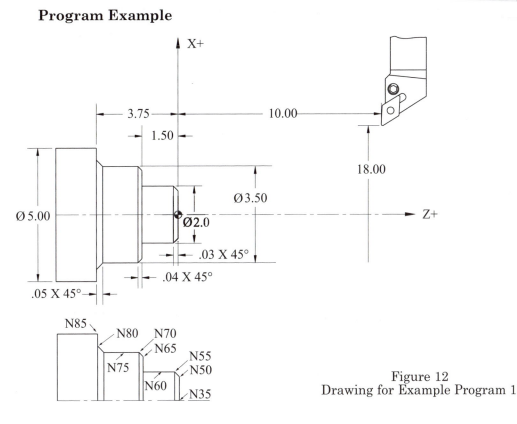

Figure 12
Drawing for Example Program 1

Example Program 1

For this example, only a finish cutting pass is programmed along the part geometry. The tool nose radius (for the purpose of simplification only) in this example is programmed as if to be zero.

```
O0001
N10G50X18.0Z10.0S1800
N15T0100M42
N20G96S600M04
N25G00X2.1Z.1T0101M08
N30G01Z0.0F.02
N35X-.03F.008
N40G00Z.1
N45X1.94
N50G01Z0.0F.02
N55X2.0Z-.03F.004
N60Z-1.5F.008
N65X3.42
N70X3.5W-.04F.004
```

N75Z-3.7F.008

N80U.1Z-3.75F.004

N85X5.1F.008

N90G00Z.1M09

N95X18.0Z10.0T0100M05

N100M30

The following explanations are given for the individual parts of the program above.

O0001 is the program number.

N10 through N100 are sequence or block numbers.

N10G50X18.0Z10.0S1800

Block N10 contains very important information about Workpiece Zero. This information must be provided by the operator, programmer or a setup person and entered at the beginning of the program with function G50. The input value $X = 18.0$ refers to the diameter the tool tip is set at, in the Machine Home position. Input value $Z = 10.0$ refers to the distance between the position of the tool tip and the Workpiece Zero position of the coordinate system on the Z axis. In this case, it is the face of the workpiece. Please refer to "Coordinate System Setting (G50)" in this section and "Turning Center Tool Offsets" in Part 2 for specific techniques for setting and detailed explanation. The program word, S1800 refers to a maximum spindle speed of 1800 r/min (because it is specified simultaneously in the G50 block) applicable during machining for tool No. 1 (T0100).

N15T0100M42

Tool function T0100 commands that tool No. 1 is in the work position. If the turret is in another position, it will rotate tool No. 1 into the ready position. The number 00 cancels any wear offset so that no offset compensation is used for this tool, at this time. Miscellaneous function M42 provides the machine with the information that the highest spindle speed range is applicable to this tool.

N20G96S600M04

Function G96 commands that the machining process will take place with a constant cutting speed of 600 ft/min, and the spindle speed will be adjusted automatically, based on the diameter of the workpiece up to a maximum of 1800 r/min. M04 refers to the direction of spindle rotation, which in this case is clockwise and also activates the rotation of the spindle. Please note the spindle direction in this example is pertinent to the tool orientation in the drawing provided in Figure 12. With most modern Slant Bed style CNC Turning Centers the tool is mounted with the insert facing downward which would require clockwise rotation (M03) for cutting.

N25G00X2.1Z.1T0101M08

The information contained in this block activates the execution of many of the commands related to tool motion and the flood coolant system, etc. First, the tool turret and carrier advances to a position specified by the coordinates X2.1 and Z.1 at rapid traverse, while simultaneously, the flood coolant system is activated. In addition, an applicable wear offset (coded by 01 after T01) will affect the actual path taken by the tool.

N30G01Z0.0F.02

Function G01 in block N30 refers to the linear interpolation with respect to the Z axis, with the ending coordinate of $Z = 0.0$ and a value of feed .02 in/rev.

> **N35X-.03F.008**
>
> **N40G00Z.1**
>
> **N45X1.94**
>
> **N50G01Z0.0F.02**
>
> **N55X2.0Z-.03F.004**
>
> **N60Z-1.5F.008**
>
> **N65X3.42;**
>
> **N70X3.5W-0.04F.004**
>
> **N75Z-3.7F.008;**
>
> **N80U.1Z-3.75F.004**
>
> **N85X5.1F.008**

The information found in blocks N35 to N85 refer to a change of position in the programmed points of the workpiece coordinate system which machine the part profile.

N90G00Z.109M09

In this block, the tool will advance at rapid traverse to Z.1 and the flood coolant system will be turned off.

N95X18.0Z10.0T0100M05

Block N95 is also very important in the program. It commands the tool to return to its starting point identified by the G50 in block N10, cancel the wear offset activated for tool No. 1 and stop the rotation of the spindle.

N100M30

Miscellaneous function M30 ends the program and resets the program to its beginning point or head.

Preparatory Functions for turning Centers (G-Codes)
Rapid Traverse Positioning (G00)

G00 is the rapid traverse positioning; it is used for position changes without machining. G00 can be interconnected with the M, S, or T functions. G00 is a modal G-Code and will remain in effect until replaced by another command from the same group.

For this example, a finish cutting pass is programmed along the part geometry. The tool nose radius (for the purpose of simplification only) in this example is programmed as if to be zero.

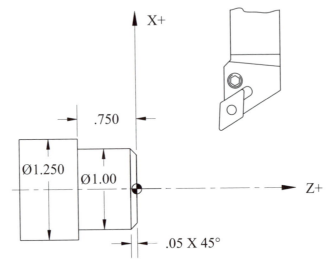

Figure 13 Drawing for Rapid Traverse Function (G00)

Example: Program using the Rapid Traverse Function.

O2345

N10G50X15.Z10.S2000

N15T0100M42

N20G96S500M03

N25G00X1.1Z.2T0101

N30G01Z0F.02

N35X0F.01

N40G00X.9Z.03

N45G01Z0.F.02

N50X1.W-.05F.01

N55Z-.75

N60X1.25

N65G00X15.Z10.0T0100M05

N70M30

Caution: If G00 rapid positioning is programmed in the direction of both axes, note that the tool will not advance to a specified point following the shortest possible path. The tool path is determined by the speed of rapid traverse with respect to each axis. In most cases, the speed for the X axis is much greater than that for the Z axis. The axis that must travel the least distance will be reached first.

Figure 14 gives a comparison between rapid and feed traverse movements when the axes are commanded simultaneously. The rapid traverse moves are indicated by dashed lines and feed traverse by solid lines.

N25 = a rapid traverse move toward the material (workpiece).

N40 = a rapid traverse move to the starting point of the chamfer

N50 = a linear feed traverse move (G01) to create the .05 X 45° chamfer

N65 = a rapid traverse move to return to the reference position identified by G50 in line N10.

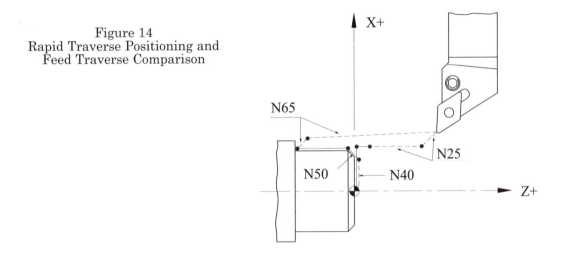

Figure 14
Rapid Traverse Positioning and
Feed Traverse Comparison

Workpiece holding devices (chuck jaws) quite often extend beyond the workpiece holding equipment (chuck). Therefore, careful consideration should be given to the position of work holding devices (including tailstock centers and quill extension position), so that rapid traverse paths of the tool do not interfere with them and cause a crash that may damage the machine, clamping device, or part.

LINEAR INTERPOLATION (G01)

Linear interpolation is programmed by using the G01 function and may be applied simultaneously for both axes. The G01 function commands the movement of the tool from a given position to the position with the assigned coordinates having feed rate specified by the F-Word. The block format for linear interpolation is given as follows. G01 is a modal command that stays in effect until replaced by another command from the same group.

G01X(U) . . .Z(W) . . .F . . .

The interpolator in the control system calculates various speeds for the motion axis so that the resulting speed is equivalent to the programmed feed rate. For this example, only a finish cutting pass is programmed along the part geometry. The tool nose radius (for the purpose of simplification only) in this example is programmed as if to be zero.

The following is an example using linear interpolation.

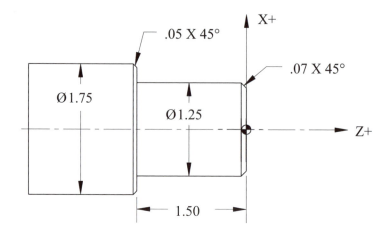

Figure 15 Drawing for Linear Interpolation (G01) Example

O3456

N10G50X10.Z5.S2500

N15T0200M42

N20G96S500M03

N25G00X1.35Z.2T0202M08

N30G01Z0F.01

N35X0

N40G00X1.11W.03

N45G01Z0

N50X1.25Z-.07

N55Z-1.5

N60X1.650

N65X1.75Z-1.55

N70X1.8

N75G00X10.Z5.T0200M05

N80M30

In block N30 a Linear Interpolation (feed) move is commanded to the work face along the Z axis at a rate of .01 in/rev.

In block N35 a Linear feed move is made along the X axis to the centerline of the part at the same feed rate as commanded in block N30.

In block N40 a Rapid Traverse move is made to the X and Z axes start point in front of the contour.

In blocks N45 through N65 Linear Interpolation moves are made along the finish contour of the part.

CIRCULAR INTERPOLATION (G02 AND G03)

Circular interpolation allows programmed tool movements along an arc. In order to define the circular interpolation function, the following conditions must be fulfilled.

1. Selection of the direction of interpolation:

G02 = Clockwise

G03 = Counterclockwise

2. Position coordinates of the ending or point of the arc. The ending point coordinates may be omitted, if they correspond to the coordinates of the starting point (half circle, circle).

3. The dimension corresponding to the distance between the center of the tool nose radius and the center of the arc, from the starting point of each axis must be given.

The incremental distance for the X axis is defined by the value of letter address I, and the Z axis is defined by the value of the letter address K. Values for I and K may be omitted from the program, if they are equal to zero. The block format for circular interpolation is given as follows.

G02X(U) ... Z(W) ... I ... K ... F ...

G03X(U) ... Z(W) ... I ... K ... F ...

The following figure (Figure 16) graphically identifies all of the components necessary for programming arcs and their descriptions.

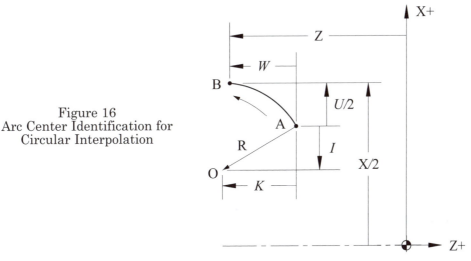

Figure 16
Arc Center Identification for
Circular Interpolation

A = Arc starting point

B = Arc ending point

I = Incremental distance from start point (A) to the arc center (O) along the X axis

K = Incremental distance from start point (A) to the arc center (O) along the Z axis

O = Arc center

R = Arc radius (a negative signed value will produce a concave arc)

U/2 = Incremental distance from the arc starting point to the ending point along the X axis

W = Incremental distance to the arc end point along the Z axis

X/2 = Absolute coordinate for the ending point of the arc along the X axis

Z = Absolute coordinate for the ending point of the arc along the Z axis

In order to establish signs for *I* and *K*, consider the following directions: Imagine a line is drawn from the arc starting point to the arc center point with a direction vector toward the center of the arc. Next, project this vector onto the axes of the coordinate system with the origin at the arc start point. If the resulting projections of the vector are oriented in the same direction as the corresponding axis of the absolute coordinate system, the sign plus (+) is applicable. Otherwise, the sign minus (-) is applicable. If no sign is given in the coordinate entry, the control assumes the sign is positive.

The following figure (Figure 17) is a pictorial representation showing how to determine the sign for vectors of *I* and *K,* while taking into account the tool nose radius.

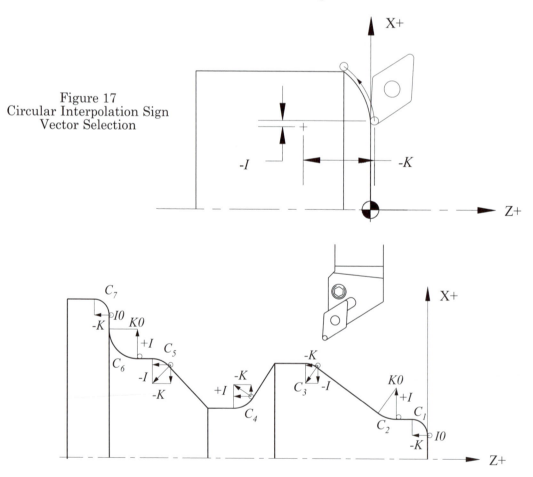

Figure 17
Circular Interpolation Sign
Vector Selection

Figure 18 Circular Interpolation Direction Vector Selection

Figure 18 shows how to determine the direction of the circular interpolation and signs for vectors I and K. (The tool tip in the drawing is represented by a circle for easier pictorial presentation.)

$$C_1 = G03,\ I0,\ -K$$
$$C_2 = G02,\ +I,\ K0$$
$$C_3 = G03,\ -I,\ -K$$
$$C_4 = G02,\ +I,\ -K$$
$$C_5 = G03,\ -I,\ -K$$
$$C_6 = G02,\ +I,\ K0$$
$$C_7 = G03,\ I0,\ -K$$

Modern CNC controls include an additional capability to use R in place of I, and K. R is the distance from the center of the tool radius to the center of the following arc. If an arc is smaller than or equal to 180°, then R assumes a positive sign; if it is greater than 180°, then R assumes a negative sign. The block format for circular interpolation using R is given as follows.

G02X(U) . . .   Z(W) . . .   R . . .   F . . .
G03X(U) . . .   Z(W) . . .   R . . .   F . . .

Circular interpolation can be performed these two different ways. The application of address R, however, is less difficult.

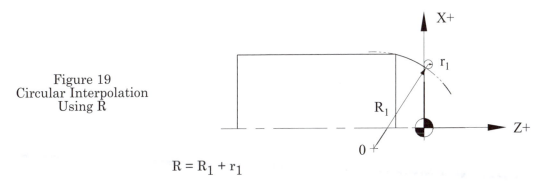

Figure 19
Circular Interpolation
Using R

$$R = R_1 + r_1$$

If I and K are applied, then machine control is fed with precise information about the position of the center of the radius of the performed arc. In this case, the coordinates of the arc ending point must correspond to a position on the programmed circle. If, however, the given values of the coordinates are incorrect, the tool will not respond by following an arc. If the second method using R is employed, the tool will follow an arc, even if the values of the coordinates are incorrect. (This is true, of course, if the ending point of the arc falls within the area of the diameter of the circle.)

The following program is an example of circular interpolation utilizing I and K. Once again, in this example, only a finish cutting pass is programmed along the part geometry. The tool nose radius (for the purpose of simplification only) in this example is programmed as if to be zero.

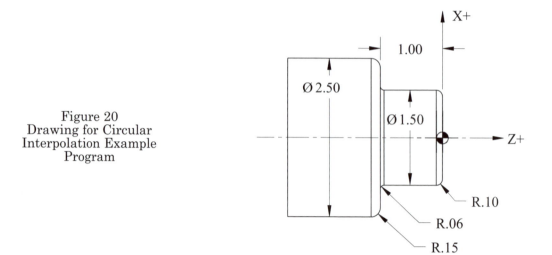

Figure 20
Drawing for Circular
Interpolation Example
Program

Example of Circular Interpolation (G02 or G03), Program 2

        O0002

        N10G50X15.Z5.S2000

        N15T0100M42

        N20G96S500M03

        N25G00X1.6Z.2T0101M08

        N30G01Z0F.02

        N35X0

        N40G00X1.3W.03

        N45G01Z0

        N50G03X1.5Z-.1K-.1I0

        N55G01Z-0.94

        N60G02U.12Z-1.0K0.I.06

        N65G01X2.2

        N70G03X2.5W-.15I0K-.15

        N75G00X15.Z5.T0100M05

        N80M30

The following is the same example as above using *R*.

        O0002

        N10G50X15.Z5.S2000

        N15T0100M42

        N20G96S500M03

        N25G00X1.6Z.2T0101M08

        N30G01Z0.F.02

N35X0

N40G00X1.3W.03

N45G01Z0

N50G03X1.5Z-1.0R.1

N55G01Z-0.94

N60G02U.12Z-1.0R.06

N65G01X2.2

N70G03X2.5W-.15R.15

N75G00X15.Z5.T0100M05

N80M30

DWELL (G04)

Dwell is initiated by use of function G04, and the length of time for the dwell is specified by P, X or U, (depending on the control type) as follows:

G04P . . . (in milliseconds)

G04X . . . (in milliseconds)

G04P . . . (in milliseconds)

Examples:

G04P2500

G04X2.5

G04U2.5

In the examples above, the dwell time values are the equivalent of 2 and 1/2 seconds. Also note that when using P to address the amount of time for dwell, a decimal point may not be used. The value of time is measured in milliseconds (ms), 1000 ms = 1 second. Function G04 is a "one shot" command and is active only in the block in which

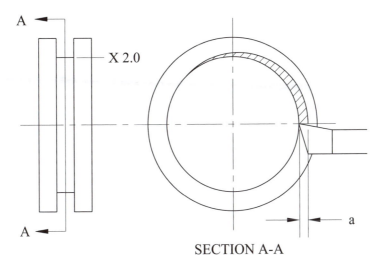

SECTION A-A

Figure 21 Drawing for Dwell (G04) Example

it is called. The dwell is activated at the end of the feed move and should be the only contents of the block. Dwell is sometimes indicated by the number of revolutions as opposed to the amount of time (dependant on a parameter setting). Study the manufacturer programming manual specific to the equipment to be sure of the exact method used. A common use for dwell is in the process of machining internal or external grooves, as shown in the following figure and described in the next paragraph.

Examples:

G01X2.0F.008

G04U.25 (or G04X0.25 or G04P250)

G00X2.5

In the process of making the groove, you must remove a layer of material with a thickness corresponding to the depth of the cut and equivalent to the feed per revolution. In order to avoid an egg-shaped workpiece, a certain amount of time must be allowed for the cut to be completed for the full circumference as the tool tip reaches the diameter indicated in the program. If function G00 follows function G01, the resulting shape of the workpiece will be that of an egg because the tool is removed from the groove before all of the material can be cut. In the above example, the programmed dwell is 1/4 second and this allows a sufficient pause necessary to clean up the bottom of the groove before retracting from it.

AUTOMATIC REFERENCE POINT RETURN CHECK (G27)

Function G27 is an automatic return function. This command positions the tool, at rapid traverse, to a reference position indicated by the coordinates of X (U) and Z (W) in the block. It serves the purpose of confirmation of the return to the reference position. The reference position does not have to be a position of Machine Zero, but it commonly is, because of the safety it insures. The block format for automatic reference point return check (G27) is given as follows.

G27X(U) . . .Z(W) . . .T . . .00

Application of this function is shown in the following program example:

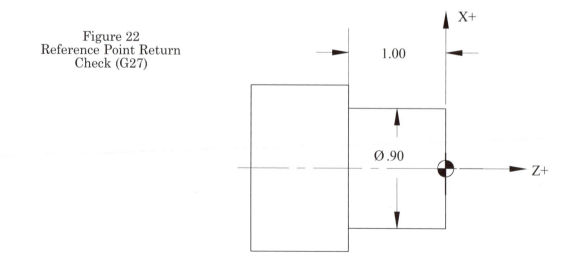

Figure 22
Reference Point Return
Check (G27)

Example of Automatic Reference Point Return Check, Program 3

O0003

N10G50X15.Z5.0S800

N15T0100M42

N20G96S500M03

N25G00X1.0Z.1T0101M08

N30G01Z0F.01

N35X0

N40G00X.9Z.03

N45G01Z-1.0

N50G27X15.Z5.0T0100M05

N55M30

Notes:

1. *The block that contains function G27 must also include cancellation of the wear offset.*

2. *Function G27 is only valid for function G50's axis input.*

3. *The time delay for the confirmation of the reference point return check by the control is approximately between 0 and 3 seconds.*

4. *A correct return to the reference position will be confirmed by a visual inspection of the control LED's being lit for the given axes. If the return position does not check accurately, an alarm will be displayed and the machine will stop until the error is corrected.*

At first glance, it may seem senseless to use this function; however, a closer look provides sufficient reasons. One good example is for tool nose radius compensation. If the programmer inputs functions G41 and G42, but for some reason omits cancellation with function G40, then the tool will return to its reference position with a displacement equal to the offset. Without the use of G27, in this case, the next cycle may result in the under or overcutting of the workpiece.

AUTOMATIC REFERENCE POINT RETURN (G28)

One of the ways to command a tool to return to the reference point after completion of an operation is through the application of function G28. The block format for Automatic Reference Point Return (G28) is given as follows.

G28X(U). . .Z(W). . .

In the above block, the values in X (U) and Z (W) are the coordinates for an intermediate point through which the tool will pass, on its way to Machine Zero. The tool will position at a rapid traverse rate in nonlinear form. Therefore, it is recommended that an escape move be programmed, so that the tool is clear of the part before commanding G28. When using G28 tool offsets should be cancelled in a prior block.

T0100 = Cancellation of tool offset

Example of Automatic Reference Point Return, Program 4

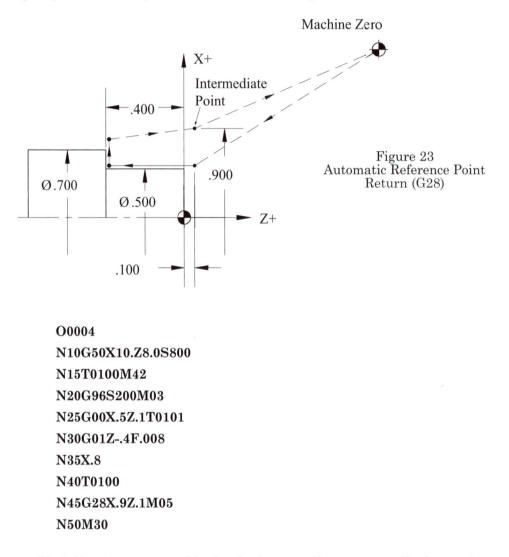

Figure 23
Automatic Reference Point
Return (G28)

O0004

N10G50X10.Z8.0S800

N15T0100M42

N20G96S200M03

N25G00X.5Z.1T0101

N30G01Z-.4F.008

N35X.8

N40T0100

N45G28X.9Z.1M05

N50M30

Block N45 is programmed in the absolute coordinate system. To change the command into the incremental system, follow this format:

N45G28U.1W.5M05

If no obstacles appear on the tool's path as the tool proceeds to Machine Zero return, then most often the intermediate point is merely a directional point.

N45G28X.8Z-.4M05 or

N45G28U0W0M05

Return from Reference Point (G29)

Function G29 follows G28 and <u>MUST NOT</u> be programmed without using G28 first. Using this command returns the tool, at rapid traverse, to a programmed point, by way of the same intermediate point as given in the G28 command. The block format used for G29 is as follows:

G29X(U). . .Z(W). . .

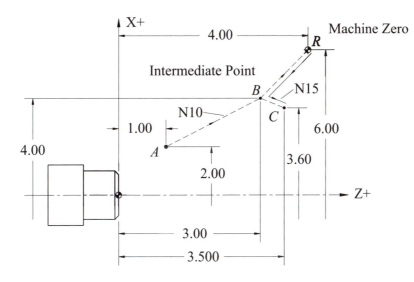

Figure 24 Return from Reference Point (G29)

N10G28U2.0W2.0

N15G29U-0.4W0.5

As shown in the drawing, the tool has moved from point A to point B (the intermediate point) and then on to point R (Machine Zero), all of which is determined in the first block with function G28. In block N15, the tool automatically returns from point R toward point B, and programming is limited to the calculation of the distance between points A and C. Point R is the Machine Zero position. Point C refers to the first position after tool change.

THREAD CUTTING (G32)

One of the oldest systems used (seldom) to program thread cutting is with the use of function G32. When it is used however, it offers absolute control of the threading if needed. The system uses a block-by-block programming technique to accomplish the desired thread. In other words, each pass of the threading tool must be programmed independently. The block format used for G32 is as follows:

1. Straight thread

G32Z(W). . . F (or E . . . for the Fanuc 6T control)

where

In the case of a straight thread, the diameter is determined by the previous block, which includes the X coordinate.

Z(W) = Length of thread

F, or E are equal to the thread pitch, 1/number of threads per inch.

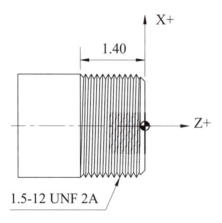

Figure 25 Thread Cutting (G32)

2. Taper thread

G32X(U)). . . Z(W). . . F (or E . . . for the Fanuc 6T control)

where

X(U) refers to the dimension-determining diameter of the thread at the end points, as in the case of a taper thread. In the case of a straight thread, X(U) may be omitted.

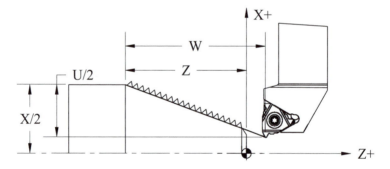

Figure 26 Taper Thread Cutting (G32)

3. Face thread

G32X(U). . . F (or E . . . for the Fanuc 6T control)

The feed rate for threading is programmed by using address F (or E and the value of feed are equivalent to the lead of the thread. Thread cutting using function G32 is rarely performed today because it is cumbersome to program each pass. Instead, function G92 is often used because it is easier to program, and function G76 is used most in these cases.

Thread Cutting Considerations

1. Changing the spindle speed or the feed rate override, while within the threading cycles, is not effective.

2. A "dry run" condition is applicable and effective.

3. The use of constant rotational speed programming, G97 S . . . is required!

Following is an example of straight thread cutting using function G32 for the part shown in Figure 25:

Example of Thread Cutting, Program 5

      O0005

      N10G50S1500

      N15T0100M42

      N20G97S600M03

      N25G00X1.58Z.5T0101M08

      N30G00X1.45

      N35G32Z-1.4F.0833

      N40G00X1.58

      N45Z.5

      N50X1.42

      N55G32Z-1.4

      N60G00X1.58

      N65Z.5

      N70X1.4

      N75G32Z-1.4

      N80G00X1.58

      N85Z.5

      N90G28U0W0T0100M05

      N95M30

THREAD CUTTING CYCLE (G92)

By using G92 a cycle of four individual movements of the tool can be obtained in one block of information, these movements are:

1. Rapid traverse to a given diameter.

2. Thread cutting with programmed feed rate.

3. Rapid withdrawal.

4. Rapid traverse return to the starting point.

Types of threads available are as follows.

1. Straight thread (cylindrical)

      G92X(U). . .Z(W). . . F or E. . .

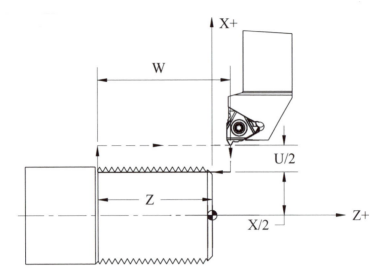

Figure 27 Thread Cutting Cycle (G92)

2. Taper thread

G92X(U)...Z(W)...I...F or E...

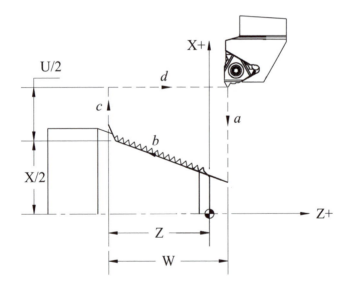

Figure 28 Thread Cutting Cycle (G92) for Tapered Threads

By activating SINGLE BLOCK, the introduced movements of a, b, c, and d can be executed by pressing CYCLE START.

Note: By using function G92, it is possible to end the thread with a 45° chamfer or simply at a right angle.

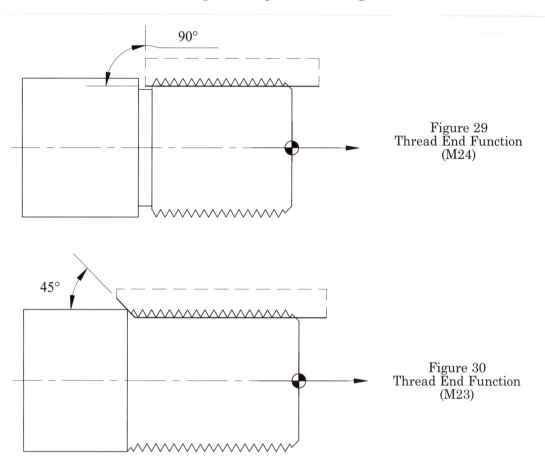

Figure 29
Thread End Function
(M24)

Figure 30
Thread End Function
(M23)

As shown in Figure 29, a 90° thread end was obtained by using function M24, and a 45° chamfer is obtained by using function M23 as in Figure 30.

At the end of the threading routine, the G92 function must be cancelled by a G00 move. If it is not, the next programmed movement will continue as if it were still threading.

On some control models other than the Fanuc 16T, function G92 is a position register setting code. In these cases, when G92 is used for threading the G50 position register command must be used.

The previous notes related to threading using function G32 are also valid for G92.

The following is an example illustrating the application of function G92 for a straight thread, see figure on next page:

Example of Thread Cutting Cycle, Program 6

      O0006

      N10G50S2000M42

      N15T0100

N20G97S800M03

N25G00X1.1Z.5T0101M08

N30G92X.97Z-2.0F0.083333

N35X.95

N40X.92

N45X.90

N50G28U0W0T0100

N55M30

Note: Using this cycle with work requiring repetitive threading and cutting opera-tions is very convenient, especially for large thread sizes. In such cases, after executing the block containing function G92, a certain number of consecutive blocks may be omitted and instead only the last few blocks may be executed.

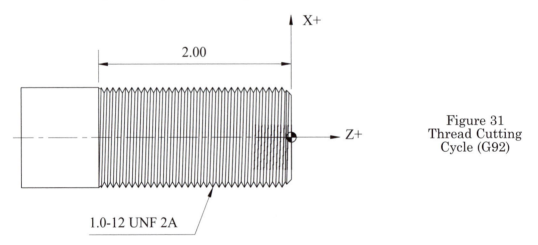

2.00

X+

Z+

1.0-12 UNF 2A

Figure 31
Thread Cutting
Cycle (G92)

It is convenient to use function G92 in programs that result in much shorter length. For the above program, blocks N35, N40, and N45, deal with diameters for par-ticular passes, while the preceding data that form block N30 are valid until block N50. The value of feed F (or E for some types of controls), may be expressed as shown in the preceding example with programmable accuracy reaching .000001 inch. In this example, the value for Z in block N25 is .5. As far as the programmer is concerned, thread cutting may be initiated from a point positioned much closer to the material.

Threading could start as close as $Z = 0$. Practically speaking, however, it is not possible for the tool to begin the operation with the feed rate given in a program. Thus, part of the tool path is followed by the tool, with acceleration, until the tool reaches the value of the feed rate equivalent to thread lead. A similar situation occurs at the end of the threading process when the tool decelerates. A certain distance is traveled by the tool after some delay.

The lengths of L_1 and L_2, in Figure 32, depend on the thread pitch and rotation-al speed of the spindle. Theoretical analysis of the mathematical formulas applied in cal-culating such lengths is rather complicated. For this reason, we will limit our explanation to the more simple calculations that are based on the following equations.

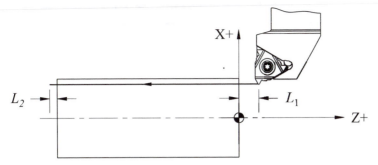

Figure 32 Threading Acceleration and Deceleration

$$L_2 = \frac{P \times S}{1800}(\text{in})$$

where

P = lead (in)

S = revolutions per minute

$$L_1 = L_2 \times K \quad K = \ln\frac{1}{a} - 1 \quad a = \frac{\Delta P}{P}$$

where

ΔP = lead error

P = lead

The amount of lead error P depends on the design of the machine, as well as its servo motor systems. Values of the coefficient $K,$ for a few specified error allowances, are contained in the table below.

a	K
.02	2.91
.015	3.2
.01	3.605
.005	4.298

Part 3 Chart 3

Example:

Given $\quad P = 0.0833, S = 1200, \quad$ and $\quad a = 0.01,$

$$L_2 = \frac{P \times S}{1800} = \frac{0.0833 \times 1200}{1800} = 0.0555$$

$$L_1 = L_2 \times K = 0.0555 \times 3.605 = 0.200$$

In most cases, these values are based on experience.

TAPERED THREAD CUTTING USING CYCLE (G92)

When cutting a tapered thread using function G92 use the block format shown in item two above. The value of I is the radius or difference per side between the thread diameter at the end of the cut to the thread diameter at the start of the cut. Depending on the sign following the taper threading command of I, the cutting tool moves as follows:

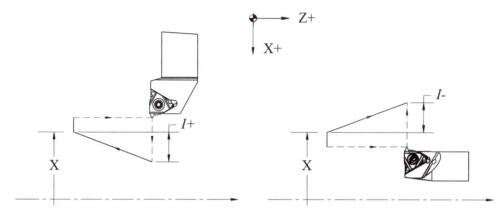

Figure 33 Tapered Thread Cutting Using Cycle (G92)

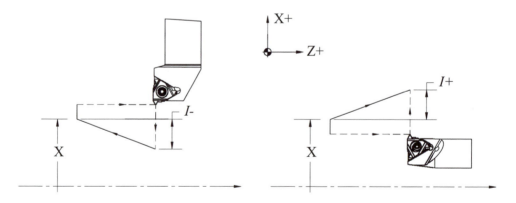

Figure 34 Tapered Thread Cutting Using Cycle (G92)

OUTER/INNER DIAMETER TURNING CYCLE (G90)

The outer/inner diameter turning cycle is a cylindrical cutting function (to cut diameters). The block format used for G90 is as follows:

G90X(U) . . . Z(W). . . F . . .

Using function G90 in a program is a convenience. However, using function G90 will result in some loss of time because, after each pass, the tool returns by the motion d, as shown in the Figure 35.

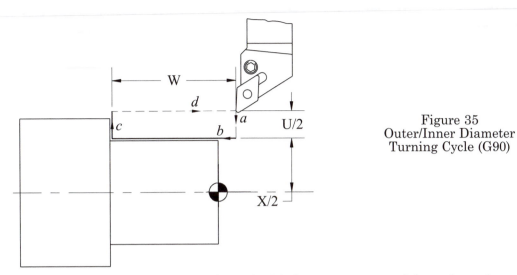

Figure 35
Outer/Inner Diameter
Turning Cycle (G90)

The cycle execution is performed with four movements of the tool, as shown in the above Figure 35, the same as it is with function G92.

Example of Outer/Inner Diameter Turning Cycle, Program 7

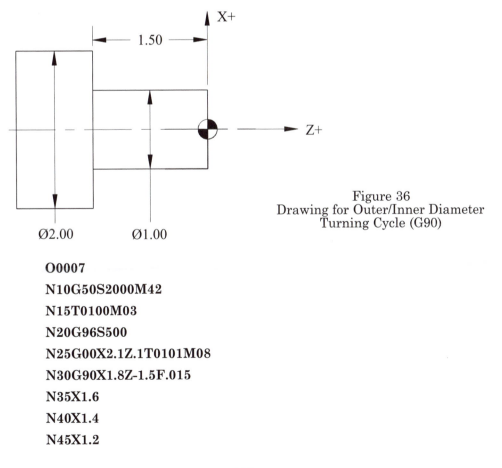

Figure 36
Drawing for Outer/Inner Diameter
Turning Cycle (G90)

O0007

N10G50S2000M42

N15T0100M03

N20G96S500

N25G00X2.1Z.1T0101M08

N30G90X1.8Z-1.5F.015

N35X1.6

N40X1.4

N45X1.2

N50X1.0

N55G28U0.W0.T0100

N60M30

Note: Depending on the control type, tapered cuts may be programmed for function G90 by inclusion of letter address I or R. Where either is input, there is a radial value of the difference between the starting and ending diameters.

FACE CUTTING CYCLE (G94)

G94X(U) . . . Z(W) . . . F . . .

The face cutting cycle is a function similar in application to function G90 except that it is used for facing.

As with function G90, the tool always returns after each pass to the starting point by motion *d*, as shown in Figure 37. This is one reason why both methods of programming are not widely used in practice today. Multiple repetitive cycles are a much better choice and are discussed next.

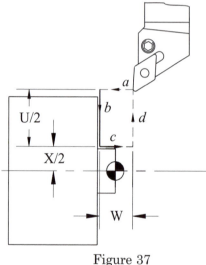

Figure 37
Face Cutting Cycle (G94)

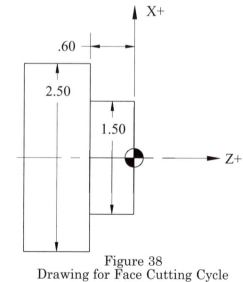

Figure 38
Drawing for Face Cutting Cycle
(G94)

Example of Face Cutting Cycle, Program 8

    O0008

    N10G50S2000M42

    N15T0100M03

    N20G96S500

    N25G00X2.6Z.05T0101M08

    N30G94X1.5Z-.1F.015

    N35Z-.2

N40Z-.3

N45Z-.4

N50Z-.5

N55Z-.6

N60G28U0W0T0100

N65M30

Note: Tapered cuts may be programmed for function G94 by inclusion of address K, where input is a radial value of the difference between the starting and ending diameters.

MULTIPLE REPETITIVE CYCLES
STOCK REMOVAL TURNING CYCLE (G71)

Function G71 is the stock removal cycle for turning that removes metal along the direction of the Z axis. In a case where there is a lot of material to be removed, this cycle provides an easy method for programming.

In Figure 39, the dotted lines refer to the initial shape of the workpiece, while the solid line refers to the final product. In the programming described so far, it has been necessary to employ many blocks of information to perform all the individual cuts for roughing. By using function G71, programming of the final shape of the workpiece is defined. Material is removed automatically, in each pass.

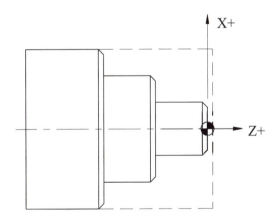

Figure 39 Stock Removal Turning Cycle (G71)

There are two types of program format for stock removal using function G71: single block and double block. The CNC control model used determines which type will be needed. Consult the manufacturer programming manual specific to the machine to determine the required method.

Function G71, Double Block

There are two program blocks required for function G71 when using this method. The finished profile of the part, as shown in Figure 40, is machined starting at point *a*, and proceeding to points *b* and *c*. The metal removal amount along the *X* axis is defined by the parameter U (depth of cut), in the first program block. A finish allowance for the X axis is defined by the parameter U in the second block. Be careful not to get the two confused, as they do different things. The finish allowance for the Z axis is defined by W in the second block.

The following is the block format for function G71:

G71U. . .R . . .

G71P. . .Q. . .U. . .W. . .F. . .S. . .

where

***in block one*:**

U = the depth of each roughing cut per side, to be used in consecutive passes (no sign)

R = the amount of retract, along the X axis, for each cut

***in block two*:**

P = the sequence number of the first block in the program, which defines the finish profile

Q = the sequence number of the last block of the program, which defines the finish profile

U = the stock allowance to be left for a finishing pass in the X axis direction (referred to as diameter; sign is + or –)

W = the stock allowance to be left for finishing in the Z axis direction (sign is + or –)

F = Cutting feed rate (in/rev or mm/rev) for blocks defined from P to Q

S = Spindle speed (ft/min or m/min) for blocks defined from P to Q

The signs attached to symbols U and W may have negative or positive values, depending on the orientation of the coordinate system and the direction in which the allowance is assumed.

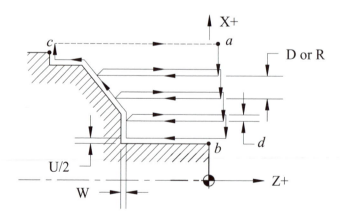

Figure 40 Stock Removal Turning Cycle (G71) Function Diagram

In Figure 40, *d* represents the amount of *X* axis retract, programmed for clearance by R in the first program block. This amount may also be set by parameter.

For Figure 40:

a = the starting point of the given cycle

b = the sequence number of the first programmed point for the finish contour, which corresponds with P number of the second G71 block

c = the sequence number of the last programmed point for the finish contour

Notes for the double block command:

1. Changes in the feed between blocks P and Q will not be ignored in G71. Only feed F, indicated by function G71, is valid.

2. The first tool path move of the programmed cycle from point a to point b cannot include any displacement in the direction of the Z axis.

3. The tool path between point b and c must be a steadily increasing, or decreasing pattern in both axes.

4. Both linear and circular interpolation is allowed.

5. The value for R must be noted as in the following examples:

R2000

R2500

R1500

6. For some controls, use of a decimal point may be allowed.

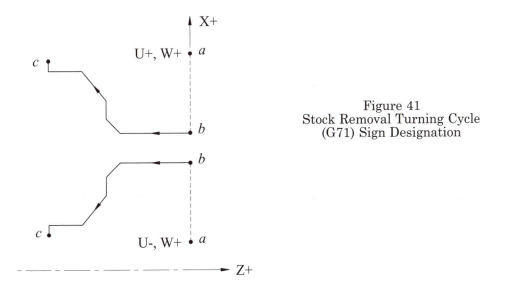

Figure 41
Stock Removal Turning Cycle
(G71) Sign Designation

7. Figure 41 illustrates that if the allowance for finishing is located on the positive side (in the direction of the X and Z axes), with respect to the programmed contour, no sign is used; if on the negative side, the negative sign (–) is used.

Function G71, Single Block

There is only one program block required for function G71 when using this method. Much more freedom is allowed in regards to programmable shapes. In this case, it is not necessary to program a steadily increasing or decreasing pattern in both axes, it is only required along the Z axis, and up to ten concave figures are allowed.

Noteworthy differences between double and single block function G71:

1. Block format for a single block is as follows:

G71P . . . Q . . . I . . . K . . . U . . . W . . . D . . . F . . . S . . .

where

All parameters are identical as stated above except:

I = Radial distance and direction of the rough cut along the X axis

K = Distance and direction of the rough cut along the Z axis

D = Depth of each roughing cut

2. In the single block format, two axes may be programmed in the first block identified by the parameter P.

3. If single block format programming is used and the first block does not include any Z displacement, 0 must be input for parameter W in the G71 block.

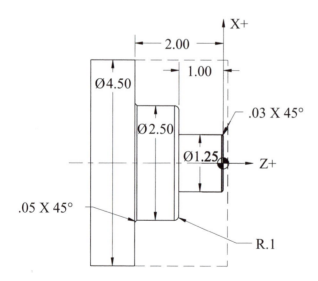

Figure 42
Drawing for Stock Removal
Turning Cycle (G71)

Example of Stock Removal Turning Cycle, Program 9

The following is a program example for a double block call of G71:

O0009

N10G50S2000

N15T0100M42

N20G96S600M03

N25G00X4.75Z.2T0101M08

N30G71U.12R.05

N35G71P40Q80U.03W.015F.019

N40G00X1.19

N45G01Z0

N50X1.25Z-.03

N55Z-1.0

N60X2.3

N65G02X2.5W-.1I0K-.1

N70G01Z-1.95

N75X2.6W-.05

N80X4.75

N85G28U0W0T0100M09

N90M05

N95M30

The following is a program example for a single block call of G71:

O0009

N10G50S2000

N15T0100M42

N20G96S600M03

N25G00X4.75Z.2T0101M08

N30G71P35Q75U.03W.003D1200F0.019

N35G00X1.19

N40G01Z0

N45X1.25Z-.03

N50Z-1.0

N55X2.3

N60G02X2.5W-.1I0K-.1

N65G01Z-1.95

N70X2.6W-.05

N75X4.75

N80G28U0W0T0100M09

N85M05

N90M30

Comments: *The block N35 first appears right after the first block containing function G71. Despite the fact that the last programmed cycle block N80 refers to the location of the tool tip Z-2.0, the tool will automatically return to the starting point indicated as Z.2 in the block N25.*

STOCK REMOVAL FACING CYCLE (G72)

The properties for function G72 are similar to G71. The only difference is in the change of the cutting direction to facing. The following is a function block diagram:

Function G72, Single Block:

G72P... Q ... I ... K ...U...W... D... F... S...

Function G72, Double Block:

G72W... R...

G72P... Q... U... W... D... F... S...

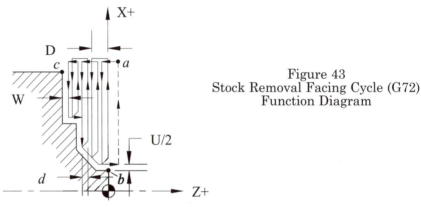

Figure 43
Stock Removal Facing Cycle (G72)
Function Diagram

The parameters for this function have the same meaning as those described for function G71.

Notes:

1. *The first block of the programmed cycle should not include any displacement in direction of the X axis.*

2. *The remaining notes for this function are identical to those for function G71.*

3. *The principle defining the choice between a positive or negative sign for U and W is identical to function G71.*

Figure 44
Drawing for Stock Removal Facing
Cycle (G72)

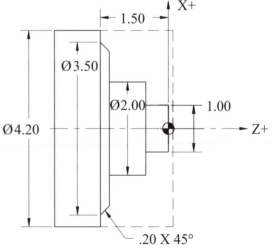

Example of Stock Removal Facing Cycle, Program 10

> **O0010**
>
> **N10G50S2000**
>
> **N15T0100M42**
>
> **N20G96S500M03**
>
> **N25G00X4.4Z.2T0101M08**
>
> **N30G72P35Q60U.03W.012D1000F.012**
>
> **N35G00Z-1.5**
>
> **N40G01X3.5**
>
> **N45X3.1Z-1.3**
>
> **N50X2.0**
>
> **N55Z-0.5**
>
> **N60X1.0**
>
> **N65G28U0W0T0100M09**
>
> **N70M05**
>
> **N75M30**

PATTERN REPEATING CYCLE (G73)

Function G73 permits the repeated cutting of a fixed pattern, with displacement of the axes position by an amount determined by the total material to be removed, divided by the number of passes desired. This cycle is well suited for previously formed castings, forgings or rough machined materials. This machining method assumes that an equal amount of material is to be removed from all surfaces. It can still be used if the amounts are not equal, but caution should be applied concerning excessive depths of cut and there may also be occasions of air cutting. Basically, the best scenario is when the finished contour closely matches the casting, forging, or rough material shape.

The following is the block diagram for function G73 using the double block format:

> G73U...W...R...
>
> G73P ... Q ... U ... W ...F ... S ...
>
> where
>
> *In block one*
>
> U = total displacement in the direction of the X axis (sign + or -)
>
> W = total displacement in the direction of the Z axis (sign + or -)
>
> R = the number of rough cutting passes
>
> *In block two*
>
> P = number of the first block of the finished profile (given in the following figure as position "*b*")
>
> Q = number of the last block of the finished profile (given in the following figure as position "*c*")

U = finish stock allowance in the direction of the X axis (sign + or -), referred to the diameter

W = finish stock allowance in the direction of the Z axis (sign + or -)

F = feed rate, effective for blocks P through Q

S = spindle speed, effective for blocks P through Q

Notes on function G73 using the double block format:

1. *Do not confuse the function of U and W with those in the single block format.*
2. *There is no use of address, D, in the double block format. Depth of cut is automatically calculated by the control based on U, and W, stock removal amount in X and Z axes and R, the number of cutting passes.*

The following is a block diagram for function G73 using a single block:

G73P . . . Q . . . U . . . I . . . K . . . U . . . W . . . D . . . F . . . S . . .

where

P = number of the first block of the finished profile (given in the following figure as position "*b*")

Q = number of the last block of the finished profile (given in the following figure as position "*c*")

I = total displacement in the direction of the X axis (sign + or -)

K = total displacement in the direction of the Z axis (sign + or -)

U = finish stock allowance in the direction of the X axis (sign + or -), referred to stock left on the diameter

W = finish stock allowance in the direction of the Z axis (sign + or -)

D = the number of rough cutting passes

F = feed rate, effective for blocks P through Q

S = spindle speed, effective for blocks P through Q

Point *a* on the drawing is the starting point. In executing this cycle, the tool trav-

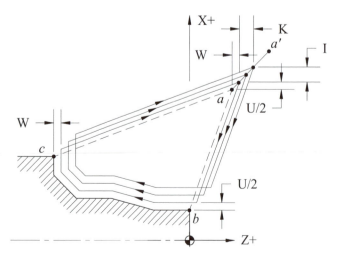

Figure 45
Pattern Repeating Cycle
(G73) Function Diagram

els from point a to point a', where displacements are defined by the values of I and K, as well as allowances for U and W, and, the cycle begins equivalent cutting passes with the number of passes, D. At the end of the cycle, the tool automatically returns to point a. Points b to c define the finished profile to be machined by function G70 as described in the next section.

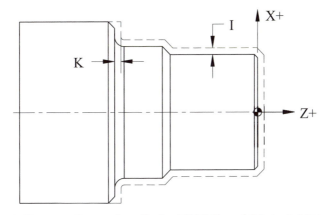

Figure 46 Pattern Repeating Cycle (G73) Equal Material Diagram

Address I and K are measured on the machined workpiece. The principle defining choice of signs is similar to that for U and W.

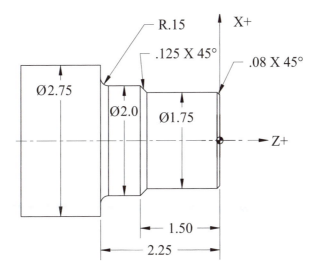

Figure 47 Drawing for Pattern Repeating Cycle (G73)

Example of Pattern Repeating Cycle, Program 11

    O0011

    N10G50S2000

    N15T0100M42

N20G96S500M03

N25G00X2.2Z.3T0101M08

N30G01Z.01F.03

N35X0.F.012

N40G00X3.0Z.2

N45G73P50Q85I.168K.169U.04W.02D3F.012

N50G00X1.59

N55G01Z0

N60X1.75Z-.08

N65Z-1.375

N70X2.0W-.125

N75Z-2.1

N80G03U.3Z-2.25I-.15K0.F.004

N85G01X2.85

N90G28U0W0T0100

N95M30

FINISHING CYCLE (G70)

Stock allowances left for finishing (U, W) may be removed by the same tool used in rough cutting. However, it is a common practice to use a different tool for the finishing pass. Application of function G70 allows for removal of the remaining stock allowance (with the previously applied cycles G71, G72, and G73, without repetitive passes, along the contour).

The following is a block diagram for function G70:

G70P . . . Q . . . F . . . S . . .

where

P = number of the first block of the finished profile (given in the above figure as position "*b*")

Q = number of the last block of the finished profile (given in the above figures as position "*c*")

F = feed rate, effective for blocks P through Q

S = spindle speed, effective for blocks P through Q

Notice from the block diagram, that it is only necessary to enter position coordinates of the first (b) through last block (c) of the previous rough cycle, which define the finished profile. This will cause an automatic return to the earlier part of the program for the coordinates needed for the completion of the process removing allowances U and W).

N10G50S1000

. . . .

....
N30G71P35Q65U.02W.01D1000F.014

......................................F.04

......................................F.06

......................................F.05

N50M01

N55G50S1500

....

....

N70G70P35Q65

....

N80M30

Attention: In the above program values of the feed used in information blocks following function G71 are ignored for the roughing cycle, but they are valid for function G70.

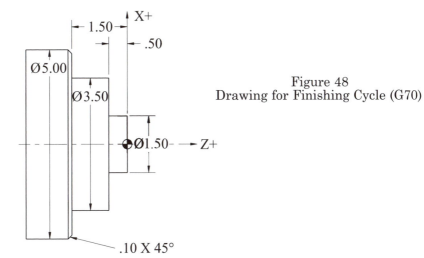

Figure 48
Drawing for Finishing Cycle (G70)

Example of Finishing Cycle, Program 12

    O0012

    N10G50S2000

    N15T0100M42

    N20G96S600M03

    N25G00X5.2Z.2T0101M08

    N30G71P35Q65U.04W.005D1000F.012

    N35G00X1.5

    N40G01Z-.5F.010

N45X3.5F.008

N50Z-1.5F.010

N55X4.8F.009

N60X5.1Z-1.7F.004

N65X5.2F.010

N70G00X7.75Z8.75T0100

N75M01

N80T0200S2000M42

N85G96S700M03

N90G00X5.2Z.2T0202M09

N95G70P35Q65

N100G28U0W0T0200

N105M09

N110T0100M05

N115M30

PECK DRILLING CYCLE (G74)

The most common use of this function is for drilling deep holes that require an interruption in the feed, in order to break long stringy chips. In spite of its name however, this function may be applied to cylindrical or face cutting of grooves that exhibit hard breaking chips as well. A block diagram of this function, as well as the movements of the tool, is illustrated below in Figure 49.

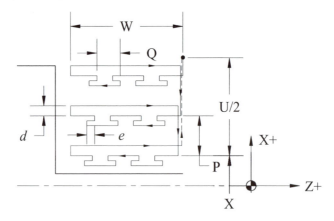

Figure 49 Peck Drilling Cycle (G74) Function Diagram

Amount of the clearance (indicated by d in the above figure), is set by a system parameter. Amount of the return (indicated by e in the above figure), is also set by a parameter.

The following is a block diagram for function G74 using a double block:

G74R . . .

G74X(U). . . Z(W). . . P. . . Q. . . R. . . F. . . S. . .

where

in the first block

R = retract amount of the tool after each cut

in the second block

X = diameter of the workpiece at the bottom of the groove

U = distance between the starting and end points (in an incremental system)

Z = final Z cut depth in the absolute system

W = Z distance from start to finish cut depth in the incremental system

P = depth of each cut for X axis (no sign)

Q = depth of each cut for Z axis (no sign)

R = retract amount of the tool at the bottom of the cutting

F = cutting feed rate

S = spindle speed

The following is a block diagram for function G74 using a single block:

G74X(U). . . Z(W). . . I. . . K. . . D. . . F. . . S. . .

X = diameter of the workpiece at the bottom of the groove

U = distance between the starting and end points along the X axis (in an incremental system)

Z = final Z cut depth in the absolute system

W = Z distance from start to finish cut depth in the incremental system

I = depth of cut per side in X direction (no sign)

K = depth of cut per side in Z direction (no sign)

D = retract amount of the tool at the bottom of the cutting

F = cutting feed rate

S = spindle speed

In the following program example, this function is applied for drilling a deep hole using the single block format.

G74Z(W). . . K. . . F. . .

Example of Peck Drilling Cycle, Program 13

    O0013

    N10G50S1000

    N15T0100M42

    N20G97S800M03

    N25G00X0.Z.2T0101M08

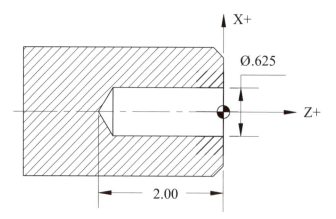

Figure 50 Drawing for Peck Drilling Cycle (G74)

N30G74Z-2.0K.550F.007

N35G28U0W0T0100M05

N40M30

At the end of drilling (i.e., block N30), the tool automatically returns to the starting position of Z.2. Advantages of the peck drilling cycle is that it includes, chip-breaking and cooling of the cutting drill tip.

GROOVE CUTTING CYCLE (G75)

Function G75, in its form, is very similar to function G74. Their difference lies in the direction of the tool movement (which is opposite that in function G74). Function G75 is used for cutting grooves that require an interrupted cut along the X axis.

The following is a block diagram for function G75 using a double block:

G75R. . .

G75X(U). . . Z(W). . . I. . . K. . . D. . . F. . . S. . .

The following is a block diagram for function G75 using a single block:

G75X(U). . . Z(W). . . I. . . K. . . D. . . F. . . S. . .

All notations assumed for this function are defined exactly as in function G74.

Example of Groove Cutting Cycle, Program 14

O0014

(RIGHT SIDE OF INSERT CUTS)

N10G50S2000M42

N15G96T0100M03

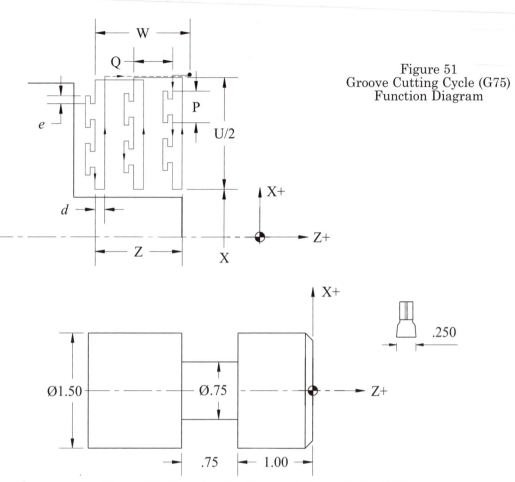

Figure 51
Groove Cutting Cycle (G75)
Function Diagram

Figure 52 Drawing for Groove Cutting Cycle (G75)

N20G00X1.55Z–1.25S400M08

N25G75X.75Z–1.75T0101K.125D0F.005

N30G00X7.75Z8.5I.150M09

N35T0100M05

N40M30

Note: At the end of the cycle, the tool returns to the starting point in both functions G74 and G75.

MULTIPLE THREAD CUTTING CYCLE (G76)

The G76 multiple thread cutting cycle is used on most modern controls, in place of the outdated G32 and G92. All of the information needed to complete the desired thread is input in either one or two blocks, depending on the control, rather than multiple blocks in the former. By inputting the appropriate data for a particular type of

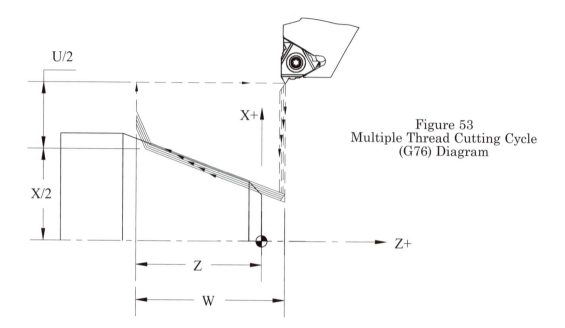

Figure 53
Multiple Thread Cutting Cycle
(G76) Diagram

thread in the program blocks, the number of cutting passes are automatically calculated by the control. Because of the limited number of blocks required, this method is very easy to program and edit.

The following is a block diagram for function G76 using a double block:

G76P. . . Q. . . R. . .

G76X. . . Z. . . R. . . P. . . Q. . . F. . .

where

in the first block

P = uses a six digit entry (P010000) of three pairs as follows:

the first two digits specify the number of finishing cuts (01 – 99)

three and four specifies the number of thread leads required for gradual pull- out (0.0 –9.9 times the lead) without a decimal point entry (00 – 99)

five and six denote the angle of the thread (only 00, 29, 30, 55, 60 and 80 degrees are allowed)

Q = the minimum cutting depth

R = finishing allowance (allows a decimal point)

in the second block

X = Final diameter of the thread

Z = Final end position of the thread along the Z axis, (can be specified as an incremental distance using address W)

R = incremental distance from the thread starting to ending, as a radial value (used for tapered threads)

P = single thread height (always a positive radial value, without a decimal point)

Q = depth of cut for the first threading pass (always a positive radial value, without a decimal point)

F = feed rate, lead of thread

The following is a block diagram for function G76 using a single block:

G76X(U). . . Z(W). . . I. . . K. . . D. . . F. . . A. . . P . . .

where

X(U) = diameter of thread core (last diameter cut)

Z(W) = full length of the cut thread (end of thread position)

I = difference of thread radius (+ or −) from start to finish (for tapered threading)

K = single thread height (always positive)

D = depth of first threading cut (always positive)

A = thread angle (matches the included angle of the threading insert and is always positive)

F = feed rate, lead of thread

Notes:

1. All notes from functions G32 and G92 are applicable to function G76.

2. The depth of a first cut D is approximately .003 to .018, depending on machining conditions.

3. With a small value of first cut, the number of passes increases and, inversely, with a greater value of the first cut, the number of passes decreases.

4. The selection of factor D depends on the type of thread, the material, and the tool tip.

Example of Multiple Thread Cutting Cycle, Program 15

> **O0015**
>
> **N10G50S2000**
>
> **N15T0100M42**
>
> **N20G97S800M03**
>
> **N25G00X1.35Z.5T0101M08**
>
> **N30G76X1.2Z−1.5I0K.055D.012A60E083333**
>
> **N35G28U0W0T0100M09**
>
> **N40M30**

Note: Depending on the value of the first cut specified by D, a certain number of passes will be obtained, as defined by the parameters that are set for the machine.

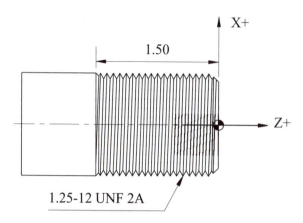

Figure 54 Drawing for Multiple
thread Cutting Cycle (G76)

Programming For the Tool nose Radius

In all of the programming examples examined thus far, the tool nose radius was considered to be zero. In reality, there is no such tool and the contour of the tool nose corresponds to the radius of its circle.

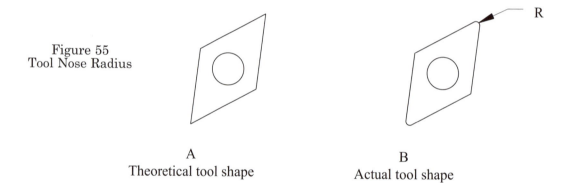

Figure 55
Tool Nose Radius

A
Theoretical tool shape

B
Actual tool shape

In the turning process on conventional lathes (in many cases), tool sharpening is performed on High Speed Steel (HSS) tool bits. On CNC lathes, HSS tools are rarely used or sharpened. Instead, indexable inserted tools tips are widely used. These (exchangeable) inserts are made from sintered carbides, and the chemical constitution of each is defined for a particular material in the form of cutting grades. The tool nose radii for inserts are standardized and can be ordered to meet the needs for most applications. The following are some of the most common radius sizes available.

Examples:

1/64 or .0156, 1/32 or .0312, 3/64 or .0469, 1/16 or .0625

When programming a tool radius that is assumed to be zero, the theoretical tool point is programmed within a given coordinate system of X and Z. However, as mentioned previously, inserts are used whose cutting tip is assigned a certain radius. Considering this fact, more complicated calculations must now be employed, in order to position the tool to compensate for the radius. The following figures demonstrate the two conditions for programming the use of tool nose radius.

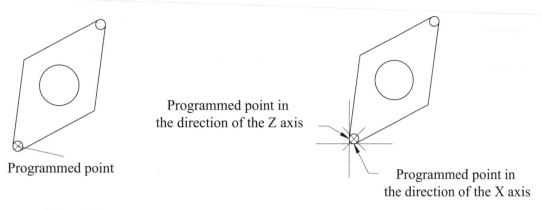

Programmed point in
the direction of the Z axis

Programmed point

Programmed point in
the direction of the X axis

Figure 56
The Center of the Tool
Nose Radius

Figure 57 The Two Initial Points

In order to gain a better understanding of these conditions, examine the following example where the tool nose radius, $r = 0$; then the above-mentioned cases can both be checked.

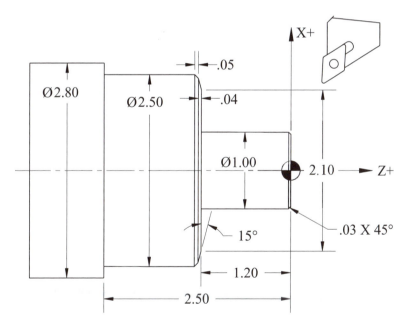

Figure 58 Drawing for Tool Nose Radius Programming

$$\tan 15° = \frac{.04}{a} \quad a = \frac{.04}{\tan 15°} = .1493$$

$$b = \frac{2.5 - (2.1 + 2 \times 0.1493)}{2} = .0507$$

$$\tan \beta = \frac{b}{.05} = \frac{.0507}{.05} = 1.0141 \quad \beta = 45.4°$$

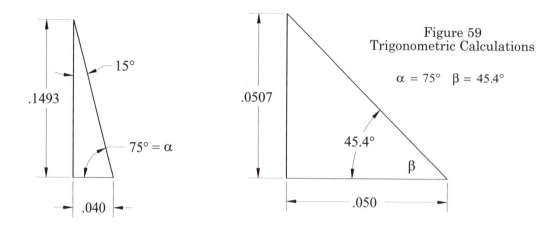

Figure 59
Trigonometric Calculations

$\alpha = 75°$ $\beta = 45.4°$

Example of Tool Nose Radius Programming, Program 16

> O0016
>
> N10G50X10.0Z8.0S2000
>
> N15T0100M42
>
> N20G96S500M03
>
> N25G00X1.1Z.2T0101M08
>
> N30G01Z0.F.02
>
> N35X0.F.01
>
> N40G00X.940Z.03
>
> N45G01Z0
>
> N50X1.0Z−.03F.005
>
> N55Z−1.2F.008
>
> N60X2.1
>
> N65X2.3986Z−1.24F.005
>
> N70X2.5Z−1.29
>
> N75Z−2.5F.008
>
> N80X2.850
>
> N85G00Z.5M09
>
> N90G28U0W0T0100M05
>
> N95M30

PROGRAMMING FOR THE CENTER OF THE TOOL NOSE RADIUS

Before proceeding with programming for the center of the circle that describes the tool nose, examine a few cases where calculations are necessary. Please note that in the drawings for the remainder of this section, the tool nose representation has been enlarged in order to make dimensional data easier to see. The tool nose radius in these cases is .0312 inch and, is therefore difficult to see otherwise.

To program the .03 X 45° chamfer as the drawing illustrates, determination of the Z_1 and X_2 coordinates are relatively simple, while X_1 and Z_2 require additional calculations.

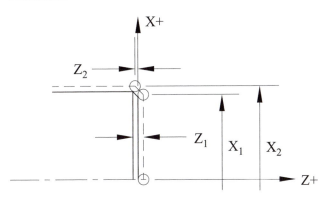

Figure 60
Programming for the Center of
the Tool Nose Radius

Calculation No. 1

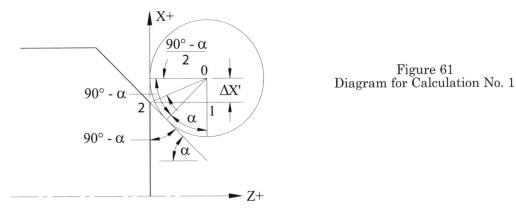

Figure 61
Diagram for Calculation No. 1

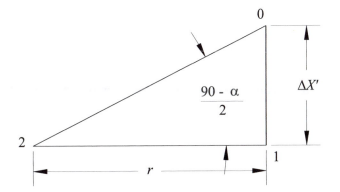

Figure 62 Trigonometric Diagram for Calculation No. 1

$$\Delta X' = r \times \tan\left(\frac{90 - \alpha}{2}\right) = r \times \tan\left(45 - \frac{\alpha}{2}\right)$$

147

Part 3 Programming CNC Turning Centers

Calculation No. 2

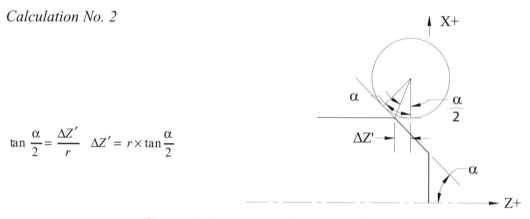

$$\tan\frac{\alpha}{2} = \frac{\Delta Z'}{r} \quad \Delta Z' = r \times \tan\frac{\alpha}{2}$$

Figure 63 Diagram for Calculation No. 2

Calculation No. 3

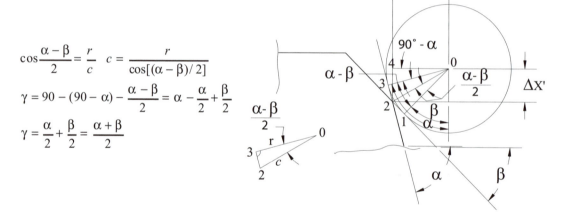

$$\cos\frac{\alpha - \beta}{2} = \frac{r}{c} \quad c = \frac{r}{\cos[(\alpha - \beta)/2]}$$

$$\gamma = 90 - (90 - \alpha) - \frac{\alpha - \beta}{2} = \alpha - \frac{\alpha}{2} + \frac{\beta}{2}$$

$$\gamma = \frac{\alpha}{2} + \frac{\beta}{2} = \frac{\alpha + \beta}{2}$$

Figure 64 Diagram for Calculation No. 3

$$\sin\gamma = \frac{\Delta z'}{c} \quad z' = c \times \sin\gamma$$

$$\Delta z' = \frac{r}{\cos[(\alpha - \beta)/2]}\sin\frac{\alpha + \beta}{2}$$

$$\Delta z' = r\frac{\sin[(\alpha + \beta)/2]}{\cos[(\alpha - \beta)/2]}$$

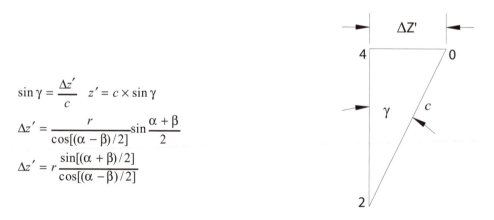

Figure 65 Trigonometric Diagram for Calculation No. 3

148

In the same manner, a formula for $\Delta X'$: can be derived

$$\Delta X' = r\frac{\cos[(\alpha + \beta)/2]}{\cos[(\alpha - \beta)/2]}$$

O0016

N10G50X10.0Z8.0S2000

N15T0100M42

N20G96S500M03

N25G00X1.1Z.2T0101M08

N30G01Z.0312F.02

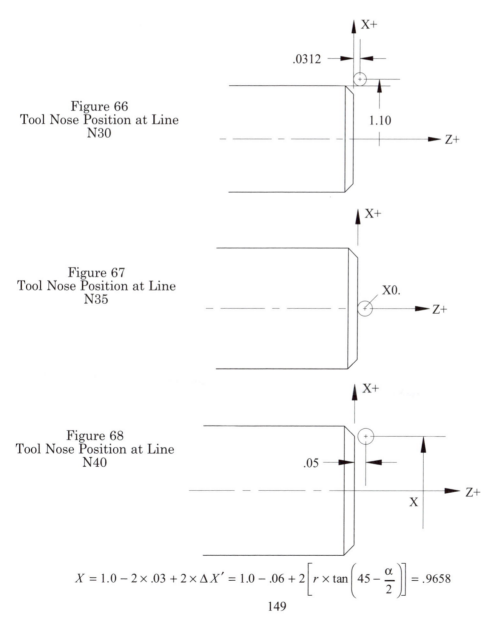

Figure 66
Tool Nose Position at Line
N30

Figure 67
Tool Nose Position at Line
N35

Figure 68
Tool Nose Position at Line
N40

$$X = 1.0 - 2 \times .03 + 2 \times \Delta X' = 1.0 - .06 + 2\left[r \times \tan\left(45 - \frac{\alpha}{2}\right)\right] = .9658$$

N35X0.F.01

N40G00X.9658W.05

N45G01Z.0312

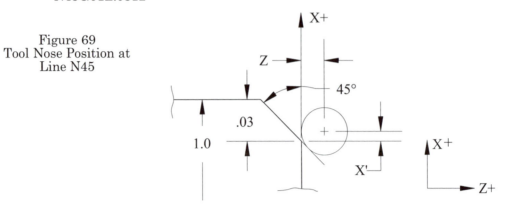

Figure 69
Tool Nose Position at
Line N45

N50X1.0624Z-.0171F.005

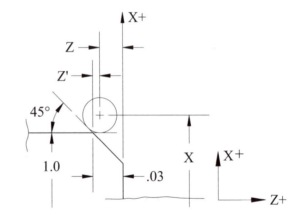

Figure 70
Tool Nose Position at
Line N50

$$X = 1.00 + 2 \times r = 1.00 + 2 \times .0312 = 1.0624$$

$$Z = 0.03 - \Delta z' = .03 - r \times \tan\frac{\alpha}{2} = .03 - .0312 \times \tan\frac{45}{2} = .0171$$

N55Z-1.1688

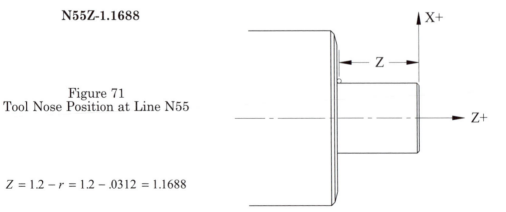

Figure 71
Tool Nose Position at Line N55

$$Z = 1.2 - r = 1.2 - .0312 = 1.1688$$

N60X2.1082

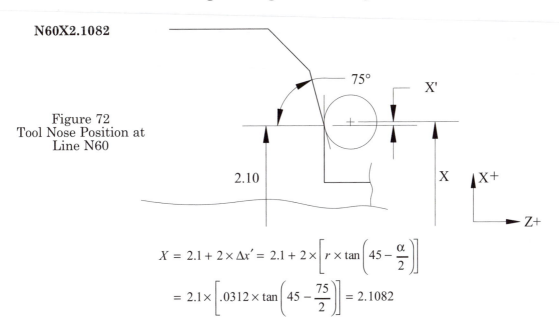

Figure 72
Tool Nose Position at
Line N60

$$X = 2.1 + 2 \times \Delta x' = 2.1 + 2 \times \left[r \times \tan\left(45 - \frac{\alpha}{2} \right) \right]$$

$$= 2.1 \times \left[.0312 \times \tan\left(45 - \frac{75}{2} \right) \right] = 2.1082$$

N65X2.4310Z-1.212F.005

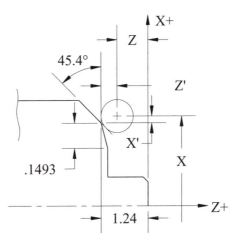

Figure 73 Tool Nose Position at Line N65

$$X = 2.1 + 2 \times .1493 + 2 \times \Delta X' = 2.1 + 0.2986 + 2 \times r \frac{\cos[(\alpha + \beta)/2]}{\cos[(\alpha - \beta)/2]}$$

$$= 2.3986 + 2 \times .0312 \frac{\cos[(75 + 45.4)/2]}{\cos[(75 - 45.4)/2]} = 2.4310$$

$$Z = 1.24 - \Delta z' = 1.24 - r \times \frac{\sin[(\alpha + \beta)/2]}{\sin[(\alpha - \beta)/2]}$$

$$= 1.24 - .0312 \frac{\sin[(75 + 45.4)/2]}{\sin[(75 - 45.4)/2]} = 1.212$$

N70X2.5624Z-1.2769

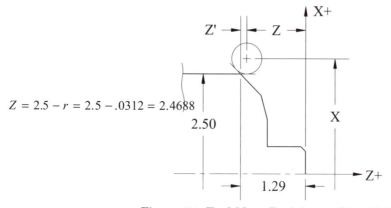

Figure 74 Tool Nose Position at Line N70

$$X = 2.5 + 2 \times r = 2.5 + 2 \times .0312 = 2.5624$$
$$Z = 1.290 - \Delta z' = 1.290 - r \times \tan \alpha$$
$$Z = 1.290 - .0312 \times \tan 45.4° = 1.2769$$

N75Z-2.4688

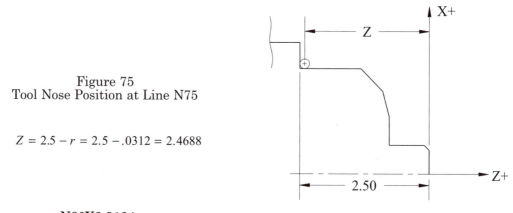

Figure 75
Tool Nose Position at Line N75

$$Z = 2.5 - r = 2.5 - .0312 = 2.4688$$

N80X2.9124

N85G00Z.5M09

N90G28U0W0T0100M05

N95M30

The last example leads to the following conclusions:

1. The operator will have difficulty in reading the program, because all the dimensions in the direction of the Z axis differ by the value of the tool radius, whereas, the diameters differ by double the value of the radius making program changes virtually impossible.

2. When using Coordinate Systems Setting (G50) values of X and Z are assigned to

function G50, facing is performed with point No. 1 on the tool, whereas, the diameter is turned with point No. 2.

Figure 76 Tool Nose Contact
    Points No. 1 and No. 2

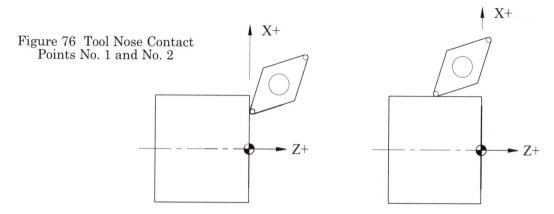

It is clear, therefore, that to the resulting value of Z, the value of the tool radius must be added, whereas to the resulting value of X, double the value of the tool radius must be added.

PROGRAMMING USING THE TWO INITIAL POINTS

The second method of programming refers to the description of the movements of a tool in two directions of motion: along the X and Z axes (assuming two different reference points). Point No. 1 on the tool is a reference point in the direction of the Z axis, and point No. 2 on the tool is a reference point in the direction of the X axis (see Figure 57). In order to make these changes, the formulas for X and Z must be rewritten as follows:

$$\Delta X = r - \Delta X' = r - r \times \tan\left(45 - \frac{\alpha}{2}\right)$$

$$\Delta X = r \times \left[1 - \tan\left(45 - \frac{\alpha}{2}\right)\right]$$

$$\Delta Z = r - \Delta Z'$$

$$\Delta Z = r - r \times \tan\frac{\alpha}{2}$$

$$\Delta Z = r \times \left(1 - \tan\frac{\alpha}{2}\right)$$

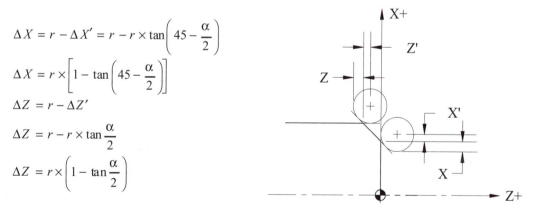

Figure 77 Diagram for the 45° Chamfer

Rewrite the formulas for the double chamfer as follows:

$$\Delta X = r - \Delta X' = r - r\frac{\cos[(\alpha + \beta)/2]}{\cos[(\alpha - \beta)/2]} = r\left(1 - \frac{\cos[(\alpha + \beta)/2]}{\cos[(\alpha - \beta)/2]}\right)$$

$$\Delta Z = r - \Delta Z' = r - r\frac{\sin[(\alpha + \beta)/2]}{\cos[(\alpha - \beta)/2]} = r\left(1 - \frac{\sin[(\alpha + \beta)/2]}{\cos[(\alpha - \beta)/2]}\right)$$

O0016

N10G50X10.0Z8.0S2000

N15T0100M42

N20G96S500M03

N25G00X1.1Z.2T0101M08

N30G01Z0F.02

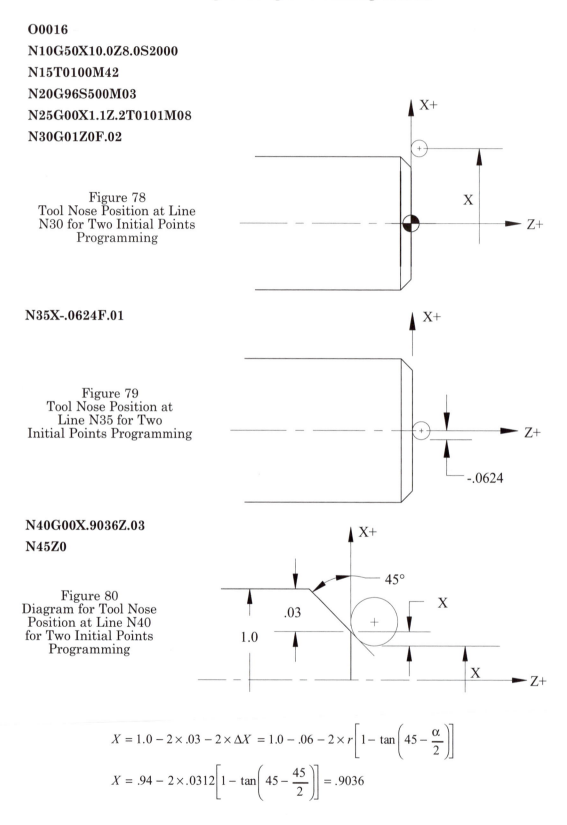

Figure 78
Tool Nose Position at Line
N30 for Two Initial Points
Programming

N35X-.0624F.01

Figure 79
Tool Nose Position at
Line N35 for Two
Initial Points Programming

N40G00X.9036Z.03

N45Z0

Figure 80
Diagram for Tool Nose
Position at Line N40
for Two Initial Points
Programming

$$X = 1.0 - 2 \times .03 - 2 \times \Delta X = 1.0 - .06 - 2 \times r\left[1 - \tan\left(45 - \frac{\alpha}{2}\right)\right]$$

$$X = .94 - 2 \times .0312\left[1 - \tan\left(45 - \frac{45}{2}\right)\right] = .9036$$

N50X1.0Z-.0483F.005

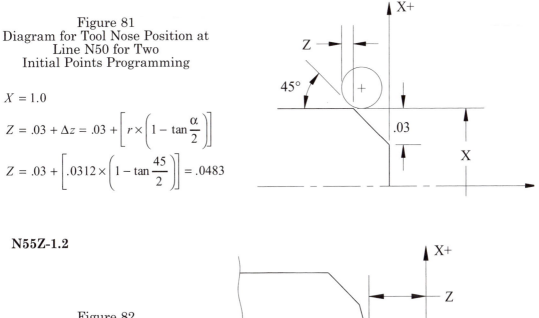

Figure 81
Diagram for Tool Nose Position at
Line N50 for Two
Initial Points Programming

$X = 1.0$

$$Z = .03 + \Delta z = .03 + \left[r \times \left(1 - \tan\frac{\alpha}{2} \right) \right]$$

$$Z = .03 + \left[.0312 \times \left(1 - \tan\frac{45}{2} \right) \right] = .0483$$

N55Z-1.2

Figure 82
Diagram for Tool Nose Position at
Line N55 for Two Initial
Points Programming

N60X2.0458

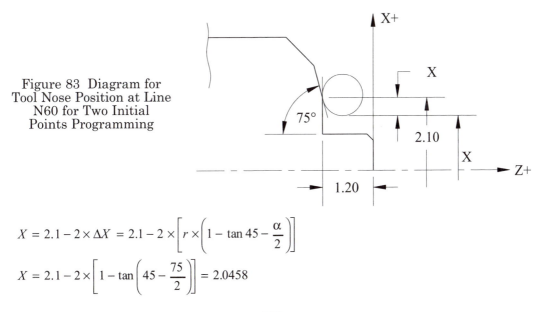

Figure 83 Diagram for
Tool Nose Position at Line
N60 for Two Initial
Points Programming

$$X = 2.1 - 2 \times \Delta X = 2.1 - 2 \times \left[r \times \left(1 - \tan 45 - \frac{\alpha}{2} \right) \right]$$

$$X = 2.1 - 2 \times \left[1 - \tan \left(45 - \frac{75}{2} \right) \right] = 2.0458$$

Figures 87 and 88 detail the necessary information to select the proper setting for the Tool Tip Orientation (T), in the offset register, for a rear turret lathe.

In the Figure 87, the tool tip is represented and the contact point is identified. The drawing should help with the selection of the appropriate Tool Tip Orientation (T), based on the direction the contact point is pointing. In Figure 88, the same concept is superimposed onto the actual tool nose and the arrows indicate the contact point and direction.

Thus, the selection of the Tool Tip Orientation number is determined by the direction the tool nose is pointing. Use number 0 or 9 for programming the center of the tool nose.

CALLING G41 OR G42 IN THE PROGRAM
Selecting which code to use

Selection of either G41 or G42 is based on the side of the part profile that the tool needs to be on, to create the desired results. Think of the part profile as the centerline of a highway and, then, based on the direction of travel, decide which side of the road to drive on. For the left side, G41 is selected and, for the right side, G42 is used. This procedure may be applied whether the cut is internal, or external.

Initiating Tool Nose Radius Compensation with G41 or G42

To initiate the use of tool nose radius compensation, the G41 or G42 should occur on a G00 rapid positioning move that is at least .100 of an inch away from the part profile. This move need only be in one axis direction, but it can include both. Please note how the G42 is initiated in the following example program for the part shown in Figure 58.

Ending use of G41 or G42 with function G40

In order to end the use of G41 or G42, TNRC may be cancelled by using function G40. When the machine is first started, the G40 command is active by default. To program the cancellation of Tool Nose Radius Compensation, the command is generally input on a move that is in departing vector from the machined profile. This move may be either G00 or G01 and cancellation may be initiated with the G28 command, where the compensation will be cancelled upon reaching the intermediate point. Please note how the G40 is used to cancel tool nose radius compensation in the following example program for the part shown in Figure 58.

Example of Tool Nose Radius Compensation (TNRC) Using G41 or G42, Program 161

```
O0161
N10G50S2000
N15T0100M42
N20G96S500M03
N25G00G41X1.1Z.2T0101M08
N30G01Z0.F.02
N35X-.04F.01
```

158

N40G00Z.03

N45G42X.940

N50G01Z0

N55X1.0Z−.03F.005

N60Z−1.2F.008

N65X2.1

N70X2.3984Z−1.24F.005

N75X2.5Z−1.29

N80Z−2.5F.008

N85X2.850

N90G00G40X4.Z.5M09

N95G28U0W0T0100M05

N100M30

Notes on using G41 and G42:

If the values for Tool Nose Radius (R) and the Tool Tip Orientation (T) are omitted in the offset register, the desired results will not be obtained (0 radius is assumed in this case). Note the change from G41 for facing, to G42 for the profile.

PROGRAMMING EXAMPLES FOR TURNING CENTER
CUTTING TOOL DESCRIPTIONS FOR TURNING CENTER
PROGRAMMING EXAMPLES

The following descriptions are for the tools used in the remaining Turning Center program examples:

Tool number 01 = an outside diameter rough turning tool with an 80° diamond insert with a .031 tool nose radius.

Figure 89
Rough Turning Tool

Tool number 02 = an outside diameter finish turning tool with a 55° diamond insert with a .015 tool nose radius.

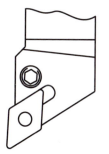

Figure 90
Finish Turning Tool

Tool number 03 = an inside diameter rough boring bar with an 80° diamond insert with a .031 tool nose radius.

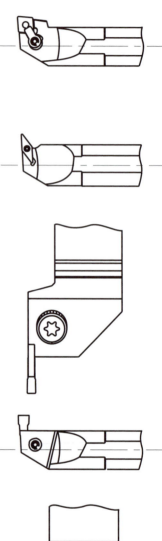

Figure 91
Rough Boring Bar

Tool number 04 = an inside diameter finish boring bar with a 55° diamond insert with a .015 tool nose radius.

Figure 92
Finish Boring Bar

Tool number 05 = an outside diameter, U style, grooving tool with a .005 tool nose radius.

Figure 93
Outside Diameter Grooving Tool

Tool number 06 = an inside diameter, U style, grooving tool with a .005 tool nose radius.

Figure 94
Inside Diameter Grooving Tool

Tool number 07 = an outside diameter 60° threading tool.

Figure 95
Outside Diameter Threading Tool

Tool number 08 = an inside diameter 60° threading tool.

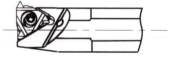

Figure 96
Inside Diameter Threading Tool

Tool number 09 = a #5 High Speed Steel center drill.

Figure 97
Center Drill

Tool number 10 = a standard drill.

Figure 98 Drill Bit

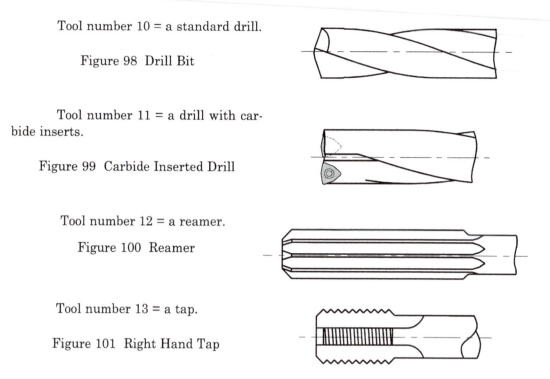

Tool number 11 = a drill with carbide inserts.

Figure 99 Carbide Inserted Drill

Tool number 12 = a reamer.

Figure 100 Reamer

Tool number 13 = a tap.

Figure 101 Right Hand Tap

Note: Using a 12 position turret tool number 13 in the list above would replace number 12.

APPLICATION OF G00 AND G01 IN BOTH ABSOLUTE AND INCREMENTAL SYSTEMS

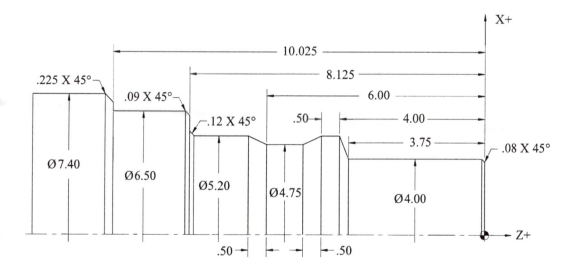

Figure 102 Drawing for Program Example 17

Part 3 Programming CNC Turning Centers

Preliminary Considerations:

1. For the following example, the workpiece is rough turned with an allowance given for finishing; therefore, the use of only one finishing tool is programmed.

2. The part profile is programmed as if the tool nose radius is equal to zero.

3. The left side of the workpiece was faced in a previous operation.

Example Application of G0 and G01 Using the Absolute System Coordinate System, Program 17

O0017

N10G50S1500

N15T0100M42

N20G96S600M03

N25G00X4.25Z.1T0101M08

N30G01Z0.F.03

N35X-.05F.01

N40G00Z.03

N45X3.84

N50G01Z0

N55X4.0Z-.08

N60Z-3.75

N65X5.2Z-4.0

N70Z-4.5

N75X4.75Z-5.0

N80Z-6.0

N85X5.2Z-6.5

N90Z-8.005

N95X5.44Z-8.125

N100X6.32

N105X6.5Z-8.215

N110Z-10.025

N115X6.95

N120X7.5Z-10.35

N125G00Z.1M09

N130G28U0W0T0100M05

N135M30

Example Application of G0 and G01 Using the Incremental System Coordinate System, Program 171

    O0171

    N10G50S1500

    N15T0100M41

    N20G96S600M03

    N25G00X4.25Z.1T0101M08

    N30G01W−.1F.03

    N35U−4.3F.01

    N40G00W.03

    N45U3.89

    N50G01W−.03

    N55U.16W−.08

    N60W−3.67

    N65U1.2W−.25

    N70W−.5

    N75U−.45W−.5

    N80W−1.0

    N85U.45W−.5

    N90W−1.505

    N100U.24W−.12

    N105U.88

    N110U.18W−.09

    N115W−1.81

    N120U.45

    N125U.55W−.325

    N130G00U12.5W10.4M09

    N135W14.9T0100M05

    N140M30

Example Application of G0 and G01 Using Both Absolute and Incremental Coordinate Systems Combined, Program 172

    O0172

    N10G50S1500

    N15T0100M41

    N20G96S600M03

    N25G00X4.25Z.1T0101M08

    N30G01Z0.F.03

N35X0.F.01

N40G00X3.84W.03

N45G01Z0

N50X4.0W-.08

N55Z-3.75

N60X5.2W-.25

N65Z-4.5

N70X4.75W-.5

N75Z-6.0

N80X5.2W-.500

N85Z-8.005

N90U.24W-.12

N95X6.32

N100X6.5W-.09

N105Z-10.025

N110X6.95

N115X7.5W-.325

N120G00Z.1M09

N125X20.0Z15.T0100M05

N130M30

Eliminating Taper in Turning

Although CNC lathes are superior in accuracy to manually operated lathes, CNC lathes are far from perfect. Older machines are adversely affected by temperature changes, which in turn may influence the performance of electronic circuits. In addition, the temperature of the coolant may change the dimensions of turned parts. Even tool wear can cause similar variations. Such influential factors, during cylindrical turning, may cause differences in the values of the diameters for two points of the same part, as shown in the following figure. In the following example, Tool # 1 (shown in Figure 90) is used and the insert has a .031 Tool Nose Radius.

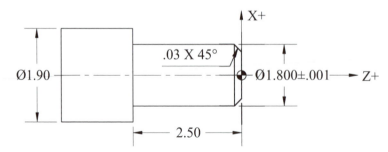

Figure 103 Drawing for Program Example 18

Example of Eliminating Taper in Turning, Program 18

O0018

N10G50S1800

N15T0100M42

N20G96S600M03

N25G00G41X2.Z.1T0101M08

N30G01Z.0F.03

N35X–.065F.008

N40W.1

N45G00G42X1.7036W.03

N50G01Z0

N55X1.8W–.0482

N60Z–2.5

N65X2.0

N70G00Z.2M09

N75G28U0W0T0100M05

N80M30

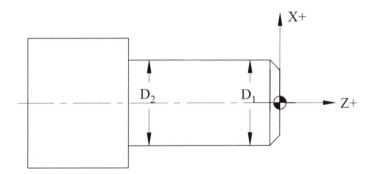

Figure 104 Drawing for Eliminating Taper in Turning Details

The measured diameters for points D_1 and D_2 are

For D_1 = 1.8 and D_2 = 1.802

The correction offset can be assigned to block N60 as follows:

N60U–.002 Z–2.5

For D_1 = 1.8 and D_2 = 1.798

N60 U.002 Z–2.5

For D_1 = 1.802 and D_2 = 1.8

N60 U.002 Z–2.5

Another common method of eliminating the taper that is less error prone (because the operator is adjusting the offset page rather than the program) is to use two offsets for the same tool. In this case the value is adjusted in the X axis wear offset register. The initial positioning move uses the original offset called in block N25 (T0101).

The correction wear offset can be assigned to block N60 as follows:

N60T0109Z–2.5

For D_1 = 1.8 D_2 = 1.798

In this case the offset amount in the #9 X axis wear compensation register needs to be a negative (-.002) to correct the error.

For D_1 = 1.802 D_2 = 1.8

In this case the offset amount in the #9 X axis wear compensation register needs to be positive (.002) to correct the error.

After eliminating the taper, the diameter will be equal for the entire turned length. To reduce the diameter, change the value in the program for U or the wear offset in a negative direction by a value of $-.002$ in.

Eliminating Taper in Threading

Just as the statement implies, the situation of taper in threading is very similar to that of turning, but the actual method for elimination differs as described below.

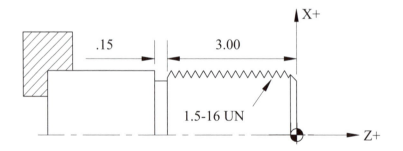

Figure 105 Drawing for Program Example 19

Example of Eliminating Taper in Threading, Program 19

O0019

N10G50S1800M08

N15T0700M42

N20G97S764M03

N25G00X1.6Z.5T0707M24

N30G76X1.4234Z–3.075I0K.0383D01F0.0625

N35G28U0W0T0700M23

N40M09

N45M30

As a reminder, look again at a part of the taper thread drawing, which first appeared as Figure 28 of this section.

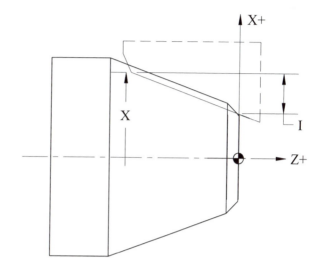

Figure 106
Taper Thread Drawing

Notice that X (the minor diameter) is the end point of threading, whereas magnitude I, appears at the beginning. By comparing this situation to straight turning, note another difference that results from the fact that I is calculated for one side only, unlike U.

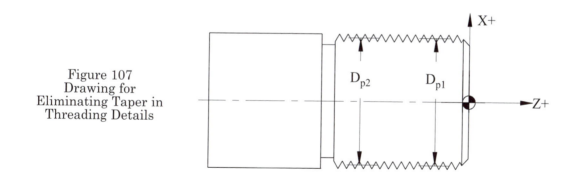

Figure 107
Drawing for
Eliminating Taper in
Threading Details

If the D_{p2} pitch diameter is correct, but D_{p1} is smaller by 0.003, then in block N30, assign –.0015 for the value of I. If the D_{p2} pitch diameter is smaller than D_{p1} by 0.003, then I should read –.0015, and the wear offset value of .003 should be assigned in the positive direction for the X axis.

Subprogram Application

By using subprograms we can easily program parts that have identical repetitive features. An advantage of subprograms is that fewer lines of programming code are needed. The following is a simple example of how a subprogram might be used for cutting multiple grooves:

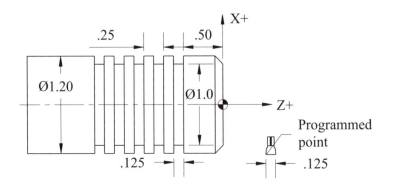

Figure 108 Drawing for Program Example 20

Example of Subprogram Application, Program 20

>O0020
>
>N10G501600
>
>N15T0500M03
>
>N20G96S400M42
>
>N25G00X1.25Z–.375T0505M08
>
>N30M98P0002L5
>
>N35G28U0W0T0500M09
>
>N40M30

Subprogram for Example of Subprogram Application, Program 20

>O0002
>
>N10W-.25
>
>N15G01X1.0F.004
>
>N20G00X1.25
>
>N25M99

In block N30 of the main program the subprogram O0002 is called and executed 5 times. When N25 of the subprogram is reached, the execution returns to the main program at block N35 and proceeds to its end.

EXAMPLE OF MAKING A TAPER PIPE THREAD

For this example a 3/4 inch National Pipe Thread (NPT) will be programmed on a 4140 steel part. Before programming can begin, several calculations are necessary as shown on the next page:

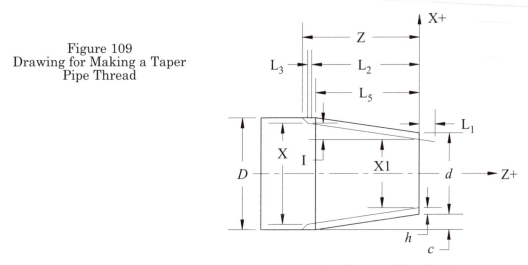

Figure 109
Drawing for Making a Taper
Pipe Thread

The thread dimensional data are drawn from *Machinery's Handbook*.

D = 1.05, diameter of pipe

L_2 = .5457, effective thread length

L_5 = .4029, length from the end of the pipe to the intersection of the taper and the outside diameter of the straight turned portion

h = .0571, height of the thread

Lead = 14

The angle of the thread is equivalent to 60°.

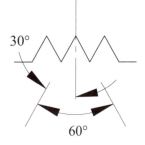

Figure 110
Thread Angle

Taper of the thread on the diameter is 3/4 of an inch per foot.

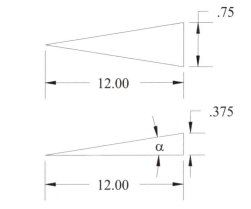

Figure 111
Taper per Foot

$$\tan \alpha = \frac{.375}{12.0} = .03125$$
$$\alpha = 1.7899° = 1°47'$$

Calculating diameter d,

$d = D - 2_c$

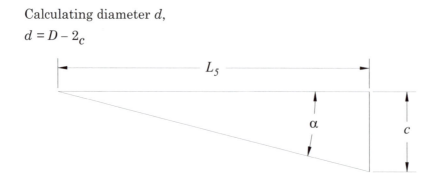

Figure 112 Small Diameter Taper Calculation

$$\tan \alpha = \frac{c}{L_5}$$

$$c = L_5 \times \tan \alpha = .4029 \times .03125 = .0126$$

$$d = 1.5 - 2 \times .0126 = 1.0248$$

The following is a calculation of the resulting length of "L", for the programmed section during threading.

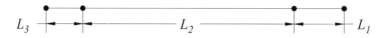

Figure 113 Tapered Length Calculation

$$L = L_1 + L_2 + L_3$$

$$L_3 = \frac{S \times P}{1800} = \frac{727 \times .0714}{1800} = .028$$

$$a = 3.2$$

$$L_1 = L_3 \times a = .028 \times 3.2 = .0896 \approx .090$$

$$L = .09 + .5457 + .028 = .6637$$

Calculating the value of I.

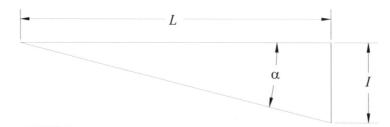

Figure 114 Drawing for the Calculation of I

$$\tan \alpha = \frac{I}{L}$$

$$I = L \times tg\,\alpha = .6637 \times .03125 = .0207$$

$$I = .0207$$

Calculating the diameter "X".

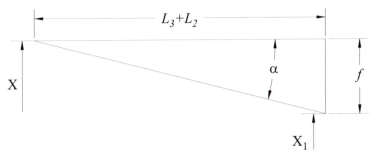

Figure 115 Drawing for the Calculation of X

$$X = X_1 + 2f$$
$$X_1 = d - 2h = 1.0248 = 2 \times .0571 = .9106$$
$$\tan\alpha = \frac{f}{L_2 + L_3}$$
$$f = (L_2 + L_3) \times \tan\alpha = (.5457 + .028) \times .03125 = .0179$$
$$X = X_1 + 2f = .9106 + 2 \times .0179 = .9464$$
$$X = .9464$$

Example of Making a Taper Pipe Thread, Program 21

In the examples that follow the CNC Setup Sheet is used (as described in Part I of this text). This provides and aid in identifying multiple tools and other specific setup information.

Machine: Turning Center	**Program Number: O0021**

Workpiece Zero = X, <u>Centerline</u> Z, <u>Part Face</u>

Setup Description: Material = 4140 Alloy Steel AISI 1300

Tool # and Offset #	**Tool Orientation #**	**Description**	**Insert Specification**	**Comments**
T0101	3	O.D. Turning Tool	80° Diamond 1/32 nose radius	400 SFPM
T0707	8	O.D. Threading Tool	60° V Thread forming insert	300 SFPM

O0021

(OD TURNING TOOL)

N10G50S1500M08

N15T0101M42

N20G96S400M03

N25G00G41X1.2Z.2

N30G01Z0.F.03

N35X-.06F.008

N40G00W.03

N45G42X1.0266

N50G01Z0

N55X1.050Z-.4336F.007

N60G00U.06Z.2M09

N65G28G40U0W0T0100

N70M01

(OD THREADING TOOL)

N75G50S1000M08

N80T0707M42

N85G97S1000M03

N90G00G42X1.15Z.09

N95G76X.9464Z-.5737I-.0207K.0571A60D.012F.071428

N100G28G40U0W0T0700M09

N105M30

Notes: An offset number was introduced in block N15. In this case, the tool will advance rapidly to the position (X, Z), whose values are determined by the offset register. As described in previous comments, the resulting motion is defined as the movement to given coordinates with a displacement equivalent to the given values of the offsets.

Function G00 is used in block N10 and is valid until block N30 where it is replaced by a linear feed move.

EXAMPLE OF TURNING BAR STOCK

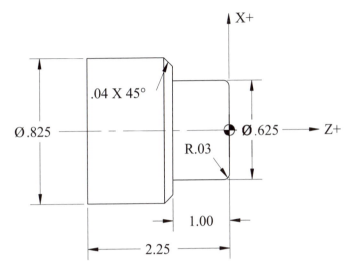

Figure 116 Drawing for Turning Bar Stock Example 22

Example of Turning Bar Stock, Program 22

Machine: Turning Center	Program Number: O0022

Workpiece Zero = X, Centerline Z, Part Face
Setup Description: Operation No. 1
Material: 1" diameter Carbon Steel 1100 bar stock

Tool # and Offset #	Tool Orientation #	Description	Insert Specification	Comments
T0101	3	O.D. Turning Tool	80° Diamond 1/32 nose radius	600 SFPM
T0505	3	O.D. Grooving Tool	.125 wide 1/64 corner radius	300 SFPM

O0022

(OD TURNING TOOL)

N10G50S4000M08

N15T0100M42

N20G96S600M03

N25G00G41X1.1Z.2T0101

N30G01Z0F3.0

N35X-.06F7.0

N40G00W.030

N45G42X1.1Z.2

N50G71P55Q85U.03W.003D0800F0.016

N55X.565Z0

N60G03X.625Z-.03I0K-.03

N65G01Z-1.0

N70X.745

N75X.825Z-1.04

N80Z-2.375

N85X1.1

N90G70P55Q85F.007

N95G28G40U0W0T0100

N100M01

(OD GROOVING TOOL)

N105G50S3000

N110T0500M42

N115G96S300M03

N120G00X1.1Z.2T0505M08

N125Z-2.3750

N130G01X.10F.012

N135X0

N140M05

N145G00X1.1

N150G28U0Z2.0T0500

N155M30

EXAMPLE OF LONG SHAFT TURNING

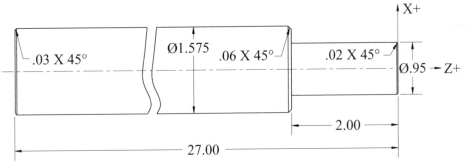

Figure 117 Drawing for Example of Long Shaft Turning

The sequence of operations for turning the shaft in the Figure 117 are: (1) face and chamfer the shaft from the left side; (2) the program stops when using function M00; (3) turn the shaft around 180°; and, (4) turn the remaining shaft end from the other side.

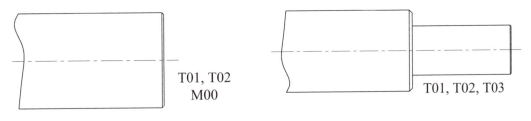

Figure 118
Drawing for Operation 1
Example of Long Shaft Turning

Figure 119
Drawing of Operation 1 Finished
Example of Long Shaft Turning

Example of Long Shaft Turning, Operation 1, Program 23

Machine: Turning Center	Program Number: O0023

Workpiece Zero = X, Centerline Z, Part Face

Setup Description: Operation No. 1

Material: 1.625" diameter Tool Steel bar stock

Tool # and Offset #	Tool Orientation #	Description	Insert Specification	Comments
T0101 T0111	3	O.D. Turning Tool	80° Diamond 1/32 nose radius	350 SFPM
T0909		#5 HSS Center Drill		35 SFPM
T0303	3	O.D. Turning Tool	55° Diamond 1/32 nose radius	400 SFPM

Program for the first operation

O0023

(ROUGH OD TURNING TOOL)

N10G50S1800

N15T0100M41

N20G96S350M03

N25G00G41X1.7Z.2T0101M08

N30G01Z0.F.02

N35X-.06F.01

N40G00W.03

N45G42X1.455

N50G01X1.635W-.06F.004

N55G00Z.2M09

N60G28G40U0W0T0100

N65M01

(#5 CENTER DRILL)

N70G50S35M03

N75T0900M41

N80G00X0Z.1T0909M08

N85G01Z-.45F.006

N90G00Z.2M09

N95X25.0Z8.0T0900

N100M00

(ROUGH OD TURNING TOOL SECOND OFFSET)

N105G50S1800

At block N105, the value of Z is greater by .07 inch than the value of Z in block N10, because after the turning process of one end is finished, the shaft length needs to be 27.07 inch; whereas, after the turning process of the other end is finished, the shaft length willbe 27.0 inches.

N110T0100M41

N115G96S350M03

N120G00G41X1.7Z.2T0111M08

N125G01Z.01F.02

N130X-.06F.014

N135G00W.03

N140G40X1.375

N145G01Z-1.99

N150G00U.06Z.1

N155X1.175

N160G01Z-1.99

N165G00U.06Z.1

N170X.975

N175G01Z-1.99

N180X1.7

N185G00Z.2M09

N190G28U0W0T0100

N195M01

(FINISH OD TURNING TOOL)

In block N120, a new offset number (11) for tool number 1 is introduced, in order to maintain independent control of the length of both sides of the shaft.

N200G50S2000

N210T0300M42

N215G96S400M03

N220G00G41X1.1Z.1T0303M08

N225G01Z0.F.02

N230X-.06F.008

N235G00W.03

N240G42X.870

N245X.95Z-.02F.004

N250Z2.0F.008

N255X1.455

N260X1.695W-.120F.004

N265G00Z.2M09

N270G28G40U0W0T0300M05

N275T0100

N280M30

In block N275, tool 1 is called, in order to prepare this tool for the turning of the next operation.

The Second Operation

In the second operation, the cylindrical portion having a diameter equal to 1.575 inch is machined. The holding arrangement of the workpiece is shown below.

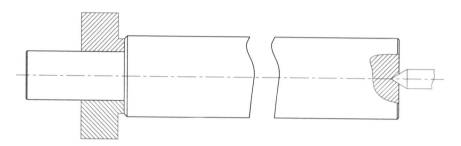

Figure 120 Drawing for Operation 2 Example of Long Shaft Turning

Example of Long Shaft Turning Operation 2, Program 231

Machine: Turning Center	Program Number: O0231

Workpiece Zero = X, Centerline Z, Part Face
Setup Description: Operation No. 2
Material: Tool Steel

Tool # and Offset #	Tool Orientation #	Description	Insert Specification	Comments
T0101	3	O.D. Turning Tool	80° Diamond 1/32 nose radius	350 SFPM
T0202	3	O.D. Turning Tool	55° Diamond 1/32 nose radius	400 SFPM

Program for the Second Operation
O0231
(ROUGH OD TURNING TOOL)
N10G50S1800
N15T0100M41
N20G96S350M03
N25G00X1.59Z.1T0101M08

N30G01Z-25.1F.012
N35G00U.06Z.1M09
N40G28U0W0T0100
N45M01
(FINISH OD TURNING TOOL)
N50G50S2000
N55T0200M42
N60G96S400M03
N65G00G42X1.6Z.1T0202M08
N70G01X1.575F.02
N75Z-25.1F.008
N80G00U.06Z.1M09
N85G28G40U0W0T0200M05
N90M30

PROGRAMMING EXAMPLE FOR MAKING A BUSHING

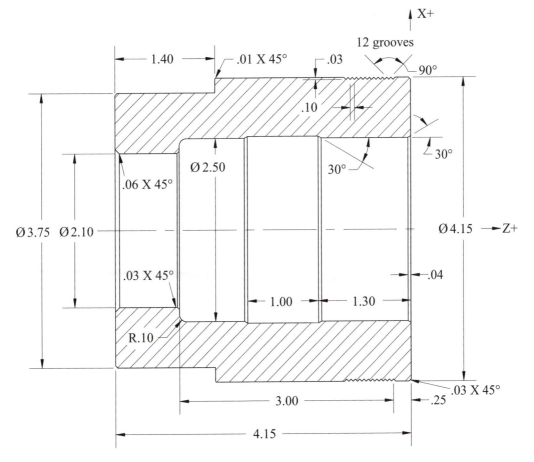

Figure 121 Drawing for Making a Bushing Example

First operation

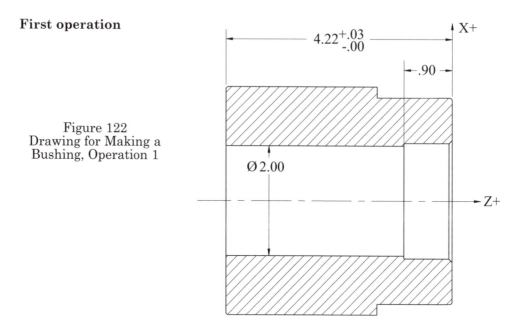

Figure 122
Drawing for Making a
Bushing, Operation 1

In the first operation, the left side of the part including the hole with a 2.0 diameter and the length of the 4.22 inch is machined. The length is finish machined in the next operation.

Example of Making a Bushing, Operation 1, Program 24

Machine: Turning Center	Program Number: O0024

Workpiece Zero = X, <u>Centerline</u> Z, <u>Part Face</u>
Setup Description: Operation No. 1
Material: Aluminum

Tool # and Offset #	Tool Orientation #	Description	Insert Specification	Comments
T0101	3	O.D. Turning Tool	80° Diamond 1/32 nose radius	800 SFPM
T0909		#5 HSS Center Drill		200 SFPM
T1010		2.0 Diameter HSS Drill		200 SFPM Min. of 4.5 length
T0202	3	O.D. Turning Tool	55° Diamond nose radius	1200 SFPM
T0303	8	I.D. Boring Bar	80° Diamond 1/32 nose radius	800 SFPM

Program for the First Operation

O0024

(ROUGH OD TURNING TOOL)

N10G50S1000

N15T0100M41

N20G96S800M03

N25G00G41X4.4Z.2T0101M08

N30G01Z.02F.02

N35X-.06F.014

N40G00W.05

N45G00G40X4.W.05

N50G01Z-1.39

N55G00U.06Z.1

N60X3.78

N65Z-1.39

N70X4.35

N75G28U0W0T0100M09

N80M01

(#5 CENTER DRILL)

N85T0900M42

N90G50S1830M03

N95G00X0.Z.2T0909M08

N100G01Z-.6F.006

N105G00Z1.M09

N110G28U0W0T0900

N115M01

(2-INCH DIAMETER DRILL)

N120T1000M41

N125G50S400M03

N130G00X0.Z.2T1010M08

N135G74Z-4.8K1.125F.008

N140G00Z1.M09

N145G28U0W0T1000

N150M01

(FINISH OD TURNING TOOL)

N155T0200M42

N160G96S1200M03

N165G00G41X3.85Z.1T0202M08

N170G01Z0.F.02

N175X1.8F.01

N180G00W.03

N185G42X3.63

N190X3.75W-.03F.004

N195Z-1.4F.001

N200X4.13

N205X4.15W-.01F.004

N210U-.03W-.03

N215G28G40U0Z1.T0200M09

N220M01

(ROUGH ID BORING BAR)

N225T0300M41

N230G96S800M03

N235G00G42X-2.22Z.1T0303M08

N240X-2.1W-.06F.004

N245Z-1.

N250G00U.06Z1.M09

N255G28G40U0W0T0300M05

N260M30

Second operation

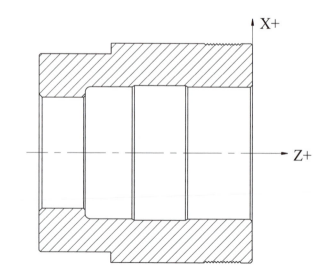

Figure 123 Drawing for Making a Bushing, Operation 2

In the second operation, the remaining portion of the bushing will be machined from the opposite end.

Example of Making a Bushing, Operation 2, Program 25

Machine: Turning Center	Program Number: O0025

Workpiece Zero = X, <u>Centerline</u> Z, <u>Part Face</u>
Setup Description: Operation No. 2
Material: Aluminum

Tool # and Offset #	Tool Orientation #	Description	Insert Specification	Comments
T0101	3	O.D. Turning Tool	80° Diamond 1/32 nose radius	800 SFPM
T0303	8	I.D. Boring Bar	80° Diamond 1/32 nose radius	800 SFPM
T0202	3	O.D. Turning Tool	55° Diamond 1/32 nose radius	1200 SFPM
T0404	8	I.D. Boring Bar	55° Diamond 1/32 nose radius	1200 SFPM
T0707	8	Special 90° V-Form Tool		300 SFPM

Program for the Second Operation

O0025

(ROUGH OD TURNING TOOL)

N10T0100M42

N15G96S800M03

N20G00X4.4Z.2T0101M08

N25G01Z.02F.03

N30X1.8F.01

N35G00X4.16W.05

N40G01Z-2.85

N45G00U.06Z1.T0100M09

N50G28U0W0

N55M01

(ROUGH ID BORING BAR)

N60T0300M42

N65G96S800M03

N70G00X-2.2Z.2T0303M08

N75G01Z-3.24F.01

N80G00U.06Z.1

N85X-2.4

N90G01Z-3.24

N95G00U.06Z.1

N100X-2.47

N105G01Z-3.14

N110U.2Z-3.24

N115X-2.

N120G28U0Z.2T0300M09

N125M01

(FINISH OD TURNING TOOL)

N130T0200M42

N135G96S1200M03

N140G00G41X4.25Z.1T0202M08

N145G01Z0.F.03

N150X2.35F.008

N155G00W.03

N160G42G00X4.03

N165X4.15W-.03F.004

N170Z-2.85F.008

N175G00X4.5Z.1

N180G28G40U.0Z1.T0200M09

N185M01

(FINISH ID BORING BAR)

N190T0400M42

N195G96S1200M03

N200G00X-2.5723Z.1T0404M08

N205G01Z0F.012

N210X-2.5W-.0628F.004

N215Z-1.2963F.008

N220X-2.55W-.0433F.004

N225Z-2.3228F.008

N230X-2.5W-.0433F.004

N235Z-3.1812F.008

N240G03U.1376Z-3.25I.0688K0.F.003

N245G01X-2.1964F.008

N250X-2.05W-.0732F.004

N255G00Z1.M09

N260G28U0W0T0400

N265M01

(SPECIAL V-FORM TOOL)

N270T0700M42

N275G96S300M03

N280G00X4.2Z-.25T0707M08

N285M98P5L12

N290G00Z1.M09

N295G28U0W0T0700M05

N300M30

Subprogram for Operation 2

O0005

N10G01X-4.09F.003

N15G00X-4.2

N20W-.1

N25M99

Notes for the second operation of the bushing program

The following notes pertain to the calculations necessary for programming the external and internal chamfers and tapers, and indicate the specific block number where the resulting values are input. Once again, in the figures that follow, the tool nose radius representation is scaled up in order to aid in the visualization.

1. N200G00X-2.5723Z.1T0404M08

X-2.5723 is the value that requires additional calculations.

Figure 124
Diagram for Making a Bushing,
Calculation 1

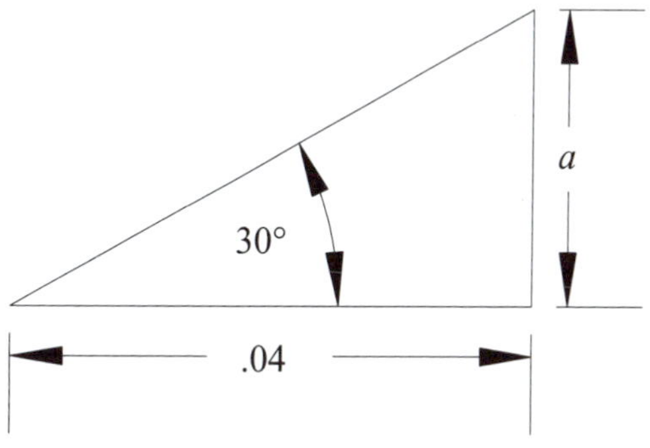

Based on the drawing, the given angle and the length of the chamfer enable us to calculate length a, of the triangle.

$$\tan 30° = \frac{a}{.04} \quad a = .04 \times \tan 30° = .0231$$

The expected value is given by a formula:

$$X = 2.5 + 2 \times a + 2 \times \Delta X$$

$$\Delta X = r \times \left[1 - \tan \left(45 - \frac{\alpha}{2} \right) \right]$$

where

r = radius of the tool =1/32

a = acute angle of chamfer, with the symmetry axis at 30°

$$\Delta X = .0312 \times \left[1 - \tan \left(45 - \frac{30°}{2} \right) \right] = .0312 \times (1 - .5774)$$

$$= .0312 \times .4226 = .01318$$

$$X = 2.5 + 2 \times .02 + 2 \times .01318 = 2.5723$$

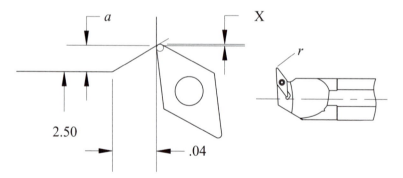

2.50

.04

Figure 125 Drawing for Making a Bushing, Calculation 1

Figure 126
Drawing for Making a Bushing,
Calculation 2

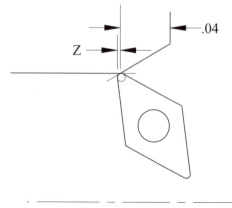

$$W = .04 + \Delta z$$

$$\Delta z = r \times \left(1 - \tan \frac{\alpha}{2} \right)$$

$$r = .0312 \quad \alpha = 30°$$

$$\Delta z = .0312 \times (1 - \tan 15°) = .02284$$

$$W = .04 + .0228 = .0628$$

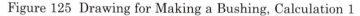

2. N210X-2.5W-.0628F.004

Look at the displacement in the direction of the Z axis, represented by W−.0628.

3. N215Z-1.2963F.008

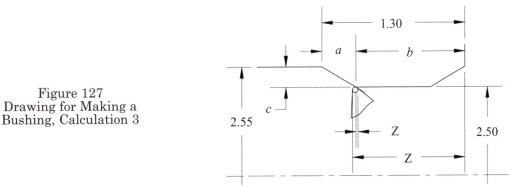

Figure 127
Drawing for Making a
Bushing, Calculation 3

Data, obtained from the drawing, refer to the dimensions 2.55, 2.5, and 1.3 and an angle of 30°

$$z = b + 2 \times r - \Delta z$$
$$b = 1.3 - a$$

Figure 128
Diagram 1 for Making a Bushing,
Calculation 3

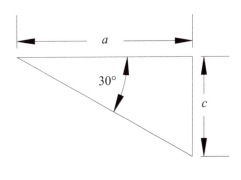

$$\tan 30° = \frac{c}{a} \quad c = \frac{2.55 - 2.5}{2} = .025$$

$$a = \frac{c}{\tan 30°} = \frac{.025}{\tan 30°} = .0433$$

$$b = 1.3 - a = 1.3 - .0433 = 1.2567$$

Figure 129
Diagram 2 for Making a Bushing,
Calculation 3

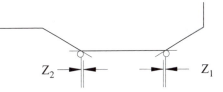

Since Z_1 is equal to Z_2 the same formula can be used:

$$\Delta z = r \times \left(1 - \tan \frac{\alpha}{2}\right)$$

$$\Delta z = .0312 \times \left(1 - \tan \frac{30}{2}\right) = .0228$$

$$z = b + 2 \times r - \Delta z = 1.2567 + .0624 - .0228$$

$$= 1.2963$$

4. N220X-2.55W-.0433F.004

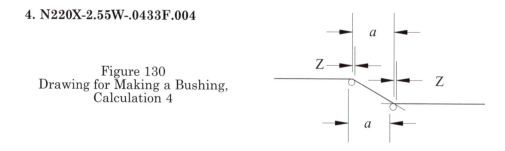

Figure 130
Drawing for Making a Bushing,
Calculation 4

As the drawing indicates, the previously calculated chamfer length a is equal to the tool displacement in the direction of the Z axis in the process of making such a chamfer. A similar situation in block N245 occurs.

5. N225Z-2.3228F.008

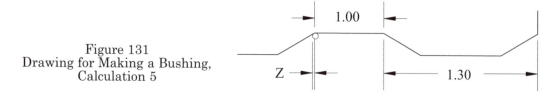

Figure 131
Drawing for Making a Bushing,
Calculation 5

$$Z = 1.3 + 1.0 + \Delta z = 1.3 + 1.0 + .0228 = 2.3228$$

6. N235Z-3.1812F.008

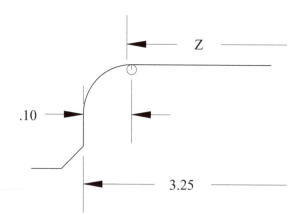

Figure 132 Drawing for Making a Bushing, Calculation 6

$$Z = 3.25 - .1 + .0312 = 3.1812$$

7. N240G03U.1376Z-3.25I.0688K0.F.003

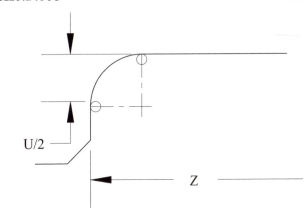

Figure 133 Drawing for Making a Bushing, Calculation 7

$$\frac{U}{2} = .1 - r \quad U = 2 \times (.1 - r) = .1376$$

EXAMPLE ILLUSTRATING THE APPLICATION OF FUNCTIONS G72 AND G75

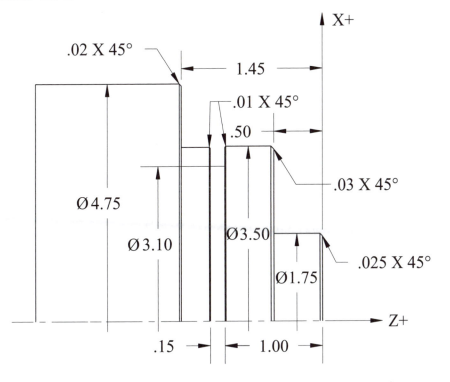

Figure 134 Drawing for Example Application of G72 and G75

Example of Application of G72 and G75, Program 26

Machine: Turning Center	Program Number: O0026

Workpiece Zero = X, <u>Centerline</u> Z, <u>Part Face</u>
Setup Description: Operation No. 1
Material: 316 Stainless Steel

Tool # and Offset #	Tool Orientation #	Description	Insert Specification	Comments
T0101	3	O.D. Turning Tool	80° Diamond 1/32 nose radius	250 SFPM
T0202	3	O.D. Turning Tool	55° Diamond 1/32 nose radius	400 SFPM
T0505	8	O.D. Grooving Tool	.125 wide 1/64 radius	250 SFPM

Program

O0026

(ROUGH OD TURNING TOOL)

N10G50S1500

N15T0100M41

N20G96S250M03

N25G00G41X4.85Z.2T0101M08

N30G01Z.0F.04

N35X-.06F.012

N40G00Z.1

N45G42X4.8

N50G72U.04W.005P55Q95D.12F.012

N55G00Z-1.47

N60G01X4.75

N65U-.02Z-1.45F.004

N70X3.5F.008

N75Z-.53

N80U-.03Z-.5F.004

N85X1.75F.008

N90G01Z-.025F.008

N95U-.025Z0F.004

Program for the First Operation

O0027

(ROUGH OD TURNING TOOL)

N10G50S2000

N15T0100M42

N20G96S500M04

N25G00X3.2Z.2T0101M08

N30G01Z.02F.05

N35X-.0624F.014

N40G00X3.03W.03

N45G01Z-2.8

N50G00U.1Z1.0M09

N55G28U0W0T0100

N60M01

(#5 CENTER DRILL)

N65T0900M42

N70G97S1019M03

N75G00X0.Z.2T0909M08

N80G01Z-.3F.006

N85G00Z.5M09

N90G28U0W0Z10.0T0900

N95M01

(1.25 DIAMETER DRILL)

N100T1000M42

N105G97S244M03

N110G00X0.Z.1T1010M08

N115G74Z-2.360K.150F.008

N120G28U0W0T1000M09

N125M01

(ROUGH ID BORING BAR)

N130G50S2000

N135T0300M42

N140G96S400M04

N145G00X1.250Z.1T0303M08

N150G71U-.03W.005P155Q180D1000F.012

N155G00X1.5364

N160G01Z0

N100Z.1

N105G28G40U0W0T0100M09

N110M01

(FINISH OD TURNING TOOL)

N115G50S2200

N120T0200M41

N125G96S400M03

N130G00G41X2.0Z.1T0202

N135G01Z0.F.04

N140X-.06F.008

N145G00Z.1

N150G42X4.8

N155G70P55Q95

N160G28G40U0W0T0200M09

N165M01

(OD GROOVING TOOL)

N170G50S2000

N175T0500M41

N180G96S250M03

N185G00X3.55Z.2T0505M08

N190Z-1.1375

N195G75X3.11I.05F.004

N200G00X3.55

N205Z-1.1591

N210G01X3.5F.008

N215U.0382W.0191F.003

N220X3.1F.005

N225W.005

N230G00X3.55

N235Z-1.1059T0313

N240G01X3.5F.008

N245U-.0382W-.0191F.003

N250X3.1F.005

N255W-.022

N260G00X3.75

N265Z1.M09

N270G28U0W0T0500

N275M30

In block N195, when the tool reaches diameter X3.11, it automatically returns to the starting point because of cycle G75. As a result of this, consecutive block N200 (with X3.55) is actually a safety clearance block since in some older types of controls, the automatic return may not appear. In cycle G75, some symbols such as *K*, *D*, and *Z* may be omitted, and, if any (K = 0, D = 0) are equal to zero, it may be omitted. The value of Z does not change and is omitted as well

COMPLEX PROGRAM EXAMPLE

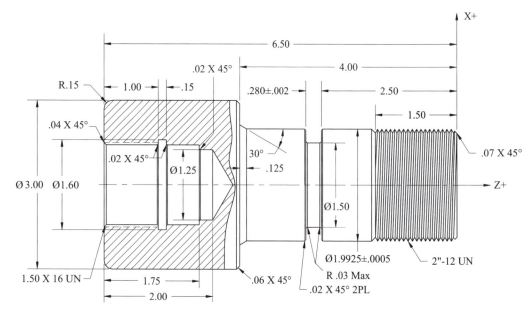

Figure 135 Drawing for Complex Program Example

First Operation

1. The order calls for 1,025 pieces of material with a 3.125 diameter and a length of 6.625.

2. Complete all the dimensions on the left side of the part, along with a length of 6.60 inch, which leaves .100 inch for the finish pass on the second operation on the opposite end.

The following are the initial program data for the first operation:

1. The center drill has a maximum diameter of .3 inch and the r/min is calculated as follows:

Cutting speed for the steel is:

$$V = 80 \text{ ft/min}$$

$$n = \frac{12 \times V}{\pi \times d} = \frac{12 \times 80}{3.14 \times .3} = 1019$$

2. The tool has a maximum diameter of 1.25 inch and the r/min for drill bit is calculated as follows:

$$V = 80 \text{ ft/min}$$

$$n = \frac{12 \times V}{\pi \times d} = 244$$

3. The cutting speeds for the boring bar, I.D groove, and the internal thread are $V = 300$ ft/min due to the unfavorable cutting conditions and the likelihood that vibrations will be present.

4. The minor diameter for the 1.5-16 internal thread is obtained from the *Machinery's Handbook* and is from 1.432 minimum to 1.446 maximum.

Example of Complex Program, Operation 1, Program 27

Machine: Turning Center	Program Number: O0027

Workpiece Zero = X, <u>Centerline</u> Z, <u>Part Face</u>
Setup Description: Operation No. 1
Material: Brass
Machine the left end including all internal work.

Tool # and Offset #	Tool Orientation #	Description	Insert Specification	Comments
T0101	3	O.D. Turning Tool	80° Diamond 1/32 nose radius	500 SFF
T0909		#5 HSS Center drill		80 SF
T1010		1.25 diameter HSS drill		80
T0303	2	I.D. Turning Tool	80° Diamond 1/32 nose radius	4
T0202	3	O.D. Turning Tool	55° Diamond 1/32 nose radius	
T0404	2	I.D. Turning Tool	55° Diamond 1/32 nose radiu	
T0606	2	I.D. Grooving Tool	.125 wide 1/64 radiu	
T0808	2	I.D. Threading Tool	1/32 ra	

N165X1.420Z-.0582F.004

N170Z-1.75F.007

N175X1.3264

N180X1.25W-.0382F.004

N185G00Z.1M09

N190G28U0W0T0300

N195M01

(FINISH OD TURNING TOOL)

N200G50S2000

N205T0200M42

N210G96S600M04

N215G00X3.1Z.1T0202M08

N220G01Z0.F.05

N225X1.0F.008

N230G00X2.6376W.03

N235G01Z0

N240G02X3.0W-.1812F.005K-.1812

N245G01Z-2.8F.008

N250G00U.1Z1.0M09

N255G28U0W0T0200

N260M01

(FINISH ID BORING BAR)

N265G50S2000

N270T0400M42

N275G96S600M04

N280G00X1.250Z.1T0404M08

N285G70P155Q180

N290G28U0W0T0400M09

N295M01

(ID GROOVING TOOL)

N300G50S2000

N305T0600M42

N310G96S300M04

N315G00X1.4Z.2T0606M08

N320Z-1.1375

N325G01X1.590F.004

N330G00X1.4

N335Z-1.179

N340G01X1.420F.008

N345U.058W.029F.004

N350X1.6F.005

N355W.005

N360G00X1.4

N370Z-1.096

N375G01X1.420F.008

N380U.058W-.029F.004

N385G01X1.6F.005

N390W-.023F.006

N395G00X1.4

N400Z1.0M09

N405G28U0W0T0600

<u>N410M01</u>

(ID THREADING TOOL)

N415T0800M42

N420G97S764M04

N425G00X1.340Z.5T0808M24

N430G76X1.5Z-1.075A60I0K.040D.0140F.0625

N435G28U0W0T0800M23

N440T0100M09

N445M30

Notes: During the process of drilling, notice that function G97 is applied, which corresponds to a constant spindle speed. It is used in blocks N70 and N105, because when a centerline tool such as a drill is used, the assigned coordinate for its position is X0. In block N20 function G96, constant cutting speed is instated. If G96 were to remain active, the spindle speed for the X position of zero would calculate to infinity and the machine would be limited to its maximum programmable r/min. Obviously, this is not practical. In this program, the value of the spindle speed is limited, with the value of S assigned to function G50.

Figure 136 Drawing for Constant Spindle r/min Setting

The drill bit, however, has a certain diameter. Therefore, the suitable r/min is calculated based on the diameter and entered in the program along with function G97.

When using function G96, there must be a programmed point located on the circumference in the direction of the X axis. This value, with respect to the X axis, must also be taken into consideration, if function G50 is assigned.

Second Operation

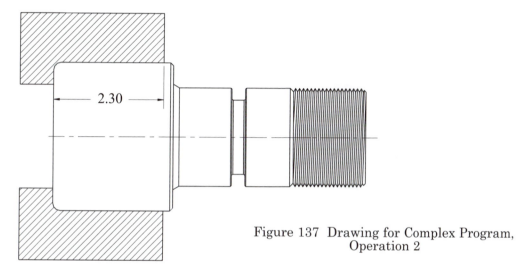

Figure 137 Drawing for Complex Program, Operation 2

Example of Complex Program, Operation 2, Program 271

Machine: Turning Center	Program Number: O0271

Workpiece Zero = X, <u>Centerline</u> Z, <u>Part Face</u>
Setup Description: Operation No. 2
Material: Brass
Machine the right end including all external work.

Tool # and Offset #	Tool Orientation #	Description	Insert Specification	Comments
T0101	3	O.D. Turning Tool	80° Diamond 1/32 nose radius	500 SFPM
T0202	3	O.D. Turning Tool	55° Diamond 1/32 nose radius	600 SFPM
T0505 T0515	8	O.D. Grooving Tool	1/4 wide 1/64radius	400 SFPM
T0707	3	O.D. Threading Tool	60° V-form threading tool	285 SFPM

Program for the Second Operation

O0271

(ROUGH OD TURNING TOOL)

N10G50S2000

N15T0100M41

N20G96S500M03

N25G00X3.2Z.2T0101M08

N30G01Z.015F.05

N35X-.0624F.014

N40G00X3.2Z.05

N45G71U.03W.005P50Q85D1500F.014

N50G00X1.8161

N55G01Z0

N60X1.9925W-.0882F.004

N65Z-3.8932F.008

N70X2.1105Z-4.0F.004

N75X2.8436F.008

N80X3.0W-.0782F.004

N85X3.2

N90G28U0W0T0100M09

N95M01

(FINISH OD TURNING TOOL)

N100G50S2000

N105T0200M41

N110G96S600M03

N115G00X2.2Z.05T0202M08

N120G01Z0.F.05

N125X-.0624F.008

N130G00X3.2Z.05

N135G70P50Q85

N140G28U0W0T0200M09

N145M01

(OD GROOVING TOOL)

N150G50S2000

N155T0500M42

N160G96S400M03

N165G00X2.1Z.2T0505M08

N170Z-2.765

N175G01X1.51F.006

N180G00X2.1

N185Z-2.809

N190G01X1.9925F.008

N200U-.058W.029F.004

N205X1.5F.006

N210W.005

N215G00X2.1

N220Z-2.721T0515

N225G01X1.9925F.008

N230U-.058W-.029F.004

N235X1.5F.006

N240W-.028

N245G00X2.1

N250G28U0W0T0500M09

<u>N255M01</u>

(OD THREADING TOOL)

In lines N165 and N220 the application of two different offsets for both sides of the groove allows a better control of its width.

Note: Values of offsets in the direction of the X axis must be equivalent.

N260T0700M42

N265G97S573M03

N270G00X2.08Z.5T0707M08

N275G76X1.896Z-1.5I0K.0485A60D140F.083333

N280G28U0W0T0700M09

N285T0100

N290M30

Example of Cutting a Three-Start Thread

The following example is given for machining a three-start 1.5 - 4 Acme thread, as illustrated in the figure on the next page:

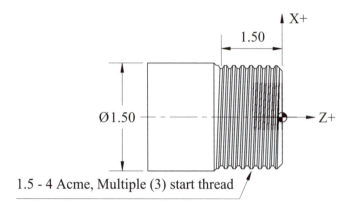

Figure 138 Drawing for a Three Start Thread Example

Example of Cutting a Three Start Thread, Program 28

        O0028
        N10G50S1500
        N15T0100M41
        N20G97S750M03
        N25G00X1.58Z.7T0101
        N30M98P1235L3
        N35G28U0W0T0100M09
        N40M30
        Subprogram for program O0028
        O1235
        N10G00W-.0833
        N15G76X1.395Z-1.5I0K.0525A60D140F.250000
        N20M99

EXAMPLE OF THREADING WITH A COMMON TAP

Figure 139 Drawing for
Threading with a Common Tap

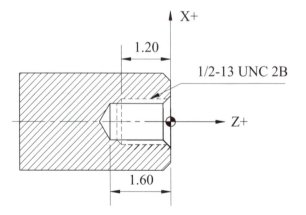

Example of Threading with a Common Tap, Program 29

Machine: Turning Center	Program Number: O0029

Workpiece Zero = X, Centerline Z, Part Face

Setup Description: Operation No. 1

Material: Carbon Steel

Tool # and Offset #	Tool Orientation #	Description	Insert Specification	Comments
T0909		#4 HSS Center Drill		100 SFPM
T1010		27/64 diameter HSS drill		100 SFPM
T1313		1/2–13 Tap		20 SFPM

Program

O0029

(#4 CENTER DRILL)

N10T0900M41

N15G97S1528M03

N20G00X0.Z.2T0909M08

N25G01Z-.250F.006

N30G00Z1.M09

N35G28U0.W0.T0900

N40M01

(.421 DIAMETER DRILL)

N45T1000M41

N50G97S907M03

N55G00X0.Z.2T1010M08

N60G01Z-1.721F.006

N65G00Z1.0M09

N70G28U0.W0.T1000

N75M01

(1/2-13 TAP)

N80T1300M41

N85G97S76M03

N90G00X0.Z.3T1313M08

N95G32Z-1.4F.0769M05

N100Z.2M04

N105G28U0W0T1300M09

N110M30

Notes:

1. The feed rate override must be set at 100% during the tapping, if function G01 is used.

2. A floating tap holder, which can float and expand forward and backward, should be used for the tapping operation.

EXAMPLE ILLUSTRATING THE APPLICATION OF TOOL NOSE RADIUS COMPENSATION (G41 AND G42)

Functions G41 and G42 refer to the tool nose radius compensation. These functions were described earlier here, in Part 3, Application of Tool Nose Radius Compensation (TNRC).

In short, when using the two initial points in programming, we are limited to programming as if the tool radius is equal to zero. When programming with tool nose radius compensation, the part profile may be programmed. Function G41 refers to the tool radius compensation in a left direction, whereas function G42 refers to the tool radius compensation in a right direction, as shown below. The block format is:

G41X(U) . . .   Z(W) . . .

G42X(U) . . .   Z(W) . . .

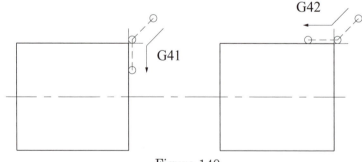

Figure 140
Tool Nose Radius Compensation for Turning

The value of the tool nose radius is entered in the offset number to which corresponds with the tool number. After pressing the OFFSET SETTING button on the control, a list of geometry offsets (including columns for R and T) appears on the screen. Under column R, enter the tool nose radius. Under column T, enter the tool orientation number as described earlier in Part 2, Turning Center Offset and Part 3, Application of Tool Nose Radius Compensation sections. After completing the turning process, cancel the tool radius compensation with function G40.

Example of Application of Tool Nose Radius Compensation,

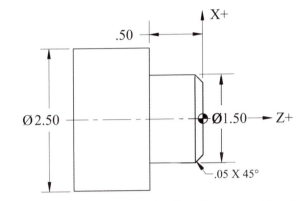

Figure 141 Drawing for Tool Nose Radius Compensation for Turning

Program 30

O0030

N10G50S2000

N15T0100M42

N20G96S500M03

N25G00G41X1.6Z.1T0101

N30G01Z0F.016

N35G00X0F.008

N40G00Z.1

N45G42X1.3

N50X1.5Z-.05

N55Z-.5

N60X2.5

N65G00G40X3.5.Z5.0

N70G28U0W0T0100M9

N75M30

Notes:Functions G41 and G42 are used in turning when tapers and radii must be closely controlled. For all straight turning and facing, initial point programming will obtain the desired results. In examination of the previous program, the only place it is necessary is along the chamfer. Furthermore, G41 applies only to facing while G42 applies for the chamfer and turning. In the following figure, the resultant path is given. Facing to the center of the part along the X axis and then, after withdrawal, perform chamfering and the turning of the remaining part of the shaft. Examine the tool motions with the application of both G41 and G42.

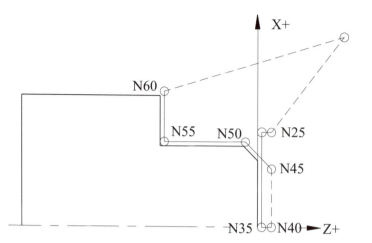

Figure 142
Tool Path for Tool Nose Radius Compensation for Turning

In the Figure 142, function G41 is activated for facing in line N25 and for turning in G42 is active in lines N45 through N60.

Part 3 Study Questions

1. In NC (tape) controlled machines, the M02 command rewinds the tape to its start. How is a CNC machine commanded to return the program to its start?

    a. M03

    b. M08

    c. M30

    d. M05

2. The term "modal commands" means that once the command is initiated it stays in effect until cancelled or replaced by another command from the same group.

    T or F

3. In programming, tool function is commanded by the four digits that follow the letter address T. (Example T0404). What do these two sets of numbers refer to?

4. Of the following choices, which is the best method for compensating for dimensional inaccuracies caused by tool deflection or wear?

    a. Geometry Offsets

    b. Wear Offsets

    c. Tool Length Offsets

    d. Absolute Position Register

5. When the rough turning cycle G71 is used, which letters identify the amount of stock to leave for finish for pass X axis and Z axis respectively?

    a. U and V

    b. X and Z

    c. P and Q

    d. U and W

6. Sequence (N) numbers in programs may be omitted entirely and the program will execute without any problem.

    T or F

7. The default (at machine start-up) feed rate on lathes is typically measured in

    a. Cutting Speed

    b. Inches per revolution in/rev

    c. Inches per Minute in/min

    d. Constant Surface Speed

Part 3 Programming CNC Turning Centers

8. The advantage of using G96 Constant Cutting Speed in turning is that as the diameter changes (position of the tool changes in relation to the centerline) the r/min increases or decreases to accomplish the programmed cutting speed.

   T or F

9. The preparatory function G50 relates to two things in programming.

   a. Absolute positioning setting and maximum spindle r/min setting

   b. Absolute position setting and constant spindle r/min

   c. Work offset and tool offsets

   d. Absolute position setting and Constant Surface Speed

10. When incremental programming is required in turning diameters and facing, which letters identify the axis movements respectively?

   a. I and J

   b. U and W

   c. I and K

   d. U and V

11. Which M Codes are used to activate and deactivate a subprogram respectively?

   a. M03 and M04

   b. M2 and M30

   c. M98 and M99

   d. M41 and 42

12. What M-Code is listed in the last line of a subprogram?

   a. M30

   b. M99

   c. M98

   d. M02

13. What two letters identify the incremental distance from the starting point to the arcs center in G02/G03 programming?

   a. I and J

   b. I and K

   c. J and K

   d. X and R

14. When programming arcs with modern CNC controllers I, J and K can be omitted and replaced by R. If the arc is greater than 180°, what must be added to the R command?

15. Fixed cutting cycle G90 is limited to orthogonal tool movements (no contouring or chamfers are allowed)

    T or F

16. When using the Multiple Repetitive Cycle, Rough Cutting Cycle, G71, U and W represent the stock allowance for finishing. What cycle is required to remove the stock allowance?

    a. G90

    b. G73

    c. G76

    d. G70

17. What is the major reason for selection of the Pattern Repeating Cycle, G73, as opposed to the Rough Cutting Cycle, G71?

    a. It is required to make the finish allowance cut on X and Z

    b. This cycle is well suited where an equal amount of material is to be removed from all surfaces.

    c. G73 is limited to orthogonal cuts while G71 can cut radius and chamfers

    d. G71 is limited to rough cutting only while G73 is required to remove stock allowances

18. When programming arc and angular cuts using tool nose radius compensation, G41 and G42, which is used for facing and turning when the tool is mounted above the part centerline.

    a. G41 facing, G42 turning

    b. G42 facing, G41 turning

    c. Neither, the tool nose radius amount must be calculated and programmed to compensate.

    d. G41 for facing if contours or angles are involved, G42 for turning if contours or angles are involved.

19. Tool tip orientation needs to be identified in the controller when TNRC is used to program functions G41 or G42. How is this information input?

    a. The number is input by R into the program

    b. The number is input by T into the offset page

    c. The number is input by R into the offset page

    d. The number is input by T into the program.

Part 3 Programming CNC Turning Centers

20. Using the tool figures 89, 91, 95 and 96. Identify the tool tip identification numbers.

21. When programming the G76 multiple thread cutting cycle, the letter address I allows for programming

    a. Tapered thread cutting

    b. Elimination of taper in thread cutting

    c. Both a and b

    d. a only

22. Name one reason that writing programs using Tool Nose Radius Compensation (TNRC) is advantageous.

PART 4

PROGRAMMING
CNC MACHINING CENTERS

Objectives:

1. Learn G-Codes associated with Machining Center Programming.

2. Learn M-Codes associated with Machining Center Programming.

3. Apply the proper use of feeds and speeds within Machining Center programs.

4. Learn how to properly use coordinate systems for programming the Machining Center.

5. Learn program structure.

6. Learn how to use Canned Cycles.

7. Learn the mathematical method for offsetting tool path to compensate for the Cutter Diameter Compensation.

8. Learn how to use Cutter Diameter Compensation (G40, G41 and G42).9.

9. Examine several practical examples of Machining Center programs.

Programming of CNC Machining Centers will be presented in this section. In Part 1 of this text, the basic steps necessary for programming were identified. These steps will be followed to create programs. Individual programming words and codes will be defined and demonstrated with examples. This section is written as if the program manuscript is being created manually, line-by-line. This experience will help you understand the programs and how they are made which will make it easier to edit existing programs. In reality though, the most common method for creating programs today is by using CAD/CAM software, which is introduced in Part 5 and Part 6 of this text. When programming one of the first things that should be dealt with in the programming process is the cutting tool.

TOOL FUNCTION (T-WORD)

Proper selection and application of cutting tools and work holding ensure that the programs you create produce desired results (refer to Part 1 "Cutting Tool Selection", *Machinery's Handbook* and the tool and insert manufacturers ordering catalogs for detailed tool information). The tool function is utilized to prepare and select the appropriate tools from the tool magazine. In order to describe the tool in the program, the address T is followed by one or more digits that refer to the pocket numbers in the tool magazine.

Example:

T05 = tool number 5

Please note that on most modern controllers, it is not necessary to use the leading zero in a tool call; thus, T5 has the same meaning as T05.

TOOL CHANGES

A tool change is specified in the program by the miscellaneous function M06. To initiate a tool change, first call for the desired tool number and then use the miscellaneous function M06 to execute the change.

Example:

N10T01 (TOOL IN THE READY POSITION)

N15M06 (ACTUAL TOOL CHANGE)

N20T02 (NEXT TOOL IN THE READY POSITION)

N25 . . .

N30M06T03

N35. . .

In block N10, the requested tool is positioned in the tool magazine to a ready state (waiting position). In block N15, tool T01 is automatically installed into the spindle and, in the mean time, tool T02 is positioned to a ready state in the tool magazine for the next tool change. In block N20, tool T01 performs programmed work. In block N30, a tool change takes place. Tool T02 is installed into the spindle, while tool T01 is returned to the tool magazine and tool T03 is positioned to a ready state in the tool magazine for the next tool change. In block N35, tool T03 performs the programmed work.

Please note that on most modern controllers it is not necessary to use the leading zero in a miscellaneous function call; thus, M6 has the same meaning as M06. Also note that on machines with umbrella style tool magazines, the tool call (T01) and tool change call (M06) must be stated together.

For example: T01M06

FEED FUNCTION (F-WORD)

The F-word is utilized to determine the work feed rates. This program word is used to establish feed rate values and precedes a numeric input for the feed amount in inches per minute (in/min), millimeters per minute (mm/min) or, inches per revolution (in/rev), millimeters per revolution (mm/rev). The value that is set by this command stays effective until changed by reentering a new value for the F-word.

Example:

F20 = a feed rate of 20 inches per minute (in/min)

F.006 = a feed rate of .006 inches per revolution (in/rev)

If the function G20 (data in inches) is active, such a notation refers to the feed speed of 20 in/min; whereas, with the function G21 (data in millimeters), the notation refers to a feed rate of 20 mm/min.

With rapid traverse G00, the machine traverses at the highest possible feed rate that is specified in control memory (actual rates depend on the design of the machine). In the case of feed rate motion, G01, the value of the feed rate must be accurately specified. The machine default setting for feed rate is inches per minute.

Examples:

F20.0 = 20.0 inches of feed per minute

F500. = 500 millimeters of feed per minute

F2.0 = 2.0 inches of feed per minute

F50 = 50 millimeters of feed per minute

F.02 = twenty thousandths inch of feed per revolution

F0.50 = millimeters of feed per minute

F.002 = two thousandths inch of feed per revolution

F0.050 = millimeters of feed per minute

The proper methods for selection and calculation formula for feed rates were discussed in Part 1 "Metal Cutting Factors". Also, *Machinery's Handbook* and the tool and insert manufacturers ordering catalogs provide excellent feed and speed data.

SPINDLE SPEED FUNCTION (S-WORD)

The letter address S is followed by a specified value in revolutions per minute (rev/min).

Example: S2100 (specifies 2100 r/min)

One or more digits following the letter address S are used for the value of the rotational speed. If S0 is input, this command deactivates the spindle rotation and leaves it in a neutral position so that the spindle can be rotated manually, depending on the machine tool. This is quite useful, especially for "dialing-in" the workpiece holder to establish Workpiece Zero coordinates. The value of S is specified in revolutions per minute (r/min).

PREPARATORY FUNCTIONS (G-CODES)

Preparatory functions are the G-codes that identify the type of activities the machine will execute. A program block may contain one or more G-codes.

The letter address G and specific numerical codes allow communication between the controller and the machine tool. This combination of letters and numerical values is commonly called G-Code. In order to perform a specific machining operation, a G-Code must be used. There are two types of G-Codes, modal and non-modal. Modal commands remain in effect, in multiple blocks until they are changed by another command from the same group; whereas, non-modal commands are only in effect for the block in which they are stated.

For example:

Group 00, are non-modal "One-Shot" commands.

Group 01, are modal commands.

There are several different groups of G-codes as indicated in column 2 of the following chart, "Preparatory Functions (G-Codes) specific to Machining Centers". One code from each group may be specified in an individual block. If two codes from the same group are used in the same block, the first will be ignored by the control and the second will be executed. There are G-Codes that are active upon startup of the machine indicated by an asterisk (*) in the chart. A Safety Block is commonly placed in the first line of the program where cancellation codes are used to cancel all G-Codes that have been in effect in prior programs (described later in this section, "Explanation of the Safety Block"). Typically, they are: (G40) cutter compensation cancellation, (G49) tool length compensation cancel and (G80) canned cycle cancellation and also (G17) *XY* plane selection. This cancellation is important because of modal commands that stay in effect until either cancelled or replaced by a command from the same group. It is also a good idea to insert this Safety Block after tool change blocks, in case of the need to rerun a single operation from within the program. By doing this, there is no chance of modal commands remaining active. The digits following the letter address G identify the action of the command for that block. (Also see the section, Explanation of the Safety Block.)

If the measurement system is changed i.e., from the G20 inch to G21 metric system, then G21 will be in effect at the next startup of the machine or until a program call of G20 is executed. On machines sold in the U.S. the parameters are set to default to the G20-inch system upon startup.

Notes for Prepatory Functions (G-Codes) specific to Machining Centers:

*The items marked with an * in Chart 1 are active upon startup of the machine or when the RESET button has been pressed.*

For G00 and G01, G90 and G91 the initial code that is active is determined by a parameter setting. These are typically set at G01 and G90 for startup condition.

PREPARATORY FUNCTIONS (G-CODES)
SPECIFIC TO MACHINING CENTERS

Code	Group	Function
*G00	01	Rapid Traverse Positioning
*G01	01	Linear Interpolation
G02	01	Circular and Helical Interpolation CW (Clockwise)
G03	01	Circular and Helical Interpolation CCW (Counterclockwise)
G04	00	Dwell
G09	00	Exact Stop
G10	00	Data Setting
*G15	17	Polar Coordinates Cancellation
G16	17	Polar Coordinates System
*G17	02	XY Plane Selection
G18	02	ZX Plane Selection
G19	02	YZ Plane Selection
G20	06	Input in Inches
G21	06	Input in Millimeters
*G22	04	Stored Stroke Limit ON
G23	04	Stored Stroke Limit OFF
G27	00	Reference Position Return Check
G28	00	Reference Position Return
G29	00	Return From Reference Position
G30	00	Return to Second, Third, and Fourth Reference Position
G33	01	Thread Cutting
G37	00	Automatic Tool Length Measurement
*G40	07	Cutter Compensation Cancel
G41	07	Cutter Compensation Left Side
G42	07	Cutter Compensation Right Side
G43	08	Tool length Offset Compensation Positive (+) Direction
G44	08	Tool Length Offset Compensation Negative (-) Direction
G45	00	Tool Offset Increase
G46	00	Tool Offset Decrease
G47	00	Tool Offset Double Increase
G48	00	Tool Offset Double Decrease
*G49	08	Tool Length Offset Compensation Cancel
*G50	11	Scaling Cancel
G51	11	Scaling
G52	00	Local Coordinate System
G53	00	Machine Coordinate System

Part 4 Chart 1 Continued on following page

PREPARATORY FUNCTIONS (G-CODES)
SPECIFIC TO MACHINING CENTERS, CONTINUED

Code Group Function

Code	Group	Function
*G54	14	Work Coordinate System 1
G55	14	Work Coordinate System 2
G56	14	Work Coordinate System 3
G57	14	Work Coordinate System 4
G58	14	Work Coordinate System 5
G59	14	Work Coordinate System 6
G60	00	Single Direction Positioning
G63	15	Tapping Mode
G68	16	Rotation of Coordinate System
*G69	16	Cancellation of Coordinate System Rotation
G73	09	High Speed Peck Drilling Cycle
G74	09	Reverse Tapping Cycle
G76	09	Fine Boring Cycle
*G80	09	Canned Cycle Cancel
G81	09	Drilling Cycle, Spot Drilling
G82	09	Drilling Cycle, Counter Boring
G83	09	Deep Hole Drilling Cycle
G84	09	Tapping Cycle
G85	09	Reaming Cycle
G86	09	Boring Cycle
G87	09	Back Boring Cycle
G88	09	Boring Cycle
G89	09	Boring Cycle
*G90	03	Absolute Programming Command
*G91	03	Incremental Programming Command
G92	00	Setting for the Work Coordinate System or Maximum Spindle r/min
*G94	05	Feed per Minute
G95	05	Feed per Revolution
G96	13	Constant Surface Speed Control
*G97	13	Constant Surface Speed Control Cancel
*G98	10	Canned Cycle Initial Level Return
G99	10	Canned Cycle R-Level Return

Part 4 Chart 1

MISCELLANEOUS FUNCTIONS (M FUNCTIONS)

Miscellaneous function or "M-Codes," control the working components that activate and deactivate coolant flow, spindle rotation, the direction of the spindle rotation and similar activities.

MISCELLANEOUS FUNCTIONS, (M-CODES) FOR MACHINING CENTERS

Code	Function
M00	Program Stop
M01	Optional Stop
M02	Program End Without Rewind
M03	Spindle ON Clockwise (CW) Rotation
M04	Spindle ON Counterclockwise (CCW) Rotation
M05	Spindle OFF Rotation Stop
M06	Tool Change
M07	Mist Coolant ON
M08	Flood Coolant ON
M09	Coolant OFF
M10	Work Table Rotation Locked
M11	Work Table Rotation Unlocked
M13	Spindle ON Clockwise and Coolant ON, Dual Command
M14	Spindle ON Counterclockwise and Coolant ON, Dual Command
M16	Change of Heavy Tools
M19	Spindle Orientation
M21	Mirror Image in the Direction of the X Axis
M22	Mirror Image in the Direction of the Y Axis
M23	Cancellation of the Mirror Image
M30	Program End with Rewind
M98	Subprogram Call
M99	Return to Main Program from Subprogram

Part 4   Chart 2

PROGRAMMING OF CNC MACHINING CENTER IN ABSOLUTE AND INCREMENTAL SYSTEMS

These two coordinate measuring systems are used to determine the values that are input into the programming code for the X, Y and/or Z program words. They can also be used in the same manner for rotary axes A, B and/or C.

ABSOLUTE COORDINATE PROGRAMMING (G90) OF THE MACHINING CENTER

In absolute programming, all coordinate values are relative to a fixed origin of the coordinate system. Axis movement in the positive direction does not require inclusion of the sign; while negative movements do require signs.

INCREMENTAL COORDINATE PROGRAMMING (G91) OF THE MACHINING CENTER

In incremental systems, every measurement refers to a previously dimensioned position (point-to-point). Incremental dimensions are the distances between two adjacent points.

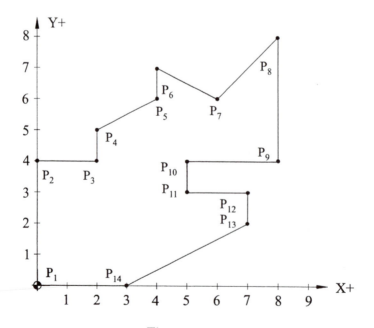

Figure 1
Absolute and Incremental Coordinates for CNC Machining Centers

The coordinate notations for the points on the drawing (in absolute and incremental systems) appear in the chart as follows:

For G90 ABSOLUTE		For G91 INCREMENTAL	
P1	G90 X0 Y0	P1	G91 X0 Y0
P2	G90 X0 Y4	P2	G91 X0 Y4
P3	G90 X2 Y4	P3	G91 X2 Y0
P4	G90 X2 Y5	P4	G91 X0 Y1
P5	G90 X4 Y6	P5	G91 X2 Y1
P6	G90 X4Y7	P6	G91 X0 Y1
P7	G90 X6 Y6	P7	G91 X2 Y-1
P8	G90 X8Y8	P8	G91 X2 Y2
P9	G90 X8 Y4	P9	G91 X0 Y-4
P10	G90 X5 Y4	P10	G91 X-3 Y0
P11	G90 X5 Y3	P11	G91 X0 Y-1
P12	G90 X7 Y3	P12	G91 X2 Y0
P13	G90 X7 Y2	P13	G91 X0 Y-1
P14	G90 X3 Y0	P14	G91 X4Y-2
P1	G90 X0 Y0	P1	G91 X-3 Y0

Absolute and Incremental Coordinates for Figure 1

WORK COORDINATE SYSTEM SETTING (G92)

Workpiece Zero is a chosen position for the origin of the coordinate system on the drawing and is determined by the programmer. Workpiece Zero, in most cases, is selected as the most suitable point on a given workpiece to simplify or eliminate coordinate calculations. For example, it may be in the center, as in Figure 2 (if the part is symmetrical), or, at intersecting edges as in Figure 3.

Example:

Figure 2
Work Coordinate System
Setting for a Symmetrical
Part

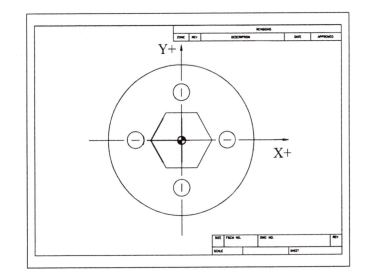

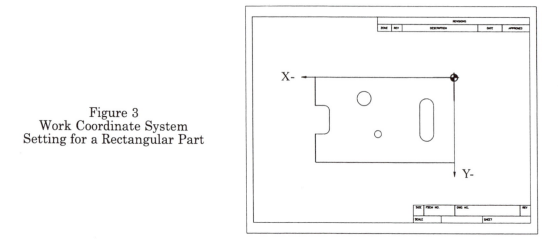

Figure 3
Work Coordinate System
Setting for a Rectangular Part

All tool motions are performed with respect to the Workpiece Zero. The G92 block format is:

G92   X...   Y...   Z...

The coordinate values input to the *X*, *Y* and *Z* program words assigned in the block with function G92, correspond with the distance from Workpiece Zero to Machine Zero for each axis.

MEASURING COORDINATES FOR SETTING G92

A way to determine the value of these coordinates, is to Home the machine for each axis and then press the "POSITION" function button on the control panel and a position readout appears on the screen. Then, press the "PAGE" button to find the corresponding Absolute Position readout. Now, check to see whether all of the values shown on the readout are equal to zero. If not, then press button "*X*" followed by the soft key labeled "ORIGIN". Follow the same pattern for the remaining axes *Y* and *Z*. Then proceed to measure the distance by aligning the center of the spindle to the part origin taking note of the position coordinates and inputting them into the *X* and *Y* values within the G92 program block.

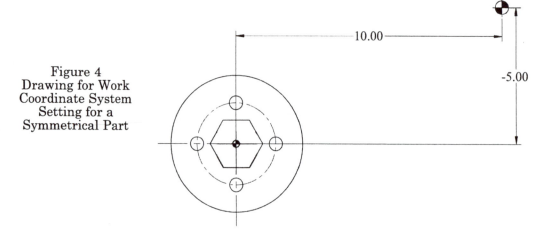

Figure 4
Drawing for Work
Coordinate System
Setting for a
Symmetrical Part

The steps for establishing the Workpiece Zero of the coordinate system for the drawing in Figure 4 are as follows:

Attach a dial indicator to the spindle of the machine then, switch to the handle mode (MPG, as described in Part 2 "CNC Machine Operation") to move the machine position until the axes are centered over the part and then dial-in the outside diameter (O.D.).

Then, input the values of X and Y that are shown on the Position readout at the beginning of the program to the function G92 as follows: G92X10.Y5. The values of X and Y correspond to the distance between Workpiece Zero and the Machine Zero.

Workpiece Zero in the example illustrated on the second drawing, Figure 3, is found in a similar manner. Manually position to Machine Zero with respect to all axes and set the ORIGIN then, install an edge finder to the spindle of the machine.

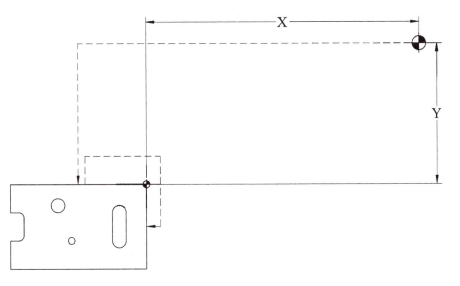

Figure 5 Drawing for Work Coordinate System Setting for a Rectangular Part

Set the r/min at 1000 and move the spindle along the path shown in Figure 5 by the dashed line. As you approach the edge, reduce the speed of the traverse in order to accurately spot the deflection of the edge finder. As soon as the first deflection of the edge finder is noticed, interrupt the feed along the Y axis and read the value of Y from the position readout. To this value, add the radius of the tip of the edge finder, and then enter this entire value under the Y assigned to the function G92.

The same procedure is applied for the edge positioned along the X axis. As soon as the edge finder deflection is noticed, read this value from the position readout. Add to that the value of the radius of the tip of the edge finder and enter it under the X axis assigned to the function G92. The values of X and Y so obtained correspond to the distance along the X and Y axes between Workpiece Zero and the Machine Zero. The value of Z assigned to function G92 is usually zero, which means that the Workpiece Zero corresponding to the Z axis is in the position of machine zero for the same axis. The values entered into the Tool Length Offsets compensate for the movements of the Z axis.

The values obtained for X, Y, and Z, assigned to function G92, are always entered at the beginning of the program and it is also advisable to enter it after each tool change, so if the program is restarted at the tool change then the coordinate system will be reestablished.

Function G92 is best used with less complicated programs in which only one Workpiece Zero is utilized. If the contour of the machined workpiece is complicated and difficult to program or if multiple coordinate systems are required on one part, it is best to use G54 through G59 to establish Workpiece Zero coordinates. This method is described later in this section in detail in, "Work Coordinate Systems". Work Coordinate Systems, functions G54-G59 are the most widely used method for assigning Workpiece Zeros today.

In order to specify multiple Workpiece Zeros for Figure 6, use the following functions:

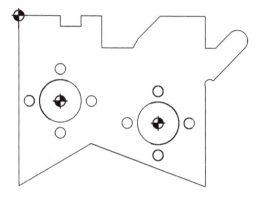

G54	first Workpiece Zero
G55	second Workpiece Zero
G56	third Workpiece Zero
G57	fourth Workpiece Zero
G58	fifth Workpiece Zero
G59	sixth Workpiece Zero

Functions G54, G55, G56, G57, G58, and G59 are described in detail later in this section and for the detailed process of setting G54 – G59 refer to Part 2 CNC Machine Operation, "Measuring Work Offsets, Machining Center".

Figure 6 Multiple Work Coordinate Systems

PROGRAM STRUCTURE FOR MACHINNG CENTERS
PROGRAM NUMBER

Each program is assigned a number. The capital letter O is reserved for the program number identification and is usually followed by four digits, which specify the actual program number. For example, to create a program with the number 1234, the programmer must input the letter address O, and then the number 1234 (O1234). All programs require this format.

Note: Refer also to Part 1, "Program Format".

Examples:

O0001 =   program number 1

O0014 =   program number 14

A common mistake made here, is to enter zero (0) instead of the letter O, which results in an alarm on the control system.

COMMENTS

Comments, to aid the operator, may be added to the program by using the parenthesis. The comments or data inside the parenthesis will not affect the execution of the program in any way. A very common place to add comments is at the program number, to identify a part number, or at a tool change or program stop to direct the operator in some way. An End-of-Block character (EOB) is typically required after the parenthesis if entered via MDI. No EOB character is needed if the data is entered via an offline text editor.

Examples:

O0001

(PN587985-B)

N9M00

(REMOVE CLAMPS FROM OUTSIDE OF PART)

BLOCK NUMBER

The letter "N" is reserved to identify the program line sequence numbers (block numbers) and precedes any other data in a program line. For each line in a program, a block number is assigned sequentially. For example, the first block of a program is labeled N10, the second N15 etc. Typically the program block numbering system is sequenced by an increment of other than one. An example of this is sequencing by five, where line one is labeled N5 and line two is labeled N10. The intent of an increment of other than one is to allow for adding additional blocks of data between the increments. Doing this is sometimes advantageous when editing programs. It is important to note that block numbering is not necessary for the program to be executed. The program blocks, even if not numbered, will be executed sequentially during automatic cycle of the machine. Removing block numbers is sometimes helpful when a program is too large to fit into the resident controller memory, as each character of a program takes memory space and large programs have block numbers into five digits. It is also worth noting that it is a common practice to place block numbers only at a tool change command. This enables the operator to restart the execution of the machining program at a specific tool change where, otherwise, the entire program may have to be executed in order to rerun a specific operation and time would be wasted. Block numbers enable movement within the program to enter offsets, verify data, or search for a block in the program. They are often referenced by the control in case of a programming error that causes an alarm enabling the programmer to search to the problem directly by block number. Block numbers are not required at all for the program to work, but they are necessary for restarting the program at a specific place.

SUBPROGRAM (M98 – M99)

If a certain part of the program can be used repeatedly within the program, assign a number for this part and list it separately, calling for it whenever it is needed. This part of the program is called a subprogram. Subprograms greatly simplify programming and decrease the amount of data that must be placed into the controller memory. A subprogram is called up from control memory by the function M98 and the letter P, both of which are entered in the main program. The letter P refers to the program number of the subroutine call. Similarly, as in the main program, in order to enter the subprogram number, you must first write the letter O and then the number. For each subprogram number, the four spaces after the letter O are reserved just as with the main program. Ordinarily, there is no workpiece coordinate system established within the subprogram because of its dependence (subordinate) on the main program.

Example:

O0002

N10

. . .

. . .

N50M99

Function M99 refers to the end of the subroutine and returns to the main program block following the block containing the subroutine call in the main program.

Example:

M98P0002

This is a call for subprogram No. 2.

Main program	Subprogram	Subprogram
O0001		
N10G54X . . .Y . . .		
N15 . . .		
N20 . . .		
. . .		
. . .		
. . .		
. . .	O0002	O0003
. . .	N10 . . .	
	N10	. . .
. . .	N15 . . .	
	. . .	
N55M98P0002	. . .	. . .
N60 . . .	. . .	
. . .		
. . .	N30M98P0003	. . .
. . .	N35 . . .	. . .
. . .	. . .	
	. . .	
N80M30	N45M99	N45M99

In the above example, enter the subroutine call in the program block N55 of the main program, and then use function M98 to call up subprogram O0002. Work is then performed according to the commands in the subprogram until block N30 is reached. The processing is then transferred to the next subprogram O0003. After the execution of subprogram O0003, the program is returned to block N35 of the subprogram O0002, which executes the remaining information blocks of this subprogram. From block N45 of this subprogram, the program is returned to block N60 of the main program, which then executes the remaining part of the main program. As many as four levels of subprograms can be linked together and is called nesting. To repeat a given subprogram twice, enter the following:

M98P0002L2

L2   repeat subprogram twice

For the controller identified in this edition (16M and 18M) the syntax is slightly different. The subprogram call works like this.

M98P00010002

Where M98 is the call for the subprogram, P0001 is the number of repeats and 0002 is the subprogram number. It is highly recommended that the programmer study the manufacturer manuals for specific instructions on using subprograms.

PROGRAM END (M30)

The difference between M02 and M30 is that M02 refers to the program end, while M30 refers to the program end and a simultaneous return to the program beginning (head). Both commands are found in the final line of the main program only.

Note: on some controls, the M02 behaves the same as M30 for compatibility with older programs.

THE LINK BETWEEN FUNCTIONS G92 AND G43

Workpiece Zero for the Z axis is placed within the machining envelope in relationship to Machine Zero. Workpiece Zero for a specific program for the Z axis is typically on the top most surface of the machined part. At first glance, a contradiction between these two statements seems to exist. However, both statements are true, because there is a direct relation between the functions G92 and Tool Length Compensation (G43) if we refer them both to the Z axis. Workpiece Zero for the Z axis is to be in the same position as it is for Machine Zero for that axis; this is applied to all the tools. Due to different tool lengths, in order to transfer Workpiece Zero along the Z axis from Machine Zero to the surface of the workpiece, you must apply function G43. Function G43 is a Tool Length Compensation function (Tool Length Offset). Offset number H . . . (H01, H02) is always assigned to function G43. In order to simplify program execution, offset number H, as a rule, is to be the same as the tool number for each corresponding tool. The measured value of offset H is entered into the offset registers in the computer memory (for example, H01 = -11.1283). The value of the Tool Length Offset for a given tool corresponds to the distance between the tool tip and the surface of the workpiece as shown in Figure 7.

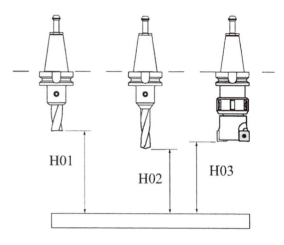

Figure 7 Tool Length Offset

In order to determine the value of offset H for a specific tool, zero the machine with respect to the Z axis with the tool to be measured in the spindle. Zero the position readout for Z; then, manually move the tool along the Z axis to the surface of the workpiece in such a way that the tool tip touches the surface of the workpiece. The value in the position register is the distance that determines the value of offset H for a given tool, and it is registered in the offset table with the number corresponding to the offset number in the program.

Function G43, with the assigned offset number H, must be entered in the program before the tool does any work. If the offset number H . . . is not entered, the tool will perform the work with the previous tool length offset. If the value of the tool length offset for the previous tool is smaller than the value of the tool length offset of the working tool, then the tool will not approach the material.

Caution: If, however, this value is greater than the value of the tool length offset of the working tool, the tool will then rapidly advance toward the workpiece and crash into it, causing damage to the tool, the workpiece, and the holding equipment.

This condition can be avoided by employing the Tool Length Compensation Cancel function (G49) in a Safety Block at the beginning of the next tool sequence. See "Explanation of the Safety Block" in this section.

Example of a CNC Machining Center Program, Program 1

Two holes, with diameters of .500 inch each, are to be drilled in a steel plate having the dimensions 9 x 4 inches. The position of the holes can be seen in the drawing below.

Machine: Machining Center	Program Number: O0001

Workpiece Zero: X= Upper Right Corner   Y= Upper Right Corner   Z= Top Surface

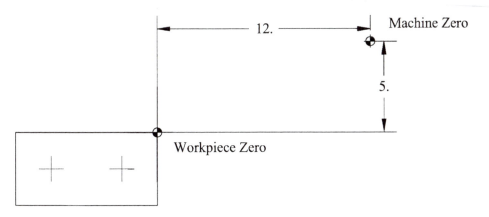

Figure 8 Machining Center Example Program 1, Setup Description

Setup Description: The plate is clamped in a vice on parallels.

Tool #	Description	Offset	Comments
1	No. 3 Center Drill		
2	.500 diameter drill		

Please note the use of the CNC Setup sheet above as described in Part 1 of this book. For the remainder of this section, it will be used for all of the examples. The top portion including the Title, Date, Prepared By, Part Name and Part Number will be omitted and only identification for each tool will be given to save space.

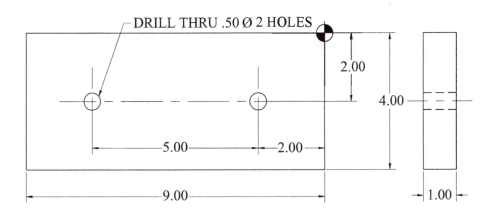

Figure 9 Drawing for Machining Center Example Program 1

Program

O0001

N10G90G20G80G40G49

(NO. 3 CENTER DRILL)

N15G92X12.Y5.Z0T02S900M03

N20G00X-2.Y-2.0

N25G43Z1.H01M08

N30G81G99Z-.219R.1F4.6

N35G00X-7.Y-2.0

N40G80Z1.M09

N45G91G28Z1.0

N50M01

N55M06

(.500 DIAMETER DRILL)

N60G92X12.Y5.Z0

N65G00G90X-2.Y-2.S620M03

N70G43Z1.H02M08

N75G81G99Z-1.25R.1F7.0

N80G00X-7.Y-2.0

N85G80Z1.M09

N90G91G28Z0M05

N95G28X0Y0

N100T01M06

N105M30

The following descriptions are given for Machining Center example program 1:

Tool one, the number 3 center drill, is mounted in the spindle prior to beginning.

In block N10, functions G90, G20, G80, and G40 are defined as follows:

G90 establishes absolute coordinate system dimensioning.

G20 establishes the inch system of measurement.

G80 cancels all canned cycle functions and is entered at the beginning of the program to ensure that all cyclic functions from the previous program are no
longer in effect.

G40 cancels Cutter Diameter Compensation functions, G41 and G42.

G49 cancels the tool length compensation functions, G43 and G44.

In block N15, distances between Workpiece Zero and Machine Zero (X12.0Y5.0) are entered into the controller memory through function G92. In addition, the spindle speed and direction of the rotation of 900 r/min (S900, M03) are defined, while the program word T02, positions tool two in the magazine for replacement in the spindle (the tool is placed in the spindle in block N55).

Block N20 positions the tool to the proper position for drilling the first hole.

In block N25, the values of tool length offset are read (center drill), while rapid traverse of the spindle along the Z axis (with a value of $Z = 1.000$) is performed to a position above the workpiece. The miscellaneous function, M08, activates the flood coolant flow.

In block N30, the first hole is center drilled with a feed equivalent to 4.6 in/min.

In block N35, the second hole is center drilled.

Block N40 cancels canned cycle function G81, raises the tool to the initial reference plane position of Z1.0, and deactivates the coolant flow with M09.

In line N45 the machine returns to zero with respect to the Z axis.

In block N50, the programmed machine work is stopped by function M01, but only if the "OPTIONAL STOP" button on the operator control panel is ON. The purpose of this block is primarily to confirm whether tool T01 has performed the work that has been assigned in the program. This block usually appears before the tool change. This stoppage of the program execution gives the operator an opportunity to perform in-process inspection of the completed operation.

In block N55, tool 1 is replaced by tool 2 in the spindle. Up to now, only one program segment for one tool has been executed. Now, by comparing the remaining part of the program, notice that it contains many similar elements for the second tool, the .500 diameter drill.

Generally speaking, the program segment consists of a few characteristic elements that play the following roles:

- Establish the absolute or incremental coordinates system, spindle r/min and rotation direction.
- Execute the tool length offset given and activate the flow of the coolant.
- Determine the drilling canned cycle.
- Determine the tool's work path to consecutive holes in the pattern to be drilled.
- Positioning of the tool to the consecutive locations.
- Cancel any canned cycles and deactivate the coolant flow.
- Command of the Z, X and Y axes to return to Machine Zero position.
- Tool change.
- Program Ending.

Example of a Machining Center Program, Program 2

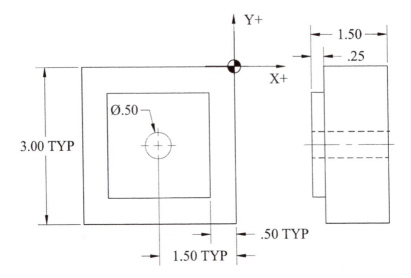

Figure 10 Drawing for Machining Center Example, Program 2

Machine: Machining Center	Program Number: O0002

Workpiece Zero X =<u>Upper Right Corner</u> Y= <u>Upper Right Corner</u> Z =<u>Top Surface</u>

Setup Description:

The workpiece material is aluminum.

Dimensions of the plate before machining are 3.0 x 3.0 x 1.5 inches.

The workpiece is mounted in the vice on parallels with the stop on the right side.

Tool 1 is placed in the spindle before the work from the program begins.

Tool #	Description	Offset	Comments
T01	Two flute 5/8 inch (.625) diameter High Speed Steel End Mill		SFM = 400
T02	No. 6 Center Drill		SFM = 400
T03	1/2 inch Drill		SFM = 400

Program

O0002

N10G90G20G80G40G49

N15G92X21.025Y6.127Z0

N20G00X.4Y-.1875S2445M03T02

N25G43Z1.M08H01

N30G01Z-.25F20.0

N35X-2.8125

N40Y-2.8125

N45X-.1875

N50Y.4

N55G28Z1.0.M09

N60M01

N65M06

N70G90G80G40G49

N71G92X21.025Y6.127Z0

N75G00X-1.5Y-1.5S3057 M03T03

N80G43Z1.H02M08

N85G81G98Z-.438R.1F12.0

N90G80M09

N95G28Z1.0

N100M01

N105M06

N110G90G80G40G49

N111G92X21.025Y6.127Z0

N115G00 X-1.5Y-1.5S3057M03T01

N120G43Z1.H03M08

N125G81G98Z-1.664R.1F18.0

N130G80Z1.0M09

N135G91G28Z0M05

N140G28X0Y0

N145M06

N150M30

EXPLANATION OF THE SAFETY BLOCK

Block N10 is called the Safety Block; the name refers to its role in the program. This Safety Block consists of G-Codes that establish the type of coordinate system used (G90 Absolute or G91 Incremental), establishes the measurement system used (G20 inch or G21 metric), cancellation of all canned cycles (G80), cancellation of cutter diameter compensation (G40), and cancellation of tool length compensation (G49). In order to more accurately define the canceling function of G80, study the Machining Center Example Program number 2. In block N85, the canned drilling cycle G81 is activated. If the canned drilling cycle is interrupted for some reason (the drill or the workpiece become damaged), the machining must be stopped, in order to exchange the tool and or the workpiece. Then by pressing the "reset" button, the program is returned to its beginning or head. However, because of the above occurrence, canceling function G80 (included in block N90) has not occurred. If machining is started from the beginning of the program, tool one (an end mill of .625 inch) will need to be returned to the spindle and the new drill returned to the tool magazine prior to starting. In the above-described procedure, if function G80 is omitted from block N10, then function G81 is still valid. Thus, all position changes of the end mill (along the X and Y axes) will be executed as if the canned drilling cycle including the assigned parameters of Z and R from block N85 are still valid. On many machines, RESET will cancel canned cycles.

To avoid such mistakes, use the Safety Block including the canceling functions in the first block of the program and immediately following all tool changes.

A similar situation in which tool radius compensation was applied is described as follows: In programming the milling process, you may establish the movement of the tool by determining the movement of the center of the tool. This approach applies to standard programming. However, you may also program the actual contour of the workpiece. From that, the computer will calculate the path of the center of the mill (or other tool) with a given radius. This means that the actual machine movements will differ from the programmed movements by the value of the offset placed in the offset register, which is the radius of the tool. Such an arrangement can be applied by use of functions G41 (cutter compensation to the left) and G42 (cutter compensation to the right). If, for any reason, the milling process with the assigned tool radius compensation is interrupted, then the next tool used, for example, a drill, will be positioned with a displacement value equal to the mill radius compensation previously used. In order to avoid such situations, employ the cancellation function G40 in the Safety Block.

The same scenario can be applied to tool length compensation values that are activated in the program by the letter address H followed by the number of the tool as shown in block N25. If this offset value is not cancelled by G49 in the safety block, it too will remain valid. Note: If function G28 is used prior to tool change as in line N55, the tool length offset is cancelled.

To avoid such mistakes, use the Safety Block, including the canceling functions, in the first block of the program and immediately following all tool changes.

PREPARATORY FUNCTIONS FOR MACHINING CENTERS (G-CODES)

These preparatory functions, often called G-Codes, are a major part of the programming puzzle. They identify to the controller, what type of machining activity is needed. For example, if a hole needs to be drilled, function G81 may be used, or if programming in the incremental coordinate system is required, then function G91 is used. These codes, along with other data, control machine motion.

The motion of the axes of a machine may be performed along a straight line, an arc or a circle.

Codes G00 and G01 allow axes movement along a straight line.

Codes G02 and G03 allow axes movement along a circular path of motion.

RAPID TRAVERSE POSITIONING (G00)

The rapid traverse function is entered to relocate the tool from position *A* to position *B* along a straight line at the fastest possible traverse. The shortest axis movement distance will be accomplished first and, therefore, you must be aware of the workpiece holding the equipment in order to avoid any collision between the tool and the holding equipment. The path traveled by the tool for G00, X30, and Y20, is as follows:

Figure 11
Rapid Traverse Tool Movement

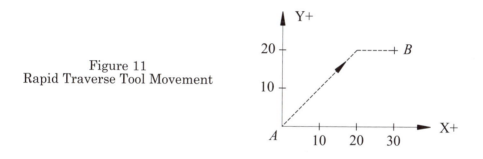

If you are uncertain of the tool path, then position the "RAPID TRAVERSE OVERRIDE" switch at 25 or 50% (see Part 2, "Rapid Traverse Override"). This reduces the traverse speed and increases the time allowance for a possible manual interruption of the motion. Programmers should position the *Z* axis to an acceptable clearance plane of 1.0 inch or any amount necessary to move over clamps or obstructions to reduce the chance of collision.

By adjusting a system parameter, the movement described in Figure 11 can be changed to a simultaneous or diagonal move of both axes. Consult the manufacturer manual for specific instructions.

Linear Interpolation (G01)

The work function G01 is used to move the tool from point *P1* to point *P2* along one or all of the axes simultaneously, along a straight line of motion and at a given feed rate specified by the F-word.

Example:

G01X10.Y20.F8.0

Linear interpolation may be performed along three axes simultaneously. The control system calculates the particular speeds for each axis so that the resulting speed is equivalent to the programmed feed rate.

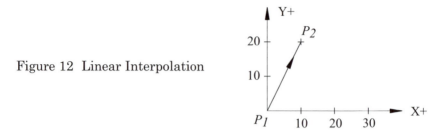

Figure 12 Linear Interpolation

Example of Linear Interpolation, Program 3

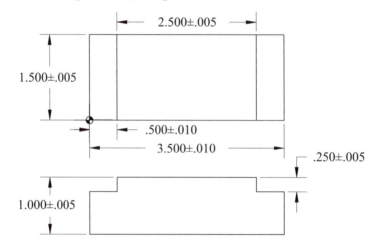

Figure 13 Drawing for Linear Interpolation Example

Machine: Machining Center	Program Number: O0003

Workpiece Zero X = <u>Lower Left Corner</u>   Y = <u>Lower Left Corner</u>   Z = <u>Top Surface</u>

Setup Description:

Material used is Aluminum 6061-T6, with plate dimensions of 3.5 x 1.5 x 1.0.

$$S = \frac{12 \times SFM}{\pi \times d} = \frac{12 \times 400}{3.14 \times .625} = 2445 \text{ rev/min}$$

$$F = 4 \times .003 \times 2445 = 29 IPM$$

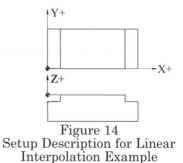

Figure 14
Setup Description for Linear
Interpolation Example

Tool #	Description	Offset	Comments
T01	Two Flute .625 Diameter HSS End Mill		SFM = 400 ft/min Feed =.003 in per tooth

Program

O0003

N10G90G20G80G40G49

N15G92X11.0Y6.0Z0

N20G00X.1875Y-.4S2445M03

N25G43Z1.0H01M08

N30G01Z-.250F50.0

N35Y1.9F29.0

N40G00Z1.0

N45X3.3125

N50G01Z-.250F50.0

N55Y-.4F29.0

N60G28Z1.0M09

N65G91G28X0Y0M05

N70M30

CIRCULAR INTERPOLATION (G02, G03)

Circular interpolation allows tool movements to be programmed to move along the arc of a circle. When applying the circular interpolation, the plane in which the arc is positioned must be determined initially. To do this, employ preparatory function G17, G18, or G19. Then, depending on the direction of the machining, select function G02 to make a clockwise movement along the arc and function G03 to make a counterclockwise movement along the arc. In order to describe the movements of the tool along the arc, apply the following two methods:

1. Determine the radius R and the values of the end point coordinates in the given plane.

2. Determine the value of the endpoint of the arc and the values of the incremental distance to the arc center in a given plane.

P_1 = the start of an arc.

P_2 = the end of an arc.

P_3 = the center of the arc.

R = radius vector.

I = the incremental distance to the arc center along the X axis.

J = the incremental distance to the arc center along the Y axis.

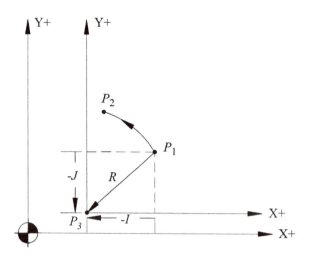

Figure 15
Required Data for Circular
Interpolation

The incremental, distance is defined as a radius projection onto a given axis. The radius incremental distance is always attached. It begins at the starting point and ends at the center of the circle. It is always directed toward the center of the circle.

- Vector projections onto the X axis are identified by use of the letter address I.
- Vector projections onto the Y axis are identified by use of the letter address J.
- Vector projections onto the Z axis are identified by use of the letter address K.

In the following example, a vector projection is illustrated. The radius is positioned at the origin of the coordinate system, as below:

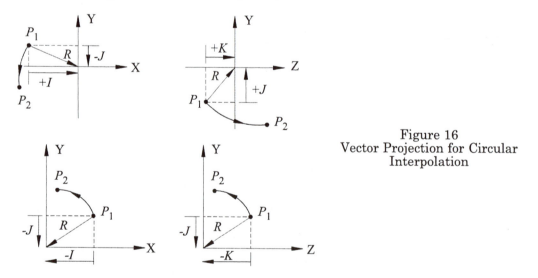

Figure 16
Vector Projection for Circular
Interpolation

Note: in the Figure 16 the signs (+, -) of the incremental distances I, J, and K depend on the position of the starting point of the arc with respect to the center of the arc, that is, with respect to the coordinate system. If the direction of the vector is consistent with the direction of the assigned axis of the coordinate system, then we apply the positive sign. If not, then we apply the negative sign.

Most modern controllers do not require the use of the positive sign. If no sign is present the value is considered to be positive.

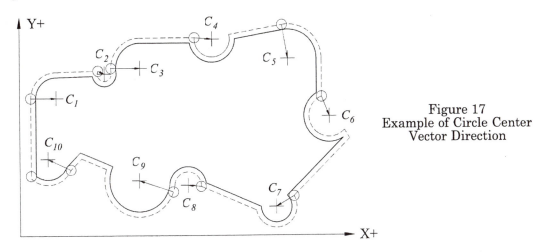

Figure 17
Example of Circle Center
Vector Direction

In the Figure 17, the tool path begins in the lower left corner of the part. Each arc is identified C_1 through C_{10} with the center point of the arc shown. The arrows indicate direction vector to the arc center. Following are the arc cutting directions and the I and J values:

C_1 = G02, I, $J0$

C_2 = G03, I, $-J$

C_3 = G02, I, $J0$

C_4 = G03, I,$-J$

C_5 = G02, I,$-J$

C_6 = G03, I, $-J$

C_7 = G02, $-I$,$-J$

C_8 = G03, $-I$, J

C_9 = G02, $-I$, J

C_{10} = G02, $-I$, J

Notes: When circular motion is described with radius function R, no sign is required if the arc is less than or equal to 180° (the system defaults to positive unless otherwise specified). Assign a negative value to R (-) if the arc is greater than 180°. The maximum rotation of an arc using R is 359.9°.

A full circle may be accomplished using R by linking two 180° arcs.

If a full circle of 360° is to be performed, then it is necessary to employ the incremental distances of the arc center points for I, J, and K, not radius R, in

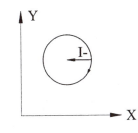

Figure 18 Programming Arcs
for Full Circles

the program.

Example:

Do not use *I*, *J*, or *K* with *R* in the same block, because if either *I*, *J*, or *K* are used, they will be ignored by the control, and the tool will follow the arc with the assigned radius of *R*.

If the value of an entered radius *R* is zero, an alarm will result.

Cutter radius compensation may be used for circular interpolation. However, it must be initiated in a G00 or G01 block preceding the G02/G03 information.

Example of Circular Interpolation, Program 4

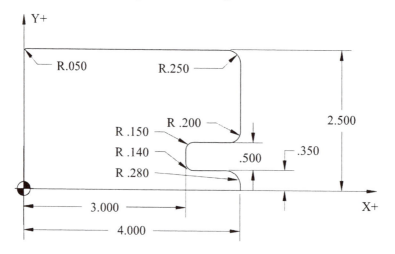

Figure 19 Drawing for Machining Center Circular Interpolation Example

Machine: Machining Center	Program Number: O0004

Workpiece Zero X = <u>Lower Left Corner</u>   Y = <u>Lower Left Corner</u>   Z = <u>Top Surface</u>

Setup Description:

Material used is Steel and is .150 thick.

The program is limited to a finishing pass.

The workpiece is mounted to a machining fixture by the existing .250 diameter holes not shown in the above print.

F = 4 x .002 x 1375 = 11 in/min

Tool #	Description	Offset	Comments
T02	Four Flute .250 Diameter HSS End Mill		SFM = 90 ft/min Feed =.002 inches per tooth

Program

O0004

N10G90G20G80G40

N15G92X10.0Y7.0Z0

N20G90G00X-.125Y-.2S1375M03

N25G43Z1.0H02M08

N30G01Z-.16F50.0

N35Y2.45 F11.0

N40G02X.05Y2.625I.175J0

N45G01X3.75

N50G02X4.125Y2.25I0.J-.375

N5G01Y1.05

N60G02X3.8Y.725I-.325J0

N65G01X3.15

N70G03X3.125Y.7I0.J-.025

N75G01Y.49

N80G03X3.140Y.475I.015J0

N85G01X3.72

N90G02X4.125Y.07I0.J-.405

N95G01Y-.125

N100X-.125

N105G28Z1.M09

N110G91G28X0Y0M05

N115M30

The following is the same program but utilizing radius R:

O0004

N10G90G20G80G40

N15G92X10.0Y7.0Z0

N20G90G00X-.125Y-.2S1375M03

N25G43Z1.0H02M08

N30G01Z-.1F50.0

N35Y2.45 F11.0

N4G02X.05Y2.625R.175

N4G01X3.75

N50G02X4.125Y2.25R.375

N55G01Y1.05

N6G02X3.8Y.725R.325

N65G01X3.15

N70G03X3.125Y.7R.025

N75G0Y.49

N80G03X3.14Y.475R.015

N85G01X3.72

N90G02X4.125Y.07R.0405

N95G01Y-.125

N100X-.125

N105G28 Z1.M09

N110G91G28X0Y0M05

N22M30

HELICAL INTERPOLATION USING G02 OR G03

Helical interpolation allows movements of the tool to be programmed so that it travels along a circular path in the *XY* plane, with a simultaneously straight-line motion along the *Z* axis. Such a combination of tool movements with respect to the three axes creates a helix contour and may be used to machine threads. For all practical purposes, this function is limited to machining threads with large diameters because of the required tool diameter.

Block Format:
G02X . . . Y . . . I . . . J . . . Z . . . F . . . (R . . . may be used in place of *I* and *J*)
G03X . . . Y . . . I . . . J . . . Z . . . F . . . (R . . . may be used in place of *I* and *J*)

Where   X = arc ending point

Y = arc ending point

I = incremental *X* axis distance of the arc center from the start point

J = incremental *Y* axis distance of the arc center from the start point

Z = arc ending point (at depth)

R = radius of the arc

F = feed rate

As stated before, function G02 refers to clockwise motion (CW); whereas, function G03 refers to counterclockwise motion (CCW).

Notes on Helical Interpolation:

This function may only be applied under the following conditions:

The minor diameter for internal threads or major diameter for external threads must be machined in a prior operation.

The XY plane is in a circular interpolation.

The Z axis is in a linear interpolation.

The tool must be positioned in the Z axis prior to the helical interpolation.

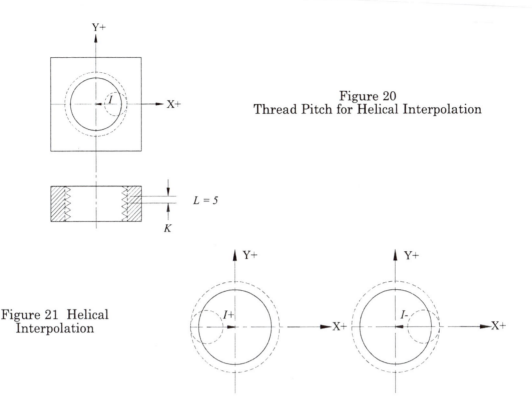

Figure 20
Thread Pitch for Helical Interpolation

Figure 21 Helical
Interpolation

The feed rate designated by the F-word is the circular interpolation for the XY plane.

Climb milling is the preferred machining method.

When one pass of the tool does not complete the full thread, adjustment of the Z axis coordinates of the starting or ending point (depending on machining starting from the bottom or top of the thread), for another pass is required.

If cutter radius compensation is used in the program, G41 or G42 may not be called in the same line as the G02 or G03. For an internal thread, it is typically called in a linear approach move from the center of the diameter moving outward and cancelled at the end of interpolation moving back towards the center. This move must be equal to, or greater than the radius of the tool used.

Figure 22 is an example of a tool that may be utilized during threading with the application of helical interpolation.

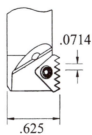

Figure 22
Helical Threading Tool

.0714

.625

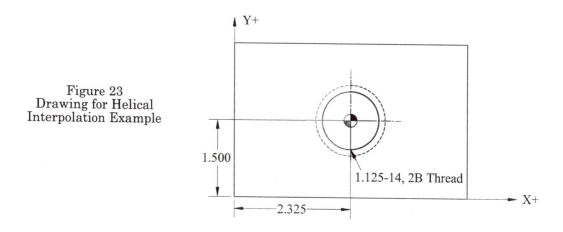

Figure 23
Drawing for Helical
Interpolation Example

1.500

2.325

1.125-14, 2B Thread

Example of Helical Interpolation, Program 5

An example for helical interpolation follows:

Machine: Machining Center   Program Number: O0005

Workpiece Zero X = Part Centerline Y = Part Centerline Z = Top Surface

Material used is .50 thick CD 12L14 steel plate.

The minor diameter was machined in a previous operation and only the threading cycle is needed.

Tool #	Description	Offset	Comments
T01	Special threading tool with a .0714 lead. See Figure above.		SFM = 70 ft/min Feed = 1.5 IPM

Program

O0005

N10G90G20G80G40G49

N15G92X10.0Y7.0Z0

N20G00X0Y0S428M03

N25G43Z1.H01M08

N30G01Z-.575F50.0

N35G01X.25F3.5

N40G03X.25Y0I-.25Z-.5036F1.5

N45I-.25Z-.4322

N50G01X0F50.0

N55G00Z1.0M09

N60G91G28Z0M05

N65G28X0Y0

N70M30

Note: The value X.25 in block N35 was calculated in the following manner:

$$X = \frac{1.125}{2} - \frac{.625}{2} = .250$$

The difference between the Z value on line N40, -.5036 and on line N45, -.4322, is equal to the lead of the thread, .0714.

Dwell (G04)

Dwell is determined by the preparatory function, G04 and by using the letter address P or X, which corresponds to the time duration of dwell.

Block Format:

    G04P . . .

    G04X . . .

Example:

    G04X1.   or   G04P1000

The length of time for the dwell is stated by X or P and, in this example, is equivalent to 1 second. The P address is programmed in milliseconds (ms) 1 second = 1000 ms.

When the letter address P is used to determine time, do not use the decimal point. Drilling or counterboring holes, with a specifically defined depth, are two examples in which to apply function G04. Time P and X, in these cases, determine a time period in which the drill, having full rotational speed, hesitates at the bottom of the hole long enough to fully and accurately remove the excess material from the bottom of the hole.

Example of Using Dwell, Program 6

Figure 24
Drawing for Dwell Example

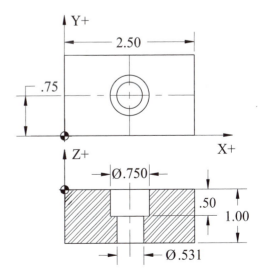

Machine: Machining Center	Program Number: O0006

Workpiece Zero X = <u>Lower Left Corner</u> Y = <u>Lower Left Corner</u> Z = <u>Top Surface</u>

Setup Description:

The material used is 4140 steel.

The part is clamped in a vise.

Tool one is mounted in the spindle prior to beginning.

Tool #	Description	Offset	Comments
T01	No. 4 HSS center drill.		SFM = 75 ft/min Feed =.001 inch per flute
T02	17/32 HSS drill		SFM = 75 ft/min Feed =.002 inch per flute
T03	.750 diameter 2 flute HSS end mill		SFM = 75 ft/min Feed =.002 inch per flute

Program

O0006

N10G90G20G80G40G49

N15G92X10.0Y7.0Z0

N20T1M06

N25G00X1.25Y.750S960M03T02

N30G43Z1.H01M08

N35G00Z.1

N40G01Z-.269F1.9

N45G91G28Z0M09

N50M01

N55M06

N60G90G80G40G49

N65G00X1.25Y.750S564M03T03

N70G43Z1.H02M08

N75Z.1

N80G01Z-1.35F2.25

N85G91G28Z0M09

N90M01

N95M06

N100G90G80G40G49

N105G00X1.25Y.750S400M03T01

N110G43Z1.H03M08

N115Z.1

N120G01Z-.5F3.2

N125G04P300

N130G00Z1.0M09

N135G9G28Z0M05

N140G28X0Y0

N145M30

In block N125, a dwell of .3 seconds is utilized to fully cut the bottom of the counterbore.

EXACT STOP (G09)

When this function is used, the machine responds by decelerating to a feed of zero in the axis commanded before executing the next line, thus, checking the programmed end point. This function is used only if you want to obtain a sharp edge around corners in the cutting feed mode. It refers only to the block to which it was assigned.

DATA SETTING (G10)

See later in section "Offset Amount Input by the Program (G10)"

POLAR COORDINATE CANCELLATION (G15)

This command is used to cancel use of the polar coordinate system called by function G16. This command must be programmed in a line, by itself.

POLAR COORDINATE SYSTEM (G16)

The polar coordinate system may be used to program the locations for holes in a bolt circle. By locating the bolt circle center, using the Local Coordinate System function G52, a rotation angle and circle radius can be programmed to locate the holes (unless the center is X0Y0). The programmed values in the X axis represent the circle radius and, in the Y, represent the angular. The angular values may be programmed as either positive, or negative. Positive rotation is counterclockwise and negative rotation is clockwise.

PLANE SELECTION (G17, G18, G19)

G17 = XY plane

G18 = XZ plane

G19 = YZ plane

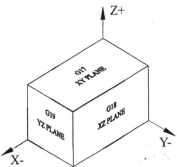

Figure 25 Plane Selection

By declaring the selection of plane G17, G18, and G19, the machine selects the given plane. The default plane on a vertical machining center is the G17 *XY* plane and, therefore, it is not necessary to input G17 but it should be part of the Safety Block especially when other planes are selected within the program.

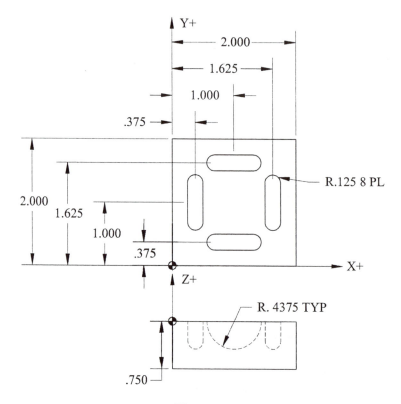

Figure 26
Drawing for Plane Selection Example

Example of Plane Selection, Program 7

Machine: Machining Center	Program Number: O0007

Workpiece Zero X = Lower Left Corner Y = Lower Left Corner Z = Top Surface
Setup Description:
Material used is Aluminum 6061-T6 , Feed per flute .001;

Tool #	Description	Offset	Comments
T01	.250 diameter HSS 2 flute ball end mill		SFM = 300 Feed =.001 inch per flute

Program

O0007

N10G90G20G17G80G40G49

N15G92X10.0Y7.0Z0

N20G90G00X1.3125Y.375S4585M03

N25G43Z1.H01M08

N30Z.2

N35G01Z-.125F18.0

N40G18G03X.6875Z-.125I-.3125 K0F12.0

N45G01Z.2F18.0

N50G00Y1.625

N5G01Z-.125

N60G02X1.3125Z-.125I.3125K0F12.0

N65G01Z.2F12.0

N70G00X1.625Y1.3125

N75G01Z-.125

N80G19G02Y.6875Z-.125J-.3125K0F12.0

N85G01Z.2F18.0

N90G00X.375

N95G01Z-.125

N100G03Y1.3125Z-.125J.3125K0F12.0

N105G01Z.2

N110G91G28Z0M09

N115G28X0Y0M05

N120M30

INPUT IN INCHES (G20) AND INPUT IN MILLIMETERS (G21)

Function G20 or G21 is entered at the beginning of the program in the Safety Block to establish the measurement system. They apply to the whole program. Functions G20 and G21 cannot be interchanged during programming. When using functions G20 or G21 the values are the same units of measure for

F = feed rate

Position of *X, Y,* and *Z*

Offset values.

The default measurement system for most American machines is inch, therefore, it is not necessary to use the above functions unless the opposite system is required. The desired system should always be stated within the Safety Block.

STORED STROKE LIMIT ON/OFF (G22, G23)

Function G22 identifies the stored stroke limit for the tool area outside the working envelope. Function G23 identifies the stored stroke limit for the tool area inside the working envelope.

The manufacturer enters these stroke limits of the axis travel into the controller memory, in order to define the work envelope for the machining center. In a case where the tool is positioned beyond the limited travel area (outside the limits of the stroke), the machine will stop and an alarm is displayed on the screen.

REFERENCE POSITION RETURN CHECK (G27)

This function checks for the accurate return to the reference point and confirms the programmed position has been reached. Entering function G27 causes rapid traverse of the tool to the reference point. If the tool is positioned accurately at the reference point, the axis LED's will light up. If the LED's do not go on, the tool is displaced from the reference point and an alarm will result. This might occur, if prior to the G27 command, cancellation of the tool radius compensation has been omitted (i.e., function G40 is omitted following the use functions G41 and G42). This will cause a position displacement of the tool equal to the value of the compensation offset and the execution of the following program block with the same displacement.

REFERENCE POSITION RETURN (G28)

This function is the automatic reference point return through an intermediate point programmed in the X, Y and/or Z axes. The machine will position at rapid traverse (G00) to the programmed intermediate point coordinate values and then to the reference point of machine zero. This function is commonly used before an Automatic Tool Change (ATC). After using function G28, the axis LED's on the control panel should light up to indicate successful return to zero for the axis. When you use function G28, you must specify the point through which the tool passes on its way to zero. If the command, G28X0Y0Z0 is entered in the incremental mode (G91), the machine will position at rapid traverse to the reference position in all three axes simultaneously. Caution must be exercised so as not to interfere with the work holding device or part. It is a good practice to use G28 with a Z axis positioning move in a prior block to ensure clearance. The following example demonstrates the best method for using function G28 by first positioning the Z axis to a level for clearance of the work holding.

Example:

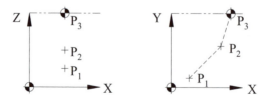

Figure 28 Reference Position Return (G28)

Function G90 is active.

    N50G28Z3.00
    N55G28X6.00Y7.00

where

X, Y, and Z = the coordinates of the point through which the tool passes on its way to Machine Zero

P_1 = present position
P_2 = point through which the tool passes
P_3 = Machine Zero position

Function G91 is activated.

    N50G91G28Z0.
    N55G28X0.Y0.

where

First in line N50, the tool will retract along the Z axis until it reaches the Machine Zero position and then in the next block N55, the machine will move to the X and Y Machine Zero coordinates.

RETURN FROM THE REFERENCE POSITION (G29)

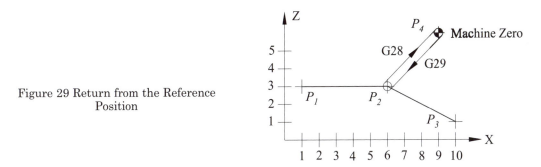

Figure 29 Return from the Reference Position

To use function G29 it must follow function G28 or G30. It can be applied after an automatic tool change so that the tool will return to the work position at rapid traverse, as it similarly does in G00. The tool will travel through the intermediate point programmed.

Example:

G91 = incremental positioning command
N50G28X5.0Y0 P_1 - P_2 - R
N55M06
N60G29X4.0Y-1.5   R - P_2 - P_3

where

P_1 = initial work point of tool 1
P_2 = end work point of tool 1
P_2 = initial work point of tool 2
P_3 = end work point of tool 2
P_4 = Machine Zero

During the execution of the block containing function G29, the tool automatically follows the return path from point P_4 to P_2 and the programming is limited only to entering the difference between points P_2 and P_3.

RETURN TO SECOND, THIRD, AND FOURTH REFERENCE POSITIONS (G30)

Usually Machine Zero coordinates are coincidental to the reference point assigned to function G28. However, if the reference point assigned to function G28 does not correspond to Machine Zero, function G30 must be used in order to place the tool in this position. (This may be the case if the tool change position is not the same point as is specified in function G28.) Before applying function G30, apply function G28. The movement determined by function G30 is rapid traverse. Functions G28, G29, and G30 can be used to program in both absolute and incremental systems. Cutter compensations (length and diameter) must be cancelled prior to this command.

CUTTER COMPENSATION (G40, G41, G42)

The use of cutter compensation allows the programmer to use the part geometry exactly as from the print for programmed coordinates. Without using compensation, the programmer must always know the cutter size and offset the programmed coordinates for the geometry by the amount of the radius. In this scenario, if a different size cutter is used the part will not be machined correctly. An added advantage for using cutter compensation is the ability to use any size cutter as long as the offset amount is input accurately into the offset register. It is also very effectively used for fine-tuning of dimensional results by minor adjustments to the amount in the offset register.

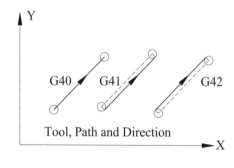

Figure 30 Cutter Compensation

G40 = Cutter Compensation Cancel
G41 = Cutter Compensation Left
G42 = Cutter Compensation Right

Cutter Compensation Cancellation (G40)

Function G40 is used to cancel cutter radius compensation initiated by G41 or G42. It should be programmed after the cut using the compensation is completed by moving away from the finished part in a linear (G01) or rapid traverse (G00) move by at least the radius of the tool. Care should be taken here because if the cancellation is on a line without movement, the cutter will move unpredictably in the opposite direction and may damage the part.

Cutter Radius Compensation Left and Right (G41 and G42)

Functions G41 and G42 offset the programmed tool position to the left (G41) or right (G42) respectively, by the value of the tool radius entered into offset registers and called in the program by the letter address D. For each tool, enter the corresponding offset amount. In the program, the letter D and the number of the offset (two digits) are input to initiate the compensation call.

The direction the tool is offset, to the left or the right is dependant upon which direction the tool is traveling. To accomplish climb cutting with right hand tools, always use G41 and for conventional cutting use G42. Consider which direction of offset is needed by facing the direction the tool is going to travel and observe which side of the part the cutter is going to be on, to the left for G41 or to the right for G42.

Procedure for Initiating Cutter Compensation

Position the tool in the X and Y axes to a point away from the required finished geometry. Then program a linear move that is larger than the radius of the cutter to feed into the part, e.g. G01 G41 or G42 offset direction X or Y absolute coordinate and D offset number. To use the cutter compensation properly, there needs to be one full line of movement to position the tool on the proper vector to cut the part. Once this is accomplished, program the part geometry per print.

The tool will not be positioned to the actual programmed point on the geometry, rather, it will be positioned to a point plus or minus the offset value of the cutter called by D, and the edge of the cutter will be aligned with finished part geometry.

Rules for Cutter Compensation Use

1. When cutter compensation is called in a program, the control looks ahead by two lines in order to set up a vector to position for the offset amount for each move.
2. Movement must be maintained once G41 or G42 commands have been called up. By following this rule, over-cutting of the part can be avoided. If two lines of non-movement commands are placed in a row after cutter compensation is called up, the control will ignore functions G41 or G42 and the part will be cut incorrectly.
3. <u>Do Not</u> start cutter compensation G41 or G42 when G02 or G03 is in effect.
4. When a change from left to right compensation, or vise versa is needed, it is recommended that cancellation of the first compensation be executed, then the second compensation may be called and; thus, the transition of the tool position vector will not conflict.
5. When machining an inside radius, the radius must be larger than the offset of the tool, or the control will stop the program and an alarm will be displayed.
6. The move used to call up cutter compensation must be larger than the radius of the tool used.

Example of Cutter Compenastion, Program 8
Offset for tool number one is D01.
Offset for tool number two is D02.

Please note in the following program example there are no Z axis moves. The intent here is to demonstrate the usage of cutter radius compensation only. The program line sequence numbers in the illustration above (Figure 31) correspond with the following program. The arrows along the tool centerline path indicate the travel direction. At each intersection point on the geometry the arrows indicate the tool offset direction.

In type B, when the angle is less than 1°:

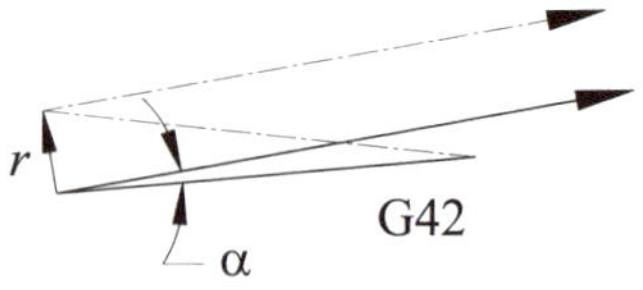

The following are examples of tool withdrawal from the workpiece when using tool radius compensation cancel function G40.

Type A: $\alpha < 90°$

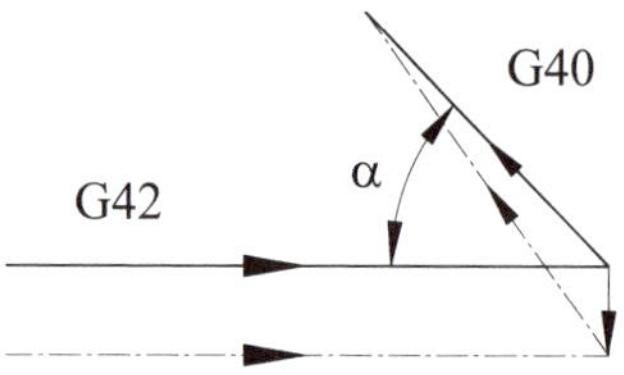

Figure 41

Figure 42
Line-Line

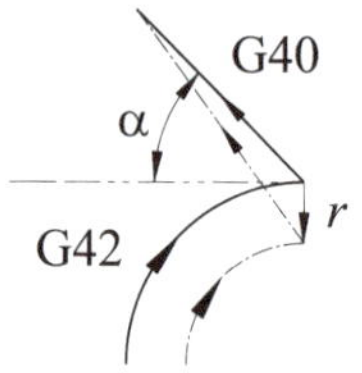

Figure 43
Line-Arc

$90° \leq \alpha \leq 180°$

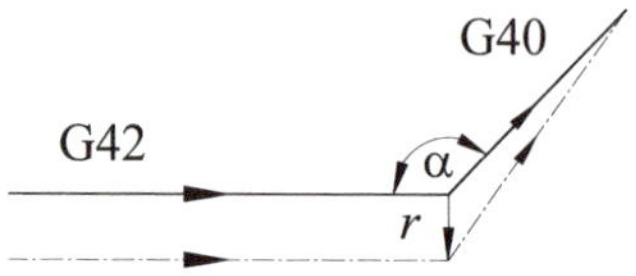

Figure 44
Line-Line

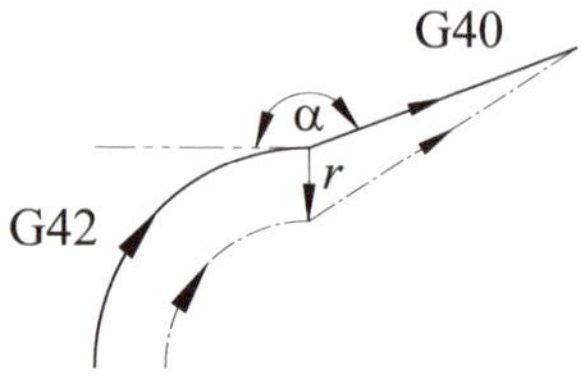

Figure 45
Arc-Line

Type B: $\alpha < 90°$

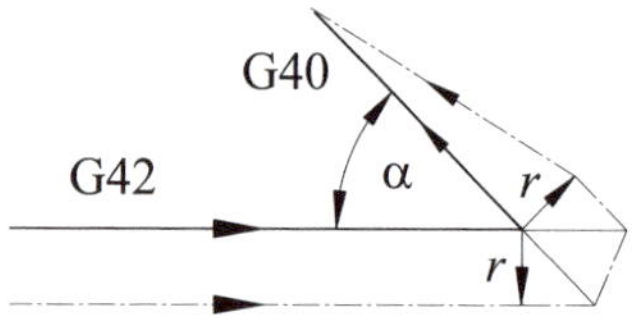

Figure 46
Line-Line

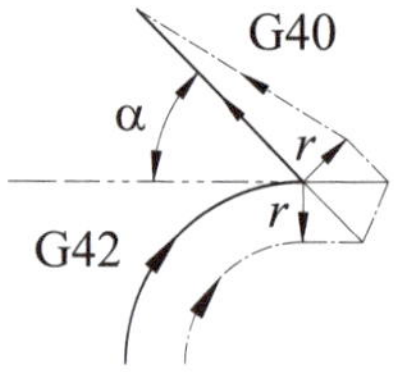

Figure 47
Arc-Line

$\alpha < 1°$

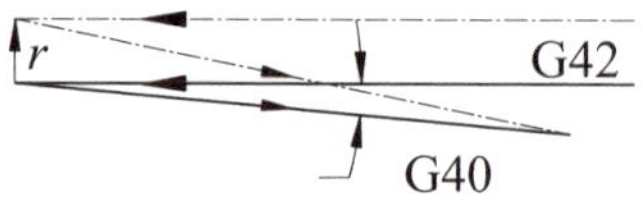

Figure 48

$90° \leq \alpha \leq 180°$

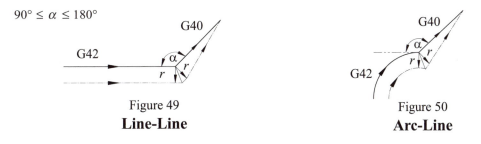

Figure 49
Line-Line

Figure 50
Arc-Line

The following is an example of a tool path configuration with reference to the program line, when executing functions G41 and G42.

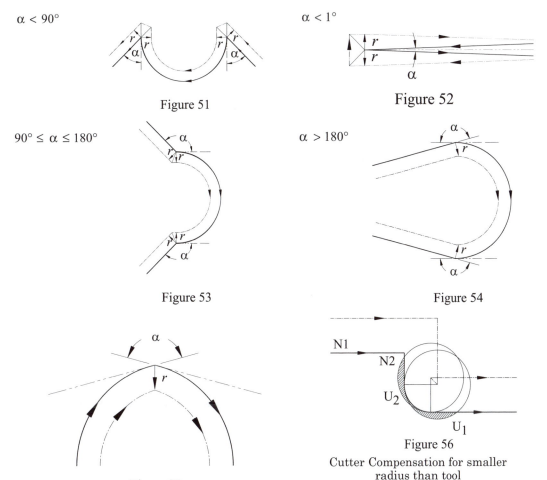

$\alpha < 90°$

Figure 51

$\alpha < 1°$

Figure 52

$90° \leq \alpha \leq 180°$

Figure 53

$\alpha > 180°$

Figure 54

Figure 55

Figure 56
Cutter Compensation for smaller
radius than tool

Special Cases:

The radius of the machined workpiece is smaller than the tool radius.

The control has a look-ahead feature that always reads one or more lines of the program in advance. For Figure 56, at the end of the execution of block N_1, an alarm will be displayed on the screen and the execution will be interrupted. If the execution of the program proceeds line by line (the SINGLE BLOCK button is ON) after the completion of operation N1, the control will not read the information contained in N_2 and thus cannot offset the tool path to compensate for the condition and will undercut the edge in U_1. In both cases a similar situation will be encountered with undercut U_2.

The width of the undercut is smaller than the tool diameter.

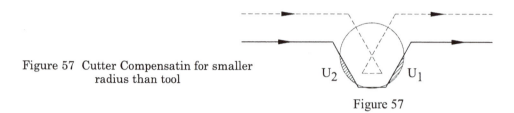

Figure 57 Cutter Compensatin for smaller radius than tool

Figure 57

Example of Cutter Compensation, Program 9

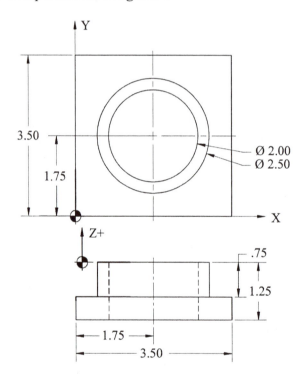

Figure 58 Drawing for Cutter Compensation Example 2

Machine: Machining Center	Program Number: O0009

Workpiece Zero X = Lower Left Corner Y = Lower Left Corner Z = Top Surface

Setup Description:

Material used is 3.5 x 3.5 x 1.25, 303 Stainless Steel.

Tool #	Description	Offset	Comments
T01	#5 HSS CTR drill		SFM = 100 Feed =.002 inch per flute
T02	1.25 diameter drill		SFM = 100 Feed =.003 inch per flute
T03	1.25 diameter, six flute, HSS roughing cutter	D60= .650	SFM = 100 Feed =.003 inch per flute
T04	1.25 diameter, six flute, HSS end mill	D61 = .625	SFM = 100 Feed =.002 inch per flute

Program
O0009
(#5 CENTER DRILL)
N10G90G20G80G40G49
N15G92X10.0 Y7.0Z0
N20T01M06
N25G00X1.75Y1.75S915M03T02
N30G43Z.1H01M08
N35G81G98Z-.382R.1F4.0
N40G00G80Z1.0M09
N45G91G28Z0
N50M01
(1.25 DIAMETER DRILL)
N55T02M06
N60G90G80G40G49
N65G00X1.75 Y1.75S320M03T03
N70G43Z.1H02M08
N75G83G98Z-1.625R.1Q.625F2.0
N80G00G80Z1.0M09
N85G91G28Z0
N90M01
(1.25 DIAMETER 6-FL ROUGHING END MILL, CUTTER COMP #D60)
N95T03M06
N100G90G80G40G49
N105G00X1.75Y1.75S320M03T04
N110G43Z.1H03M08
N1151Z-1.3F20.0

```
N120G01G42Y2.750D60F5.76
N125G02J-1.0
N130G01G40Y1.750F10.0
N135G00Z.1
N140Y-1.0
N145G01Z-.73F20.0
N150G42Y.5D60
N155G03J1.25F6.0
N160G01G40Y-1.0F20.0
N165G00Z.1M09
N170G91G28Z0
N175   M01
```
(1.25 DIAMETER 6-FL FINISHING END MILL, CUTTER COMP #D61)
```
N180T04M06
N185G90G80G40G49
N190G00X1.75Y1.75S320M03T01
N195G43Z.1H04M08
N200G41Y2.75F10.D61
N205G01Z-1.3 F50.0
N210G03J-1.0F3.84
N215G01G40Y1.75F20.0
N220G00Z.1
N225Y-1.0
N230Z-.75
N235G01G41Y0.50F10.D61
N240G02J1.25F4.0
N245G01G40  Y-1.0F20.0
N250G00Z.1M09
N255G91G28Z0M05
N260G28X0Y0
N265T01M06
N270M30
```

Example of Cutter Compensation, Program 10

The following is an example illustrating the machining of a slot through 0.5 inch thick Aluminum.

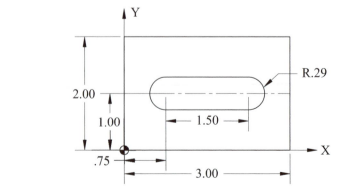

Figure 59
Drawing for Cutter
Compensation Example 3

Machine: Machining Center	Program Number: O0010

Workpiece Zero X = Lower Left Corner Y = Lower Left Corner Z = Top Surface
Setup Description:

Tool #	Description	Offset	Comments
T01	#5 HSS CTR drill		SFM = 300 Feed =.003inch per flute
T02	17/32 diameter HSS drill		SFM = 300 Feed =.003 inch per flute
T03	1/2 diameter, two flute, HSS roughing cutter.	D50 = .260	SFM = 850 Feed =.003 inch per flute
T04	1/2 diameter, two flute, HSS end mill	D51 = .250	SFM = 850 Feed =.002 inch per flute

Program
O0010
N10G90G80G20G40G49
(#5 CENTER DRILL)
N15T01M06
N20G92X10.0Y7.0Z0
N25G00X.750Y1.S2200M03T02
N30Z.1H01M08
N35G81G98Z-.3R.1F8.8
N40G80Z1.0M09
N45G91G28Z0
N50M01
(17/32 DRILL)
N55T02M06
N60G90G80G20G40G49
N65G00X.75Y1.0S2150M03T03
N70G43Z1.H02M08
N75G81G98Z-.70R.1F8.0
N80G80Z1.0M09
N85G91G28Z0
N90M01
(1/2 2-FL ROUGHING END MILL, CUTTER COMP #D50)
N95T03M06
N100G90G80G40G49
N105G00X.75Y1.0S2200M03
N110G43Z1.0H03M08
N115Z.1
N120G01Z-.52F50.0

P6 = offset register number (tool 6)
R = value of offset

Note: The function G10 is used to change the work coordinate values and for other offsets. Therefore, if you plan to use the new coordinate system, call directly for the newly entered system in the consecutive block (in the example, G59). Before G10, you should use function G90 to replace the existing offset value and G91 to increase or decrease the existing offset value by the amount input.

This G10 format varies with control models and may be a control option, consult the specific manufacturer control manuals.

WORK COORDINATE SYSTEMS (G54, G55, G56, G57, G58, G59)

In comparison to function G92 (work coordinate system setting), the application of functions G54 through G59 simplifies the programming where it is necessary to use additional Workpiece Zeros (maximum six as a standard control feature).

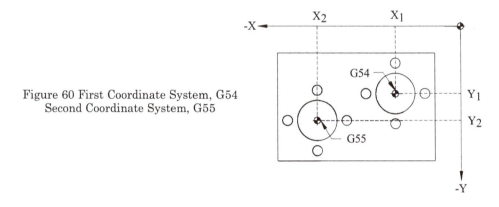

Figure 60 First Coordinate System, G54
Second Coordinate System, G55

Note: When using functions G54 through G59, avoid using function G92. Using function G92 with function G54 may cause unpredictable results.

While using function G92, the coordinates of Workpiece Zero are entered directly into the program. When using functions G54 through G59, the coordinates of the program zeros are entered into controller memory on the work offset page for the coordinate systems. Under the position 01 on the offset page, enter the coordinates of Workpiece Zero in the function G54, and so on. See specific instructions on the process for setting Workpiece Zero in Part 2, "Measuring Work Offsets, Machining Center".

The following offsets are available. See Figure 35 in Part 2.
01   G54 X. . .Y. . .Z. . .
02   G55 X. . .Y. . .Z. . .
03   G56 X. . .Y. . .Z. . .
04   G57 X. . .Y. . .Z. . .
05   G58 X. . .Y. . .Z. . .
06   G59 X. . .Y. . .Z. . .
Workpiece Zero coordinates G54 through G59 are measured in a similar manner as they are for function G92. Function G92 is measured from *X Y* Workpiece Zero to Machine Zero while G54 through G59 is measured from Machine Zero to Workpiece Zero.

The following two examples illustrate the application of the functions G92 and then G54 through G59:

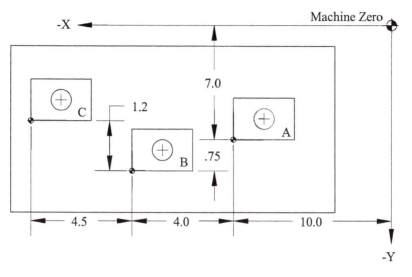

Figure 61 Work Coordinate Systems Comparison G92 to G54-59

Example of Work Coordinate Systems, Program No.11

In the following program, the tool used is a .5 diameter HSS drill and cutting speed of 80 ft/min was used to calculate the r/min and feed rate.

The following is a program utilizing function G92:

```
O0011
N10G90G80G20G40G49
N15G92X10.0Y7.0Z0 (PART A)
N20G00X.75Y.5S611M03
N25G43Z1.0M08H01
N30G00Z.1
N35G01Z-.65F3.6
N40G00Z.1
N45X0 Y0
N50G92X14.0Y7.75 (PART B)
N55G00X.75Y.5
N60G01Z-.65F3.6
N65G00Z.1
N70X0 Y0
N75G92X18.5Y6.55 (PART C)
N80G00X.75Y.75
N85G01Z-.65F3.6
N907G00Z.1
N95G28Z1.0M09
N100G91X0Y0M05
N105M30
```

The following is the same program utilizing the functions G54 through G59 instead:

Part No. 1: (A) G54   X–10.0 Y–7.0   Z0
Part No. 2: (B) G55   X–14.0 Y–7.75 Z0
Part No. 3: (C) G56   X–18.5 Y–6.55 Z0

O0011
N10G9080G20G80G40G49
N15G00G54X.75Y.5S611M03
N20G43Z1.0M08H01
N25G00Z.1
N3G01Z-.65F3.6
N35G00Z.1
N40G55G00X.75Y.5
N45G01Z-.65
N50G00Z.1
N55G56G00X.75Y.5
N60G01Z-.65
N65G00Z.1
N70G28Z1.0M09
N75G91G28X0Y0M05
N80M30

After reviewing the above examples, you will note the difference between the function G92 and the coordinate systems G54 through G59 and see the ease with which Work Coordinates can be used. A major advantage is the fact that with Work Coordinate Systems the operator inputs the coordinate values for Workpiece Zeros in the offset registers instead of within the program.

SINGLE-DIRECTION POSITIONING (G60)

Usually, function G60 is applied when accuracy is the prime factor in determining distances between points. The tool will always approach the programmed point from one direction only.

The value of the additional path is set by parameter in the control. Function G60 eliminates the undesirable influence of gear and feed-screw play (backlash) on the accurate positioning of the programmed point.

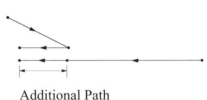

Additional Path

Figure 63
Single-Direction Positioning

CANNED CYCLE FUNCTIONS

The function of a Canned Cycle is defined as a set of operations assigned to one block and performed automatically without any possibility of interruption. Usually, it is a set of six operations, as follows:

1. Positioning of the *X* and *Y* axes at rapid traverse.
2. A rapid traverse move to an initial clearance level plane (G98).
3. The machining cycle is executed (drill, bore, etc.).
4. A dwell or other operation is executed at the bottom of the hole.

5. A rapid traverse return to the R level plane along the Z axis (G99).
6. A rapid traverse return to the initial level plane along the Z axis.

Block format is as follows:

N . . .G . . . G . . .X . . . Y . . . Z . . . R . . .Q . . . P . . . F . . . K . . . L . . .

Where
N = the block number
G = the type of cycle function
G = initial or R level return G98/G99
X, Y = the hole position (positioning is carried out by rapid traverse)
Z = the depth of the hole
R = the distance between plane R and the surface of the material

Level R refers to the horizontal plane, positioned closely above the material on which the tool tip moves (commonly .100 in. or .08 mm). The programmed value of R is valid until the new value is entered. It does not have to be included in every block. The tool will return to this level at the end of each hole drilled.

Q = the depth of cut (drilling) for individual pecks (not used for every drilling cycle)
P = the dwell time for the drill while rotating at the bottom of the hole (not used for every drilling cycle).
The dwell in seconds is for the purpose of complete removal of excess material.
F = feed rate in inches per minute (in/min).
K = the number of repeats.
L = the number of holes incrementally spaced.

When K is used with G91, L represents the number of holes incrementally spaced by the amount entered as the X or Y position coordinate (if L does not appear in the block, that means machining of only one hole, L = 1). On some modern controls, the letter address K is used in the same manner.

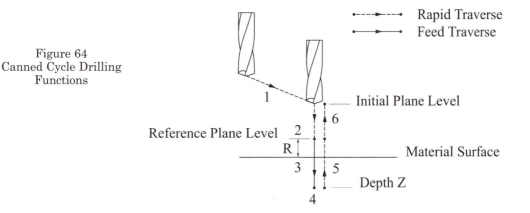

Figure 64
Canned Cycle Drilling
Functions

1. Positioning with rapid traverse
2. Rapid traverse to level R along Z axis
3. Feed traverse to Z depth
4. Operations performed on the bottom of the hole
5. Return to level R (G99)
6. Return to plane level (G98)

High Speed Peck Drilling Cycle (G73)

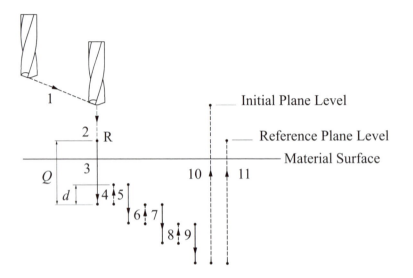

Figure 65 High Speed Peck Drilling Cycle, G73

Block format:

$$G73X \ldots Y \ldots Z \ldots R \ldots Q \ldots F \ldots$$

Figure explanations:

1. G00 — rapid traverse to a position in X or Y axes

2. G00 — rapid traverse to plane R

3, 5, 7, 9. G01 — feed traverse along Z to depth Q

4, 6, 8. G00 — rapid traverse upwards by a value specified

Established by parameter and shown here as amount "d"   .

10. G00 — rapid traverse to a plane level assigned to function G98

11. G00 — rapid traverse to plane R assigned to function G99

Q = The depth of cut for each drilling peck.

d = The value of the upward traverse that is used to accomplish momentary interruption, removal of chips, and the delivery of coolant to the bottom of the hole. The value of d is entered into the parameters of the machine's control.

Left Handed Tapping Cycle (G74)

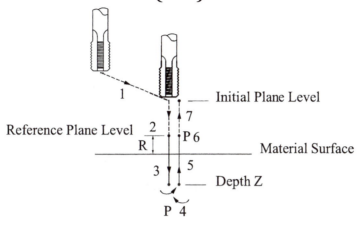

Figure 66 Left Handed tapping Cycle, G74

Block format:

G74X . . .Y. . . Z . . .R . . .P. . . F . . .

The spindle is rotating in the counterclockwise direction by using the M04 command with a value of S.

Figure explanations:

1. G00 — rapid traverse to a position in X or Y axes
2. G00 — rapid traverse to plane R
3. G01 — feed traverse along Z to depth
P. — P = dwell time in seconds at the bottom of the hole
4. — spindle direction is reversed to clockwise
5. G01 — feed traverse along the Z to the R plane
P. — P = dwell time in seconds at the R plane
6. — spindle direction is reversed to counterclockwise
7. G00 — rapid traverse to initial plane if G98 is used

Fine Boring Cycle (G76)

Block format:

G76X . . .Y . . . Z . . . R . . .Q . . . P . . . F . . .

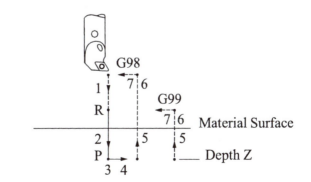

Figure 67 Fine Boring Cycle, G76

Figure explanations:
1. G00 — rapid traverse to point R.
2. G01 — feed traverse to Z depth.
P. — dwell time in seconds.
3. M19 — miscellaneous function M19

Miscellaneous function M19 initiates orientation of the spindle position and the interruption of revolutions. At function M19, the cutting edge always rotates and stops in the same position. The cutting edge stops so that it is perpendicular to the X or Y axis. This activity takes place without inserting the miscellaneous function M19 into the program. It is a part of the G76 canned cycle that is built-in.
Example:

Cutting Edge

Q

Figure 68
Spindle Orientational

4. G00 — rapid displacement along the X or Y axis
The displacement may be in the Y axis, if the cutting edge is perpendicular to the X axis, with the value of Q. The value of Q is entered into the parameters of the control. The value of Q must be known in order to avoid a collision between the tool and the back wall of the hole.
5. G00 — rapid traverse from the hole to the reference point R for function G99, or rapid traverse to the initial plane level for function G98
6. G00 — rapid displacement of the tool along the X (or Y) axis, with the value Q
7. — M03 activate clockwise spindle rotation in preparation for the boring of the next hole.

This boring cycle is usually applied for finishing in which it is intended to obtain a smooth surface, free of scratches. Because of the retract amount specified in Q, there will be no tool mark caused on the finished surface at exit from the hole.

CANNED DRILLING CYCLE CANCELLATION (G80)
This command is used to cancel all canned cycles. It should be entered at the end of each canned cycle machining sequence. This code is also entered in the Safety Block.
Notes for cycles G73 through G89:
Revolutions of the spindle (right or left), during the cycle, must be entered in the block proceeding the canned cycle block.
Never press the ORIGIN button (zero set) during the canned cycle execution because unpredictable actions may result.
In order to execute the cycle (the drilling of one hole with the SINGLE BLOCK button being ON, after the execution of each block when the machine stops), you must press the CYCLE START button three times to continue.
If feed hold is pressed during a threading canned cycle operation, the cycle will be completed before stopping.

CANNED CYCLE, SPOT DRILLING (G81)

Block format:

G81X...Y... Z...R...F...

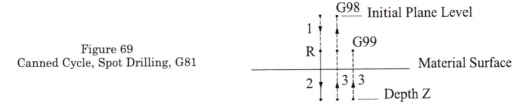

G98 Initial Plane Level

G99

Material Surface

Depth Z

Figure 69
Canned Cycle, Spot Drilling, G81

Figure explanations:
1. G00 — rapid traverse to R.
2. G01 — feed traverse to Z depth.
3. G00 — rapid traverse to R (G99) or
4. G00 — rapid traverse to initial level plane (G98).

Example of Canned Cycle, Spot Drilling, Program 12

This is an example of drilling a hole with a diameter of .25 and a chamfer with a diameter of .280 inch.

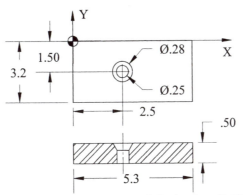

Figure 70 Drawing for Canned Cycle, Spot Drilling Example

Machine: Machining Center	Program Number: O0012

Workpiece Zero X = Upper Left Corner Y = Upper Left Corner Z = Top Surface

Setup Description:

The material is 1018 low carbon steel.

Mount the workpiece in a vice, with a part stop on the left side. Parallels are inserted below the surface of the plate on the side of the vice jaws (dotted line on the drawing).

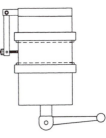

Figure 71 Setup
Diagram for Program 12

Tool #	Description	Offset	Comments
T01	#4 HSS CTR drill		SFM = 110
			Feed =.002 inch per flute
T02	.250 diameter		SFM = 110
	HSS drill		Feed =.0035 inch per flute

Program
O0012
N10G90G80G20G40G49
N15T01M06
N20G54G00X2.5Y-1.5S1408M03T02
N25G43Z1.H01M08
N30G81G98Z-.28R.1F6.0
N35G80Z1.M09
N40G91G28Z0
N45M01
N50M06
N55G90G80G40G49
N6G00 X2.5Y-1.5S160 M03
N65G43Z1.H02M08
N70G81G98Z-.58R.1F12.32
N75G80Z1.M09
N80G91G28Z0
N85G28X0Y0M05
N90M30

COUNTER BORING CYCLE (G82)

In function G82, the feed is interrupted in the drilling cycle (while the spindle is ON) at the bottom of the hole for time specified by P.
Block format:

G82X . . .Y . . . Z . . . R . . .P . . . F . . .

Figure explanations:
1. G00 — rapid traverse to reference level R.
2. G01 — feed traverse to Z depth.
3. — interruption of the feed for time duration P, in order to remove the material at the bottom of the hole.
4. G00 — rapid traverse to initial plane level for G98.
4. G00 — rapid traverse to reference plane level R for G99.

Example of Counter Boring Cycle, Program.13
In this example a counter bored (step hole) bolt hole is drilled.

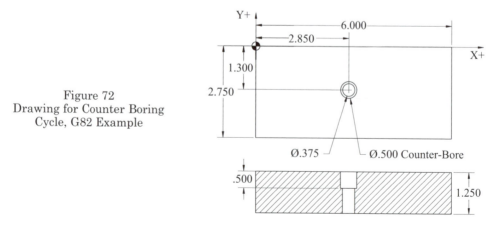

Figure 72
Drawing for Counter Boring
Cycle, G82 Example

270

	Machine: Machining Center	Program Number: O0013	

Workpiece Zero X = Upper Left Corner Y = Upper Left Corner Z = Top Surface

Setup Description:

The material is 1018 low carbon steel.

Fasten the workpiece in the same way as in the previous example.

Tool #	Description	Offset	Comments
T01	#4 HSS CTR drill		SFM = 110 Feed =.002 inch per flute
T02	.375 diameter HSS drill		SFM = 110 Feed =.0035 inch per flute
T03	.500 diameter HSS, 2 flute end mill		SFM = 110 Feed =.002 inch per flute

Program

O0013

N10G90G80G20G40G49

N15T1M06

N25G54G00X2.85Y-1.3S1408 M03T02

N30G43Z1.H01M08

N35G81G98Z-.4R.1F5.63

N40G80Z1.M09

N45G91G28Z0

N50M01

N55M06

N60G90G80G40G49

N65G00X2.85 Y-1.3S1173M03T03

N70G43Z1.H02M08

N75G81G98Z-1.4F8.21R.1

N80G80Z1.M09

N85G91G28Z0

N90M01

N95M06

N100G90G80G40G49

N105G00X2.85Y-1.3S880M03 T01

N110G43Z1.H03M08

N115G82G98Z-1.4R.1P200F7.04

N120G80Z1.M09

N125G91G28Z0

N130G28X0Y0M05

N135M30

Note: On some controls, if no value is assigned for P to function G82, its value will be automatically selected by the control. If you do enter a value for dwell P (for example, P1000 = 1 second), then the constant value included in the parameters of the machine will be ignored.

DEEP HOLE PECK DRILLING CYCLE (G83)

Block format:

G83   X . . . Y . . Z . . . R . . . Q . . . F . . .

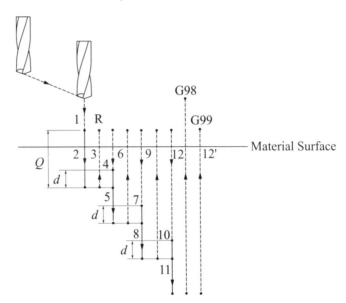

Figure 73 Deep Hole Peck Drilling Cycle, G83

Figure explanations:

1. G00 — rapid traverse to R level.

2. G01 — feed traverse with length Q (peck amount).

3, 6, 9. G00 — rapid return traverse to R.

4, 7, 10. G00 — rapid traverse to the depth previously drilled, less the value of d.

5, 8, 11 . G01— feed traverse increased by the value of d.

12. G00 — rapid traverse to initial plane level for G98.

12'. G00 — rapid traverse to reference plane level R for G99.

This drilling cycle is used to drill exceptionally deep holes. As the drill reaches the depth identified by Q, the drill then returns, at rapid traverse, to the R level point, which allows the removal of the chips and the delivery of the coolant to the bottom of the hole. Entering a given depth of Z into the control enables it to calculate the number of feed traverses necessary for Q. Q can be any incremental amount desired smaller than the total Z axis travel. The value of Q does not need to have a common factor with the dimension Z. The value of d is set by parameter.

Example of Deep Hole Peck Drilling Cycle, Program 14

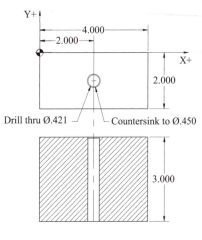

Machine: Machining Center	Program Number: O0014

Workpiece Zero X = Upper Left Corner
Y = Upper Left Corner Z = Top Surface

Setup Description:

The material is 1018 low carbon steel.

Fasten the workpiece in the same way as in the previous example.

Tool #	Description	Offset	Comments
T01	#5 HSS CTR drill		SFM = 110 Feed =.002 inch per flute
T02	.421 diameter HSS drill		SFM = 110 Feed =.0035 inch per flute

Figure 74
Drawing for Deep Hole Peck Drilling Example

Program
O0014
N10G90G80G20G40G49
N15T1M06
N20G54G00X2.Y-.1S1006M03T02
N25G43Z1.H01M08
N30G81G99Z-.38F4.03R.1
N35G80Z1.M09
N40G91G28Z0
N45M01
N50M06
N55G90G80G40G49
N60G54G00X2.Y-1.S1045M03
N65G43Z1.H02M08
N70G83G99Z-3.15R.05Q.45F7.31
N75G80Z1.M09
N80G91G28Z0
N85G28X0Y0M05
N90M30

Note: By using function G83 in block N70, the deep hole, peck drilling cycle is initiated. Starting at level R, the drill will feed by the amount of Q.45 and then, at rapid traverse, returns to the starting point R. The next move advances by the depth Q, decreased by the value of d, which is entered in the parameters of the control. This cycle repeats itself until the drill reaches a depth of Z-3.15. Also, after each feed traverse with the value Q, the tool returns to level R.

TAPPING CYCLE (G84)

Block format:

G84X . . .Y . . . Z . . . R . . . P . . . F . . .

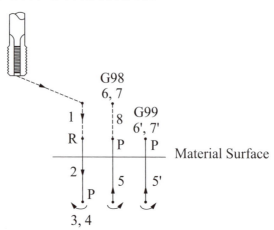

Figure 75 Tapping Cycle, G84

Figure explanations:

1. G00 — rapid traverse to R level.
2. G01 — feed traverse to Z.
3. M05 — revolutions stop.
P. — dwell time at the bottom of the hole.
4. M04 — counterclockwise revolutions are ON.
5. G01 — feed traverse to R.
6. M05 — spindle stop.
7. M03 — clockwise revolution is ON.
8. G00 — rapid traverse to a level plane for the function G98.

Example of Tapping Cycle, Program.15

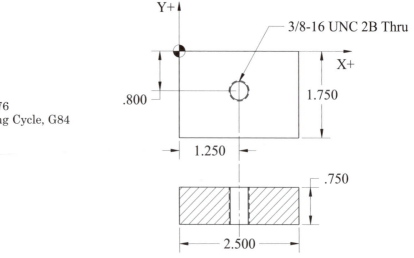

Figure 76
Drawing for Tapping Cycle, G84

Machine: Machining Center	Program Number: O0015

Workpiece Zero X = Upper Left Corner Y = Upper Left Corner Z = Top Surface

Setup Description:

The material is 1018 low carbon steel.

Fasten the workpiece in the same way as in the previous example.

Tool #	Description	Offset	Comments
T01	#4 HSS CTR drill		SFM = 110 Feed =.002 inch per flute
T02	diameter HSS drill		SFM = 110 Feed =.0035 inch per flute
T03	Tap		SFM = 15

Program
O0015
N10G90G80G20G40G49
N15T1M6
N20G00G54X1.25Y-.8S1045M03T02
N25G43Z1.H01M08
N30G81G99Z-.38R.1F4.18
N35G80Z1.M01
N40G91G28Z0
N45M01
N50M06
N55G80G40G49
N60G54G00X1.25Y-.8S1408M03T03
N65G43Z1.H02M08
N70G81G99Z-.85R.1 F9.85
N75G80Z1.M09
N80G91G28Z0
N85M01
N90M06
N95G80G40G49
N100G54G00Z1.25Y-.8S152M03T01
N105G43Z1.H03M08
N110G84G99Z-1.R.1F9.5
N115G80Z1.M09
N120G91G28Z0M05
N125G28X0Y0
N130M30

Note: The feed in block N110 (F9.5) was calculated as follows:
F = 1/16 x 152 = 9.5
        1/16 = the lead of the thread in inches
        152 = the spindle speed, expressed in revolutions per minute(r/min)

BORING CYCLES
REAMING CYCLE   (G85)
Block format:

    G85X . . . Y . . . Z . . . R . . . F . . . K . . .

Figure explanations:
1. G00 — rapid traverse to R.
2. G01 — feed traverse to Z depth.
3. G01 — feed traverse to the R level plane
    and then rapid to the level plane
    assigned to function G98.
3'. G01 — feed traverse to R for function G99.
4. & 4'. M03 — the clockwise revolution is ON.

BORING CYCLE (G86)
Block format:

    G86X . . . Y . . . Z . . . R . . . F . . .

Figure explanations:
1. G00 — rapid traverse to R.
2. G01 — feed traverse to Z depth.
3. M05 — spindle revolution is stopped.
4. G00 — rapid traverse to the level plane
    assigned to function G98.
4'.G00 — rapid traverse to the R level
plane for function G99.
5. M03 — the clockwise revolution is ON.

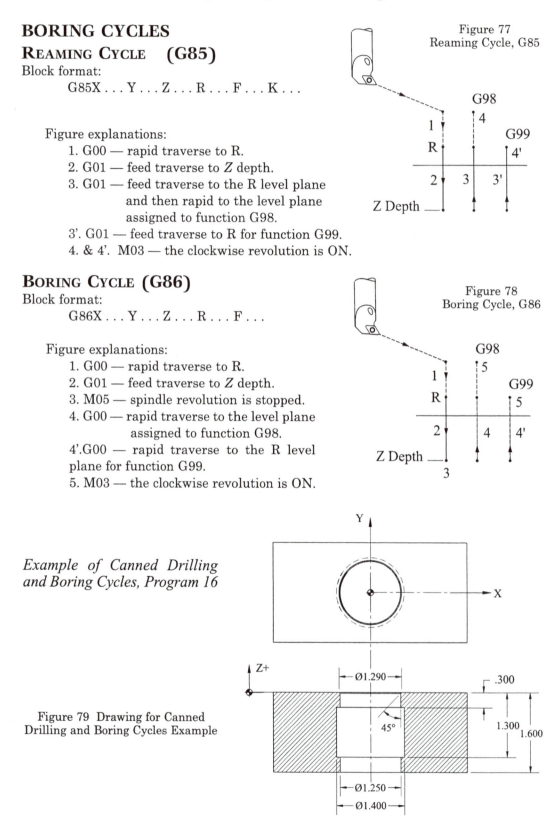

Figure 77
Reaming Cycle, G85

Figure 78
Boring Cycle, G86

*Example of Canned Drilling
and Boring Cycles, Program 16*

Figure 79 Drawing for Canned
Drilling and Boring Cycles Example

Machine: Machining Center	Program Number: O0016

Workpiece Zero X = Upper Left Corner Y = Upper Left Corner Z = Top Surface

Setup Description:

The material used is aluminum. Fasten the workpiece in the same way as in the previous
example.

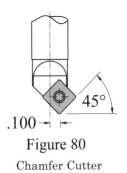

.100

Figure 80

Chamfer Cutter

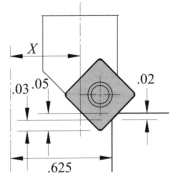

.625

Figure 81
Chamfer Cutter Detail

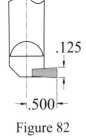

.500

Figure 82

Back Boring Bar

$$X = .625 - .3 - .05 = .545$$

Tool #	Description	Offset	Comments
T01	#6 HSS CTR drill		SFM = 300 Feed =.002 inch per flute
T02	Drill 1.125 HSS		SFM = 300 Feed =.0035 inch per flute
T03	Carbide boring bar 1.25 diameter		SFM = 800 Feed =.001 inch per flute
T04	Carbide 45∞ chamfer cutter		SFM = 800 Feed =.001 inch per flute
T05	Carbide back boring bar 1.40 diameter .125 width insert		SFM = 800 Feed =.001 inch per flute

Program
O0016
N10G90G80G20G40G49
N15T01M06
(T01 CENTER DRILL FOR BORES)
N20G54G00X0Y0S2291M03
N25G43Z1.0H01M08
N30S2291M03
N35G81G98Z-.35F9.16R.1
N40G80Z1.0M09
N45G91G28Z0
N50M01
N55T02M06
(T02 DRILL 1.125 DIAMETER FOR BORES)
N60G90G80G40G49
N65G54G00X0Y0S1182M03T03
N70G43Z1.0H02M08
N75G73G98Q.3Z-2.0F8.27R.1
N80G80Z1.0M09
N85G91G28Z0
N90M01
N95T03M06
(T03 BORE 1.25 DIAMETER HOLE THROUGH)
N100G90G80G40G49
N105G54G00X0Y0S2444M03T04
N110G43Z1.0H03M08
N115G86G98Z-1.65F2.44R.1
N120G80Z5.M09
N125G91G28Z0
N130M01
N135T04M06
(T04 CHAMFER 45° X .02 ON 1.25 DIAMETER HOLE)
N140G90G80G40G49
N145G54G00X0Y0S3086M03T05
N150G43Z1. H04M08
N155S3086M03
N160G01Z-.05F50.0
N165X.545F3.08
N170G03I-.545
N175G01X0
N180G00Z1.0M09
N185G91G28Z0
N190M01
N195T05M06
(T05 BACK BORE 1.400 DIAMETER UNDERCUT)
N200G90G80G40G49
N205G54G00X0Y0S2182M03

N210G43Z1.0H05M08
N215S2182M03
N220G01Z-.425F50.0
N225G91G41Y.7F6.5D50
(D50 = .500)
N230G03J-.7F2.18
N235M98P0002L14
N240G01G40Y-.7F50.0
N245G90G00Z5.M09
N250G91G28Z0M05
N255G28X0Y0
N260M30

The following subprogram is repeated 14 times to attain the full depth for the 1.400 diameter bore. Boring cycle G87 could be substituted in place of the subroutine. See the following description.

Subprogram
O0002
N10G01Z-.1F2.0
N15G03J-.7F3.0
N20M99

BORING CYCLE (G87)

Function G87 is used to bore that part of the hole or chamfer on the bottom of the hole (cutting is along the Z axis in the positive direction).
Block format:
   G87X...Y...Z...R...Q...P...F...

Figure explanations:

   1. G00 — rapid traverse of oriented boring bar with a value of Q.

   2. Non-programmed M19 — orientation of boring bar (same as for G76).

   3. G00 — rapid traverse to Z (bottom of the hole).

   4. G00 — rapid traverse of boring bar with a value of Q.

   5. M03 — the clockwise rotations are ON.

   6. G01 — work traverse to programmed Z value necessary to attain bore.

   7. Non-programmed M19 — orientation of boring bar.

   8. G00 — rapid traverse of boring bar with a value of Q.

   9. G00 — rapid traverse to the R level plane.

   10. G00 — rapid traverse with a value of Q.

   11. M03 — the clockwise rotation is ON.

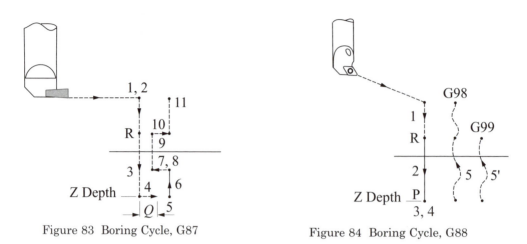

Figure 83 Boring Cycle, G87

Figure 84 Boring Cycle, G88

Boring Cycle (G88)

Block format:

G88X . . .Y . . . Z . . .R . . .P . . .F . . . K . . .

Figure explanations:

1. G00 — rapid traverse to R.

2. G01 — feed traverse to Z depth.

3. G04 — temporary interruption of feed for the time period P.

Dwell for the purpose of complete removal of material from the bottom of the hole.

4. M05 — the revolutions stop.

5., 5'. Manual or mechanical withdrawal of the tool.

After the revolutions are stopped, the tool may be removed from the holder while still at the bottom of the hole.

P. — time period of the temporary interruption is given in seconds.

Boring Cycle (G89)

Block format:

G89X . . .Y . . . Z . . .R . . .P . . .F . . .

Figure explanations:

1. G00 — rapid traverse to R.

2. G01 — feed traverse to Z depth.

3. G04 — temporary interruption of feed.

4, 4'. G01 — feed traverse to R.

5. G00 — rapid traverse to a level plane for function G98.

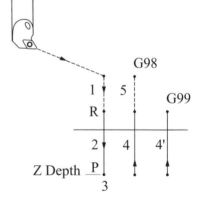

Figure 85 Boring Cycle, G89

Examples of Programming CNC Machining Centers Comparison of Absolute (G90) or Incremental (G91) Programming

In the absolute system, all dimensions are relative to the fixed origin of the workpiece coordinate system.

In the incremental system, every measurement refers to the previously dimensioned position. Incremental dimensions are distances between the adjacent points (from the current tool location).

The following example also provides application of the Initial Level Return function (G98) and R Level Return function (G99) in the Canned Drilling Cycles.

Example of Absolute or Incremental Programming and G98/G99, Program 17

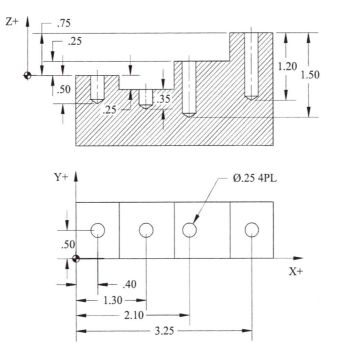

Figure 86 Drawing for Comparison of Absolute/Incremental Programming Example

Machine: Machining Center	Program Number: O0017

Workpiece Zero X = Lower Left Corner Y = Lower Left Corner Z = Top Surface
Setup Description:
Material used is steel.
Tool one is mounted in the spindle prior to beginning.

Tool #	Description	Offset	Comments
T01	Tool is drill 1/4 HSS.		SFM = 70 Feed =.0035 inch per flute

Selection of Coordinate System (G92)

G92X(A)Y(B)

For simple programs it may be easier and faster to assign the new program a new coordinate system through Work Coordinate Setting G92 to which all positioning is referred. The zero of the new workpiece coordinate system is dimensioned from Machine Zero.

The following is a program employing an absolute coordinate system.

Program
O0017
N10G90G80G20G40G49
N15G92X10.0Y-7.0Z0S1070M03
N20G00X.4Y.5
N25G43Z1.0H01M08
N30G81G98Z-.5F7.48R.1
N35X1.3Z-.6R-.15
N40X2.1Z-.75R.35
N45X3.25Z-.45R.85
N50G80Z1.0M09
N55G91G28Z0M05
N60G28X0Y0
N65M30

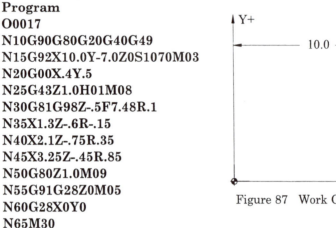

Figure 87   Work Coordinate Setting G92

The following is an example of the same program as above employing an incremental coordinate system.

O0017
N10G90G80G20G40G49
N15G92X10.0Y-7.0Z0S1070M03
N20X.4Y.5
N25G43Z1.0H01M08
N30G91G81G98Z-.6F6.4R-.9
N35Z-.45R-1.15
N40X.9Z-1.1R-.65
N45X.8Z-1.3R-.15
N50G80X1.15 Z5.0M09
N55G28Z0M05
N60G28Y0
N65X0
N70M30

Notes: Sign (+ or -) is determined with respect to the axis of the coordinate system.
Function G90 or G91 must be entered at the beginning of the program, where it remains valid until replaced by function G91 or G90. Also note the position values of the tool for Initial and R-Level Return that are applied within the Canned Drilling Cycles.

COMPLEX PROGRAM EXAMPLE 1

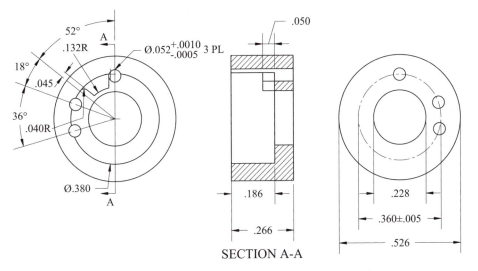

SECTION A-A

Figure 88 Drawing for Complex Program Example 1

To machine the part represented by the preceding drawing, first use a lathe to prepare the bronze bushings dimensioned, as follows in Figure 89:

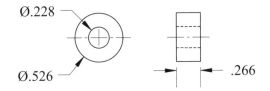

Figure 89 Bronze Bushing Blanks for Complex Program Example 1

Consecutive operations will be performed on the vertical machining center. This part will be machined in two operations. First, three holes with a diameter of .052 will be drilled, and, second, the inside of the workpiece will be milled. Machining speeds, simplicity of fastening, and maximum accuracy are the decisive factors that determine the selection of the number of operations made and the sequence they follow.

Fixture Preparation

Prepare the work holding fixture for the machined parts from a piece of rectangular steel, dimensioned as follows: 2.5 x 12.0 x .50 thick.

The fixture is fastened to the mill table with four bolts. The longer side of the part is milled with an end mill in order to establish parallelism with the part seats, which will later be machined. Additionally, this machined side will serve as a guide, in case another setup of the fixture is required. Afterwards, the 20 seats in the steel plate, in which the parts will be fastened, are machined.

The seats have the following dimensions: diameter .528, depth .100. Between the seats, drill and then tap holes for part clamping screws (not shown). For this example, the work-holding fixture has been prepared earlier.

Complex Program Example 1 Setup Description

The fixture is placed on the machining center table and is positioned so that the milled side is parallel to the X axis. To confirm this position, use a dial indicator to dial-in the setup of the fixture. As soon as parallelism with the X axis is verified, the fixture is clamped to the table with machine screws. The alignment of the fixture is checked again, after clamping, to assure that the fixture did not move during the fastening. Next, the values of X and Y for the functions G54 and G55 are determined.

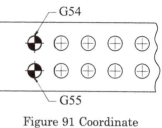

Figure 91 Coordinate Systems for Comples Example 1

Procedure for Measuring Work Offsets

Detailed descriptions for set-up procedures are given in Part 2 of this book but to clarify application to this example a brief description is given here as well. Remember that the procedures used are specific to the type of control described in this text (Fanuc 16-18). Other control types may use similar procedures. Check the specific manufacturer operation manuals for their procedures.

First of all, Home the machine with respect to the X and Y axes. By pressing the POSITION button (POS function key), the values of the X, Y, and Z values will appear on the display.

By pressing the POSITION function button, the absolute coordinate display can be found. Press the soft key labeled "ABS" to display the absolute coordinates of X, Y, and Z. This page shows the values of the coordinates of the axes positioned at Machine Zero.

After Machine Zero is established, check to verify that the absolute display values shown are zero. If not, then press button X and then press the ORIGIN soft key (X, Y ORIGIN). Now position the machine in order to determine the centers of the seat holes in the fixture, fasten an indicator in a drill chuck placed in the spindle. Use the MPG/HANDLE mode to move the axes to find the center of the first left seat-hole that is positioned in the upper row of holes. The values of X and Y are read from the absolute display and entered under the offset values of the coordinate system, of which position 01 corresponds to the function G54. The position of the left seat hole in the lower row is determined in a similar manner and the values of the coordinates corresponding to function G55 are entered under 02, in the coordinate system register. Once Workpiece Zeros are set, place the previously prepared bronze bushings in the fixture seats and fasten them with machine screws.

Procedure for Measuring Tool Length Offsets

Place the center drill in a drill chuck. Set the mode to MDI, key in T01M06; press INPUT and then CYCLE START to install the tool into the spindle. Next, change back to the MPG/HANDLE mode. By using the function button OFFSET/SETTING and the OFFSET soft key the tool-length-offsets page is found. Check the LED for the Z axis, which, for this axis, corresponds to Machine Zero. If the LED is on, the machine is at zero. The values of X, Y, or Z are shown on the bottom of the screen. Check whether the value of Z is equivalent to zero. If it is not, then press the buttons ORIGIN and then Z to zero it. To determine the tool length offset, position the tool to touch the surface of the machined part with the tool tip. Lower the tool manually using the pulse generator (MPG/HANDLE) with caution, reducing the traverse speed as the tool approaches the surface. If a touch control edge finder is used (touch sensor), then as soon as the tool tip touches the surface, the LED is lit. (Touch control provides a very accurate measurement of the tool length offset.) When the control LED lights, the value of Z must be read from the readout and entered into the tool-length-offset register under 01, which corresponds to tool 1 (T01). However, if no touch sensor is used, another common method of determining the tool length offset is given as follows:

Place a piece of paper with a known thickness between the tool tip and the surface of a machined part. While approaching the surface of the machined part with the tool tip, move the piece of paper back and forth. As soon as the first resistance to the paper is felt, stop the traverse and withdraw the piece of paper. Next, check the displayed readout and then lower the tool a distance equivalent to the thickness of the paper.

Two methods of entering the value of Z to the tool-length-offset tables are listed below:

The first method (directly): As soon as the tool tip touches the surface of a machined part, press the button Z. If the letter Z is flashing on the screen, then it is possible for the machine to transfer the value of Z directly to the offset register. By pressing the INP.C. soft key, the value of Z is transferred to the tool-length-offset register.

The second method (indirectly): The value of Z is read from the displayed readout and noted. Using the cursor move keys, the cursor is positioned at the desired offset. The value noted previously is keyed in and the INPUT button pressed. The value is of Z is transferred to a previously reserved position 01 of the tool length offset (a center drill). A similar principle is applied to determine the tool length offsets for the remaining tools (a drill and a reamer), as well as for the tools in the second operation. The tool is removed from the spindle and then transferred to the tool magazine. The machine is set at zero in the Z axis prior to each tool measurement.

The following is the main program for the first operation.

Example of Complex Program Example 1, Operation 1, Programs 18

The programs consist of two operations with tools given in the following set-up sheets:

Machine: Machining Center	Program Number: O0018, Operation 1

Workpiece Zero:
G54 X = Upper Left Seat CL Y = Upper Left Seat CL Z = Top of Part Surface
G55 X = Lower Left Seat CL Y = Lower Left Seat CL Z = Top of Part Surface
Setup Description:
Material used is bronze.

Tool #	Description	Offset	Comments
T01	#0 HSS Center drill.	H01	SFM = 70 Feed =.002 inch per flute
T02	3/64 HSS Drill	H02	SFM = 70 Feed =.002 inch per flute
T03	.052 Diameter Reamer	H03	SFM = 30 Feed =.001 inch per flute

Part 4   Programming CNC Machining Centers

The program for the first operation is specified by O0018 and the program for the second operation is specified by O0019.

O0018
N10G90G80G20G40G49
N15T01M06
N20G54G00X0Y.1787S4000M03T02
N25G43Z1.H01M08
N30G81G99Z-.06R.05F4.0
N35M98P2
N40G90G00G55X0Y.1787
N45G81G99Z-.055R.05F4.0
N50M98P2
N55G90G00G80Z5.M09
N60G91G28Z0
N65G28X0Y0
N70M01
N75M06
N80G90G80G40G49
N85G00G54X0Y.1787S4000M03T03
N90G43Z1.H02M08
N95G83G99Z-.35R.05Q.05 F4.0
N100M98P2
N105G90G00G55X0Y.1787
N110G83G99Z-.35R.05Q.05F4.0
N115M98P2
N120G80G00Z1.0M09
N125G91G28Z0
N130G28X0Y0
N135M01
N140M06
N145G90G80G40G49
N150G54G00X0Y.1787S4000M03
N155G43Z1.H03M08
N160G85G99Z-.31R.05F6.0
N165M98P2
N170G80G90G00G55X0Y.1787
N175G85G99Z-.31R.05F6.0
N180M98P2
N185G80G00Z1.0M09
N190G91G28Z0
N195G28X0Y0
N200M05
N205M30

The following is the subprogram O0002 for the first operation.

O0002
N2X-.1679Y.0611
N3X-.1718Y-.0493
N4X1.Y.1787
N5X.8321Y.0611
N6X.8282Y-.0493
N7X2.Y01787
N8X1.8321Y.0611
N9X1.8282Y-.0493
N10X3.0Y.1787
N11X2.8321Y.0611
N12X2.8282Y-.0493
N13X4.0Y.1787
N14X3.8321Y.0611
N15X3.8382Y-.0493
N16X5.0Y.1787
N17X4.8321Y.0611
N18X4.8282Y-.0493
N19X6.0Y.1787
N20X5.8321Y.0611
N21X5.8282Y-.0493
N22X7.0Y.1787
N23X6.8321Y.0611
N24X6.8321Y-.0493
N25X8.0Y.1787
N26X7.8321Y.0611
N27X7.8282Y-.0493
N28X9.0Y.1787
N29X8.8321Y.0611
N30X8.8282Y-.0493
N31M99

The following explanations are for some blocks of program O0018, operation one.

N10G90G80G20G40G49

Block N10 is the Safety Block and contains the functions G90, G80, G20, G40, and G49 where: G90 refers to programming in an absolute coordinate system (all dimensions correspond to a fixed origin which is the Workpiece Zero); G80 is entered in the beginning of the program to assure that all canned cycle functions are cancelled; G20 initiates the inch measurement system; G40 assures that all cutter diameter compensation functions are cancelled; and, G49 cancels all tool length compensations.

Example of Complex Program Example 1, Operation 2, Program 19

Machine: Machining Center	Program Number: O0019, **Operation 2**

Workpiece Zero

G54 X = Upper Left Seat CL Y = Upper Left Seat CL Z = Top of Part Surface

G55 X = Lower Left Seat CL Y = Lower Left Seat CL Z = Top of Part Surface

Setup Description:

Material used is bronze.

Tool #	Description	Offset	Comments
T04	.125 Diameter HSS Roughing End Mill	H04	SFM = 70 Feed =.002 inch per flute
T05	.0781 Diameter HSS 4-Flute Finishing End Mill	H05	SFM = 70 Feed =.002 inch per flute

The following is the main program for the second operation.

O0019
N10G90G80G20G40G49
N15T04M06
N20G54G00X0Y0S3000M03T05
N25G43Z1.H04M08
N30M98P6L10
N35G90G00G55X0Y0
N40M98P6L10
N45G00G90Z1.0M09
N50G91G28X0Y0Z0
N55M01
N60M06
N65G90G80G40G49
N70G00G54X0Y0S2300M03
N75G43Z1.H05M08
N80M98P7L10
N85G00G90G55X0Y0
N90M98P7L10
N95G00G90Z1.0M09
N100G91G28X0Y0Z0
N105M01
N110M30

Subprogram O0006 is for the second operation.

O0006
N1G90G00Z0
N2G91G01Z-.182F4.0
N3G42X-.1225D01
N4G02X.0138Y.0564I.1225

```
N5G01X.0412Y-.0369
N6G02X.088Y.0393I.0615J-.0195
N7G01Y.0608
N8G02X.1352Y-.0632I-.0265J-.0195
N9G01Z.05
N10G02X.1352Y.0632I.1087J-.1196
N11G01G40X-.0265Y-.1196
N12G90G00Z1.0
N13G91X1.0
N14M99
```

Subprogram O0007 is for the second operation.
```
O0007
N1G90G00Z0
N2G91G01Z-.184F2.0
N3G01G42X-.151D02
N4G02X.0358Y.0976I.151F.3
N5G01X.0458Y-.0357F1.0
N6G02X.0674Y.0311I.0694J-.0619F2.0
N7G01Y.058F.5
N8G02X-.1132Y-.0534I.002J-.151F2.0
N9G01Z.05F10.0
N10G02X.1132Y.0534I.1152J-.0976F2.0
N11G01G40X.002Y-.151
N12G90G00Z1.0
N13G91X1.0
N14M99
```

The Second Operation

In the second operation, the drilled bushings are milled using two end mills, first a rough pass with a tool diameter of .125, and, second, a finish pass with a tool diameter of 5/64 inch. The bushings are placed in a fixture similar to the one in the previous operation. The only difference in the second fixture is that each seat has a pin forced into the bottom with a diameter of .0515 onto which the previously reamed hole in the bushing is placed. Such positioning of the bushings assures the proper position of the hole and prevents rotation of the bushing in the seat during milling.

Main Program Number O0019

The following explanations are for some blocks of program O0019 operation two. Subprogram O0006 is for the first tool and subprogram O0007 is for the second tool.

N10G90G80G20G40G49

This is a Safety Block where G90 sets the program in the absolute coordinate system. G80 cancels all canned cycle functions. G20 initiates the inch measurement system. G40 cancels all radius compensation functions. G49 cancels all tool length compensation functions.

N15T04M06

This block commands the preparation of the tool T04 in the magazine and changes it into the spindle.

N20G54G00X0Y0S3000M03T05

Workpiece Zero coordinates are referenced by function G54; there is a rapid traverse of the tool to the position of that position of zero. The spindle is started in the clockwise direction by function M03, at a speed (S) of 3000 r/min and tool number 5 (T05) is moved to the ready position in the magazine.

N25G43Z1.H04M08

In this block, the machine is commanded to read the tool length offset for T04 from the offset register, while the tool positions at rapid traverse to 1 inch above the material along the Z axis. The flood coolant in activated by M08.

N30M98P6L10

In block N30 the subprogram, O0006 is called up and executed ten times (L10) for the 10 bushings. At this point, the tool is controlled by the subprogram. In order to simplify the program notation, the position of eight characteristics points on the drawing of the part is given in Figure 92 below.

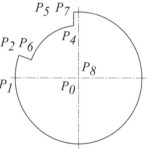

Coordinates of the individual characteristic points are:

P_0: X0Y0
P_1: X-.1255Y0
P_2: X-.1087Y.0564
P_3: X-.0615Y.0195
P_4: X.0265Y.0588
P_5: X.0265Y.1196
P_6: X-.1087Y.0564
P_7: X.0265Y.1169
P_8: X0Y0

Figure 92
Bushing Characteristic Points

The following explanations are for a few blocks of subprogram O0006 called from the main program O0019, operation two.

Subprogram O0006

N1G90G00Z0

The block refers to programming in the absolute coordinate system and traverses along the Z axis, at rapid, to position 0.

N2G91G01Z-.182F4

In this block G91 converts the coordinate system from the absolute to incremental. The tool traverses in a linear motion to Z-.182 with a feed rate of 4 in/min.

N3G01G42X-.1225D01

The tool traverses with a previously assigned feed rate to a position of X-.1225 while cutter compensation (G42) is activated, with a simultaneous reading of the radius offset (D01), from the offset register. (In this case, the value of the offset is equivalent to zero, because the traverse of the end mill is defined so that the distance from the center of the end mill to the theoretical dimension is exactly equivalent to the tool radius.) In addition, 0.005 inch of material remains for the finishing operation, in which the finishing end mill is used.

N4G02X.0138Y.0564I.1125

This block feeds the tool along a circular path in a clockwise cutting direction with a radius vector value of I.1125 (the center of the end mill is located in P_2).

N5G01X.0412Y-.0369

The tool traverses along a straight line, specified by the values of X and Y (the center of the end mill is located in P_3).

N6G02X.088Y.0393I.0615J-.0195

The tool traverses along a circular path, in a clockwise direction, specified by the values of X and Y (the center of the end mill is located in P_4).Values in I and J denote the incremental distance to the arc center point.

N7G01Y.0608

The tool traverses along, straight line, specified by the value of Y (the center of the end mill is located in P_5).

N8G02X.1352Y-.0632I-.0265J-.1196

Traverse along the circular path to P_6.

N9G01Z.050

The end mill is retracted along the Z axis, at feed rate, by a value of .050.

N10G02X.1352Y.0632I.1087J-.0564

Traverse along a circular path to P_7.

N11G01G40X-.0265Y-.1196

The active function G42 is cancelled by G40, and the tool is displaced along a straight line to P_8.

N12G90G00Z1.0

The tool is positioned, at rapid traverse, 1 inch above the material (use G90 to assure that the tool position is above the material).

N13G91G00X1.0

A rapid traverse is made along the X axis by a value of 1 in (toward the center of the second bushing).

N14M99

This block commands a return to the main program O0019. However, because function M98 L10 was entered previously in the main program, the subprogram will be executed nine more times (ten bushings will be machined in exactly the same manner). After the machining of the last bushing in the upper row is completed, control is returned to block N35 of the main program.

N35G90G00G55X0Y0

In this block, the new Workpiece Zero, G55 is introduced with its assigned coordinates given in the offset register. G00 refers to a rapid traverse to the new position of the new Workpiece Zero. (Position of the new Workpiece Zero is set to the center of the first bushing of the lower row.)

N40M98P6L10

Function M98 calls up subprogram O0006 and executes it 10 times (subprogram O0006 was described earlier). After the 10 executions of the subprogram, control is returned to block N45 of the main program for the second operation.

N45G00G90Z1.0M09

A rapid traverse movement is made to a position 1 inch above the surface of the material based on the absolute coordinate system, and the deactivation of the coolant flow (M09) occurs.

N50G91G28X0Y0Z0

Function G28 returns the machine to the Machine Zero position in the *X, Y* and *Z* axes.

N55M01

Miscellaneous command M01 provides an additional interruption of the program (only if the OPTIONAL STOP button is ON). The purpose of this interruption is to check the dimensions and condition of the tools.

N60M06

This is the tool change for T05 that was positioned in the tool magazine in block N20 and then in block N60 transfers the tool (T05) into the spindle of the machine.

N65G90G80G40G49

This is a safety block.

N70G00G54X0Y0S2300M03

A new Workpiece Zero (G54) is introduced; the tool rapidly travels to the position of the new zero and clockwise spindle rotation is initiated at 2300 r/min.

N75G43Z1.H05M08

The tool length offset H05, obtained from the offset register, is entered, allowing the traverse of the tool to a position 1 inch above the surface of the material. Flood coolant flow is also activated.

N80M98P7L10

The subprogram O0007 (P7) is called up and then executed 10 (L10) times.

Coordinates of the individual characteristic points are:

P_0: X0Y0
P_1: X-.151Y0
P_2: X-.1152Y.0976
P_3: X-.0694Y.0619
P_4: X-.002Y-.093
P_5: X-.002Y.151
P_6: X-.1152Y.0976
P_7: X-.002Y.151
P_8: X0Y0

The previous set of coordinates illustrates the position of the center of the end mill. Points P_0 through P_8 correspond to the characteristic path points of the end mill.

N1G90G00Z0

This block performs the rapid traverse movement of the end mill toward the surface of the material.

N2G91G01Z-.184F2.0

Block N2 establishes measurement in the incremental system.
The tool traverses along the Z axis with a feed rate of 2 in/min.

N3G01G42X.151D02

The radius offset is read, and there is a traverse along the X axis to the value X-.151 (point P_1).

N4G02X.0358Y.0976I.151F3.0

The traverse is along a circular path in a clockwise direction to point P_2.

N5G01X.0458Y-.0357F1.0

The traverse is along a straight line to point P_3.

N6G02X.0674Y.0311I.0694J-.0619F2.0

The traverse is along a circle in a clockwise direction to point P_4.

N7G01Y.058   F5.0

The traverse is along a straight line to point P_5.

N8G02X-.1132Y-.0534I.002J-.151F2.0

The traverse is along a circular path to point P_6.

N9G01Z.05F10.0

The tool is retracted along the Z axis by a distance of .05 in.

N10G02X.1132Y.0534I.1152J-.0976F2.0

The traverse is along a circular path to point P_7.

N11G01G40X.002Y-.151

The tool radius offset is cancelled and the traverse is along a straight line to P_8, the center of the bushing.

N12G90G00Z1.0

A rapid tool traverse is made to a position 1 inch above the surface of the material, defined in an absolute coordinate system.

N13G91X1.0

A rapid traverse is along the X axis for 1 inch (to the position of the second bushing).

N14M99

This block commands a return to the main program. Since command L10 was included in the main program, the subprogram will be executed nine more times before there is a return to block N85 of the main program.

N85G00G90G55X0Y0

A new coordinate system is assigned with function G55, and the tool moves at rapid traverse to the zero position of the new system.

N90M98P7L10

The subprogram O0007 is called up, executes the subprogram 10 times, and then returns to N95 in the main program.

N95G00G90Z1.0M09

A rapid traverse move is made to a position 1 inch above the surface of the material defined in the absolute coordinate system and the coolant will be turned off.

N100G91G28X0Y0Z0

There is a return to Machine Zero for the Y and Z axes.

N105M01

There is an additional interruption of the machine work, providing the OPTIONAL STOP switch/button is ON.

N110M30

This is the end of the program and control returns to the beginning. Completion of N110 results in a machine stop. The finished bushings must be removed and be replaced then with unfinished ones. Press the CYCLE START button and the work cycle is initiated once again.

EXAMPLE ILLUSTRATING THE APPLICATION OF MIRROR IMAGE

In the following example, use of the mirror image function is illustrated. This function makes it possible to duplicate patterns by projecting a mirror image of the pattern in the X or Y axis directions. The main purpose for this is to shorten and simplify the part program.

M21 = the mirror image in the direction of the X axis.

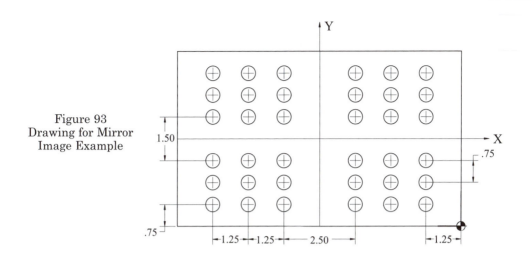

Figure 93
Drawing for Mirror
Image Example

M22 = the mirror image in the direction of the *Y* axis.

M23 = the cancellation of the mirror image.

Example of Mirror Image, Program 20

```
O0020
N10G90G80G20G40G49M23
N15G00G90G54X-1.25Y.75S1000M03
N20G43Z1.0H01M08
N25G81G98Z-.35R.1F6.0
N30M98P7
N35G00G80X0
N40M21
N45G00X-1.25Y.75
N50G81G98Z-.35F6.R.1
N55M98P7
N60G00G80X0Y0
N65M23
N70M22
N75G00X-1.25Y.75
N80G81G98Z-.35F6.R.1
N85M98P7
N90G80G00X0Y0
N95M21
N100G00X-1.25Y.75
N105G81G98Z-.35F6.R.1
N110M98P7
N115G80Z1.0M09
N120M23
N125G91G28X0Y0Z0
N130M30
```

Subprogram for Mirror Image, Program 20

O0007

N1X-2.5	N5X-1.25
N2X-3.75	N6Y2.25
N3Y1.5	N7X-2.5
N4X-2.5	N8X-3.75
	N9M99

EXAMPLE ILLUSTRATING THE APPLICATION OF AXIS ROTATION

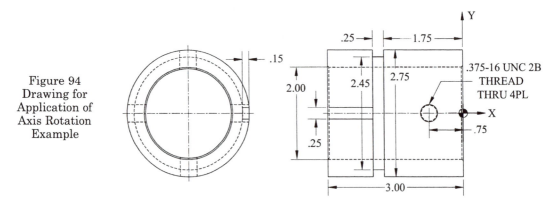

Figure 94
Drawing for
Application of
Axis Rotation
Example

Occasionally, vertical milling machines are used to perform operations on cylindrical or other workpieces that require the use of a rotational axis. In the following case, there are four drilled and tapped holes, and a slot that must be machined by using this method. Rotation about the X axis is required. This is an additional axis and is identified as the A axis.

Example of Application of Axis Rotation, Program 21

Machine: Machining Center	Program Number: O0021

Workpiece Zero
G54 X = Right End Y = Center Line Z = Top of Part Surface
Setup Description:
Material used is Steel.
Mount the part in a chuck attached to an indexing head or 4th axis.

Tool #	Description	Offset	Comments
T01	#3 HSS Center Drill		SFM = 95
			Feed =.002 inch per flute
T02	.3125 Diameter HSS Drill		SFM = 95
			Feed =.0035 inch per flute
T03	Tap 3/8 − 16		SFM = 15
T04	.250 HSS 4-flute End Mill		SFM = 95
			Feed =.002 inch per flute

Program
O0021
N10G90G80G20G40G49
N15T01M06
N20G00G54X-.75Y0A0S1520M03
N25G43Z1.H01M08
N30G98G81Z-.375R.1F6.0
N35M98P8
N40G80Z1.0M09
N45G91G28X0Y0Z0
N50M01
N55T02M06
N60G90G80G40G49
N65G54G00X-.75Y0A0S1152M03
N70G43Z1.H02M08
N75G98G81Z-.450R.1F8.0
N80M98P8
N85G80Z1.0M09
N90G91G28X0Y0Z0
N95M01
N100T03M06
N105G90G80G40G49
N110G00G54X-.75S120M03
N115G43Z1.H03M08
N120G98G84Z-.55R.2F7.5
N125M98P8
N130G80Z5.M09
N135G91G28X0Y0Z0
N140M01
N145T04M06
N150G90G80G40G49
N155G54G00X-3.3Y0A0
N160G43Z1.H04M08
N165S1520M03
N170G01Z-.15F50.0
N175X-1.875F6.0
N180A360.
N185G00Z1.0M09
N190G91G28X0Y0Z0A0
N200M30

Subprogram for Program 21
O0008
N1A90
N2A180
N3A270
N4M99

Note: If, in a given block, there is a rotation during the work traverse, then, the feed is defined in degrees per revolution.

EXAMPLE OF PROGRAMMING A HORIZONTAL MACHINING CENTER

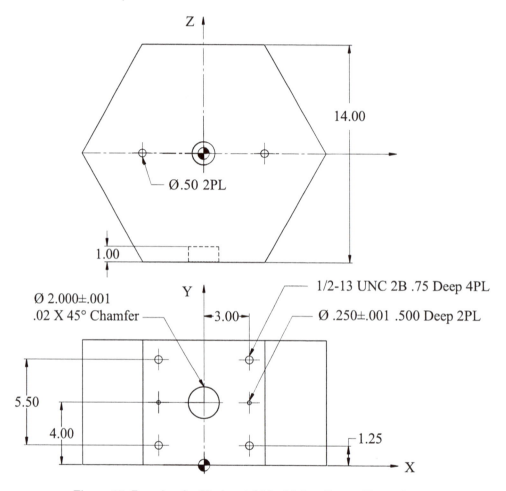

Figure 95 Drawing for Horizontal Machining Center Example

Thus far, in this part of the text, only vertical machining center programming has been described. In this example, some of the programming techniques covered earlier will be applied to a horizontal machining center. With the horizontal machine, the Z axis is horizontal. Another feature that will be useful on the illustrated part is its indexing capability.

Example of Programming a Horizontal Machining Center, Program 22

The material is 8 inch thick hexagonal plate with the following dimensions: 8 x 14 in. There are also two positioning holes in the plate having .500-in diameter.

The workpiece is placed on the fixture that has two positioning pins that are .500 inch in diameter. The workpiece is held down by a holding mechanism and a cap screw that is placed in the center of a previously made hole of 1.5 inch. To simplify programming, assume that the rotation axis of the table is aligned with the symmetry axis of the object.

Operations are performed on a four-axis horizontal milling machine.

Machine: Machining Center	Program Number: O0022

Workpiece Zero

G54 X = Center Line Y = Bottom of Part Surface Z = Center Line

Setup Description: See above

Tool #	Description	Offset	Comments
T01	#3 HSS Center Drill		
T02	.125 Diameter HSS Drill		
T03	1.0 Diameter 4-flute HSS End-Mill		
T04	27/64 (.4219) Diameter HSS Drill		
T05	1/2-13 UNC Tap		
T06	15/64 (.2344) Diameter HSS Drill		
T07	.240 HSS 4-flute End-Mill		
T08	.250 HSS Reamer		
T09	Boring Bar for 2.00 Diameter		
T10	45° Chamfer Cutter		

Program
O0022
N1G90G80G20G40G17
(T01 #3 HSS CENTER DRILL)
N2M11T01
N3M06
N4G90G00G54X0Y4.0B0S600M03
N5M10
N6G43Z8.5M08H01
N7M98P0050L6
N8G80Z10.0M09
N9M11
N10G91G28X0Y0Z0B0
N11M01
(T02 .125 DIAMETER HSS DRILL)
N12T02
N13M06
N14G90G80G20G40G17
N15G0G54X0Y4.0B0S244M03
N16M10
N17G43Z8.5M08H02
N18M98P0052L6

N19G80Z10.0M09
N20M11
N21G91G28X0Y0Z0B0
N22M01
(T03 1.0 DIAMETER 4-FLUTE HSS END-MILL)
N23T03
N24M06
N25G90G80G20G40G17
N26G00G54X0Y4.0B0S305M03
N27M10
N28G43Z8.5M08H03
N29M98P0053L6
N30G80Z10.0M09
N31M11
N32G91G28X0Y0Z0B0
N33M01
(T04 27/64 DIAMETER HSS DRILL)
N34T04
N35M06
N36G90G80G20G40G17
N37G00G54X-3.0Y6.75B0S726
N38M10
N39G43Z8.5M08H04
N40M98P0054L6
N41G80Z10.0M09
N42M11
N43G91G28X0Y0Z0B0
N44M01
(T05 1/2-13 UNC TAP)
N45T05
N46M06
N47G90G80G20G40G17
N48G00G54X-.30Y6.75B0S100M03
N49M10
N50G43Z8.5M08H05
N51M98P0055L6
N52G80Z10.0M09
N53M11
N54G91G28X0Y0Z0B0
N55M01
(T06 15/64 DIAMETER HSS DRILL)
N56T06
N57M06
N58G90G80G20G40G17
N59G00G54X3.0Y4.0B0S1300M03
N60M10
N61G43Z8.5M08H06

N62M98P0056L6
N63G80Z10.0M09
N64M11
N65G91G28X0Y0Z0B0
N66M01
(T07.240 HSS 4-FLUTE END-MILL)
N67T07
N68M06
N70G90G80G20G40G17
N71G00G54X-3.0Y4.0B0S1200M03
N72M10
N73G43Z8.5M08H07
N74M98P0057L6
N75180Z10.0M09
N76M11
N77G91G28X0Y0Z0B0
N78M01
(T08.250 HSS REAMER)
N79T08
N80M06
N81G90G80G20G40G17
N82G00G54X-3.0Y4.0B0S800M03
N83M10
N84G43Z8.5M08H08
N85M98P0058L6
N86G80Z10.0M09
N87M11
N88G91G28X0Y0Z0B0
N89M01
(T09 BORING BAR FOR 2.00 DIAMETER)
N90T09
N91M06
N92G90G80G20G40G17
N93G00G54X0Y4.0B0S500M03
N94G43Z8.5M08H09
N95G76G98Q.01Z6.0F2.0R7.2
N96B60
N97B120
N98B180
N99B240
N100B300
N101G80Z10.0M09
N102G91G28X0Y0Z0B0
N103M01
(T10 45° CHAMFER CUTTER)
N104T10
N105M06

N106G90G80G20G40G17
N107G00G54X0Y4.0S500M03
N108G43Z8.5M08H10
N109G82G98Z6.9F2.5R7.1
N110B60
N111B120
N112B180
N113B240
N114B300
N115G80Z10.0M09
N116G91G28X0Y0Z0B0M05
N117M10
N118M06
N119M30

Notes: There are two types of horizontal milling machines. On milling machines equipped with an indexing table, rotation of the table only by 1° or more is allowed. When programming the changes of angular position of the table, as a rule, a decimal point is not used (but will do no harm).

Example:

B120, B180, etc.

Special functions are used to determine the direction of rotation of the table. On a milling machine equipped with a four axis table, rapid rotation or regular feed rotation can be applied. The table can be rotated by .001°, or even more accurately. On these machines, performance of all kinds of curved contours may be accomplished while machining the object. Plus (+) or minus (-) signs determine the direction of rotation. The table should not be locked if a rotation is to be performed. At the time tools 1 to 8 perform their operations, the table is locked by function M10. When the remaining two tools perform their operations, however, the table is unlocked by function M11. Whether the table should be locked or not is determined by the conditions of machining (light cutting/table unlocked or heavy cutting/table locked).

Subprograms for example O0022

O0050
N1G81G98Z6.5F2.5R7.1
N2X-3.0Y6.75
N3M98P0051
N4X-3.0Y4.0Z6.73
N5X3.0
N6G80X0Y4.0M11
N7G91G00A60
N8G90M10
N9M99

O0051
N1X3.0
N2Y1.25
N3X-3.0
N4M99

O0052
N1G81G98Z6.0F1.5R7.1
N2G80M11
N3G91G00A60
N4G90M10
N5M99

O0053
N1G90G00Z7.1
N2G01Z6.0F3.0
N3G42Y5.0D50F2.5
N4G02J1.0
N5G01G40Y4.0
N6G00Z8.5M11
N7G91A60
N8G90M10
N9M99

O0054
N1G83G98Q.15Z5.8F4.5R7.1
N2M98P0051
N3G80X-3.0Y6.75M11
N4G91G00A60
N5G90M10
N6M99

O0055
N1G84G98Z6.0F7.2R7.3
N2M98P0051
N3G80X-3.0Y6.75M11
N4G91G00A60
N5G90M10
N6M99

O0056
N1G83G98Q.100Z6.25F6.5R7.1
N2X3.0
N3G80X-3.0Y4.0M11
N4G91G00A60
N5G90M10
N6M99

O0057
N1G81G98Z6.4F8.0R7.1
N2X3.0
N3G80X-3.0Y4.0M11
N4G91G00A60
N5G90M10
N6M99

O0058
N1G81G98Z6.47F4.8R7.1
N2X3.0
N3G80X-3.0Y4.0M11
N4G91G00A60
N5G90M10
N6M99

COMPLEX PROGRAM EXAMPLE 2

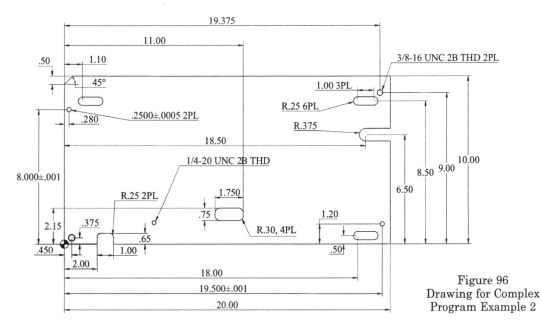

Figure 96
Drawing for Complex
Program Example 2

Part 4   Programming CNC Machining Centers

Machine: Machining Center	Program Number: O0023

Workpiece Zero

G54 X = Lower Left Corner Y = Lower Left Corner Z = Top Part Surface

Setup Description:

The material is .25 thick steel.

Tool #	Description	Offset	Comments
T01	#3 HSS Center Drill		
T02	.3125 Diameter HSS Drill		
T03	3/8-16 UNC Tap		
T04	.201 Diameter HSS Drill		
T05	1/4-20 UNC Tap		
T06	.242 Diameter HSS Drill		
T07	.250 HSS Reamer		
T08	7/16 Diameter HSS Drill		
T09	7/16 Diameter Roughing End-Mill	D51	D51 Offset = .240
T10	7/16 Diameter Finishing End-Mill	D52	D52 Offset = .2187
T11	1.0 Diameter HSS End-Mill		

Example of Complex Program 2, Program 23

        O0023
        (T01 #3 CENTER DRILL)
        N10G90G80G20G40G49
        N15T01M06
        N20G54G00X.45Y.375S611M03
        N25G43Z1.0H01M08
        N30G81G98Z-.375R.1F2.4
        N35X19.375Y9.0
        N40X19.500Y1.2Z-.26
        N45X.28Y8.0
        N50X5.5Y1.275Z-.25
        N55X1.1Y8.5Z-.437
        N60X18.0
        N65Y.5
        N70X10.5Y1.775
        N75G80Z1.0M09
        N80G91G28Z0
        N85M01
        (T02 5/16 DIAMETER TAP DRILL)
        N90T02M06
        N95G90G80G40G49
        N100G54G00X.450Y.375S733M03
        N105G43Z1.0H02M08
        N110G81G98Z-.35R.1F4.3
        N115X19.375Y9.0

N120G80Z1.0M09
N125G91G28Z0
N130M01
(T03 3/8-16 UNC 2B TAP)
N135T03M06
N140G90G80G40G49
N145G54G00X.450Y.375S150M03
N150G43Z1.0H03M08
N155G84G98Z-.45R.2F9.0
N160X19.375Y9.0
N165G80Z1.0M09
N170G91G28Z0
N175M01
(T04 .201 DIAMETER TAP DRILL)
N180T04M06
N185G90G80G40G49
N190G54G00X5.5Y1.275S1140M03
N195G43Z1.0H04M08
N200G81G98Z-.35R.1F6.8
N205G80Z1.0M09
N210G91G28Z0
N215M01
(T05 1/4-20 UNC 2B TAP)
N220T05M06
N225G90G80G40G49
N230G54G00X5.5Y1.275S220M03
N235G43Z1.0H05M08
N240G84G98Z-.42R.2F11.0
N245G80Z1.0M09
N250G91G28Z0
N255M01
(T06 .242 DIAMETER "C" DRILL)
N260T06M06
N265G90G80G40G49
N270G54G00X.28Y8.0S996M03
N275G43Z1.0H06M08
N280G81G98Z-.350R.1F5.0
N285X19.5Y1.2
N290G80Z1.0M09
N295G91G28Z0
N300M01
(T07 .250 DIAMETER REAMER)
N305T07M06
N310G90G80G40G49
N315G54G00X.28Y8.0S458M03
N320G43Z1.0H07M08
N325G85G98Z-.29R.1F2.7

N330X19.5Y1.2
N335G80Z1.0M09
N340G91G28Z0
N345M01
(T08 7/16 DRILL TO OPEN FOR SLOTS)
N350T08M06
N355G90G80G40G49
N360G54G00X1.1Y8.5S524M03
N365G43Z1.0H08M08
N370G81G98Z-.40R.1F3.1
N375X18.0
N380Y.5
N385X10.5Y1.775
N390G80Z1.0M09
N395G91G28Z0
N400M01
(T09 7/16 DIAMETER ROUGHING END-MILL)
(RADIUS COMPENSATION D51 = .240)
N405T09M06
N410G90G80G40G49
N415G54G00X1.1Y8.5S524M03
N420G43Z1.0H09M08
N425Z.2
N430M98P0060
N435X18.0Y.5
N440M98P0060
N445X18.0Y18.5
N450M98P0060
N455X10.5Y1.775
N460G01Z-.260F10.0
N465G41X11.0D51F2.0
N470Y1.85
N475G03X10.7Y2.15I-.3
N480G01X9.55
N485G03X9.25Y1.85J-.3
N490G01Y1.7
N495G03X9.55Y1.4I.3
N500G01X10.7
N505G03X11.0Y1.7J.3
N510G01Y1.775
N515G40X10.5
N520G00Z.2
N525X20.25Y6.5
N530G01Z-.26F10.0
N535G41Y6.875D51F2.0
N540X18.25
N545G03Y6.125J-.375

N550G01X20.25
N555G40Y6.5
N560G00Z.2
N565X2.5Y-.25
N570G01Z-.26F10.0
N575G41X3.0D51F2.0
N580Y.4
N585G03X2.75Y.65I-.25
N590G01X2.25
N595G03X2.0Y.4J-.25
N600G01Y-.25
N605G40X2.5
N610G00Z1.M09
N615G91G28Z0
N620M01
(T10 7/16 DIAMETER FINISHING END-MILL)
(RADIUS COMPENSATION D52 = .2187)
N625T10M06
N630G90G80G40G49
N635G54G00X1.1Y8.5T12S524M03
N640G43Z1.0H10M08
N645Z.2
N650M98P0061
N655X18.0Y8.5
N660M98P0061
N665X18.0Y.5
N670M98P0061
N675X10.5Y1.775
N680G01Z-.26F10.0
N685G41X11.0D52F2.0
N690Y1.85
N695G03X10.7Y2.15I-.3
N700G01X9.55
N705G03X9.25Y1.85J-.3
N710G01Y1.7
N715G03X9.55Y1.4I.3
N720G01X10.7
N725G03X11.0Y1.7J.3
N730G01Y1.775
N735G40X10.5
N740G00Z.2
N745X20.25Y6.5
N750G01Z-.26F10.0
N755G41Y6.875D52F2.0
N760X18.25
N765G03Y6.125J-.375
N770G01X20.25

N775G40Y6.5
N780G00Z.2
N785X2.5Y-.25
N790G01Z-.26F10.0
N795G41X3.0D52F2.0
N800Y.4
N805G03X2.75Y.65I-.25
N810G01X2.25
N815G03X2.0Y.4
N820G01Y-.25
N825G40X2.5
N830G00Z1.M09
N835G91G28Z0
N840M01

Part 4   Programming CNC Machining Centers

The following mathematical calculations are necessary to perform machining of the chamfer.

$$X = .5 - \Delta X = .5 - r \times \tan\frac{\alpha}{2} = .5 - .2071 = .2928$$

$$Y = .5 - \Delta Y = .5 - r \times \tan\left(45 - \frac{\alpha}{2}\right) = .5 - .2071 = .2928$$

(T11 1.0 DIAMETER END-MILL)
N845T11M06
N850G90G80G40G49
N855G54G00X-.5Y07072S229M03
N860G43Z1.0H11M08
N865G01Z-.26F10.0
N870X.2928Y10.5F1.3
N875G00Z1.0M09
N880G91G28Z0
N885G28X0Y0M05
N890M30

Subprograms for Complex Program Example O0023

O0060
(SUBPROGRAM OF PROGRAM O0023 FOR ROUGHING SLOTS)
N1G00Z.2
N2G91G01Z-.460F10.0
(D51 = .4375 ÷ 2 + .01)
N3G41Y.25D51F1.5
N4G03Y-.5J-.25
N5G01X.5
N6G03Y.5J.25
N7G01X-.5
N8G40Y-.25
N9G90G00Z.2
N10M99

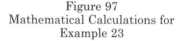

O0061
(SUBPROGRAM OF PROGRAM
O0023 FOR FINISHING SLOTS)
N1G00Z.2
N2G91G01Z-.460F10.0
(D52 = .4375 ÷ 2)
N3G41Y.25D52F1.5
N4G03Y-.5J-.25
N5G01X.5
N6G03Y.5J.25
N7G01X-.5
N8G40Y-.25
N9G90G00Z.2
N10M99

Figure 97
Mathematical Calculations for
Example 23

310

MILLING EXAMPLE

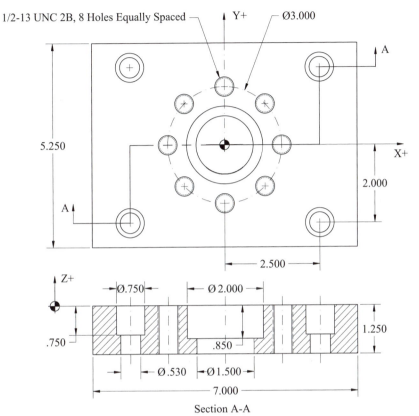

1/2-13 UNC 2B, 8 Holes Equally Spaced

Ø3.000

5.250

2.500

2.000

Z+

Ø.750

Ø 2.000

.750

.850

1.250

Ø.530

Ø1.500

7.000

Section A-A

Figure 98 Drawing for Milling Example

Example of Milling, Program 24

Machine: Machining Center	Program Number: O0024

Workpiece Zero

G54 X = Part Center Y = Part Center Z = Top Part Surface

Setup Description:

The material is steel.

Tool #	Description	Offset	Comments
T01	#6 HSS Center Drill		
T02	17/32 Diameter HSS Drill		
T03	27/64 Diameter HSS Drill		
T04	1 Diameter HSS Drill		
T05	.750 Diameter 2-flute End-Mill		
T06	1.0 Diameter 4-flutes Roughing End-Mill		
T07	1/2 -13 Tap		

Program
O0024
N10G90G80G20G40G49
N15T01M06
N20G54G00X0Y0S687M03
N25G43Z1.0H01M08
N30G81G98Z-.5F2.7R.1
N35X-2.5Y-2.0
N407M98P0070
N45X0Y-1.5
N50M98P0071
N55G80Z1.0M09
N60G91G28Z0M19
N65M01
N70M06T02
N75G90G80G40G49
N80G54G00X-2.5Y-2.0S648M03
N85G43Z1.0H02M08
N90G73G98Z-1.45R.1F3.8
N95M98P0070
N100G80Z5.0M09
N105G91G28Z0M19
N110M01
N115M06T03
N120G90G80G40G49
N125G54G00X0Y-1.5S816M03
N130G43Z1.0H03M08
N135G73G98Z-1.43R.1F4.9
N140M98P0071
N145G80Z1.0M09
N150G90G28Z0M19
N155M01
N160T04M06
N165G90G80G40G49
N170G54G00X0Y0S275M03
N175G43Z1.0H04M08
N180G73G98Z-1.65F1.65R.1
N185G80Z1.0M09
N190G91G28Z0M19
N195M01
N200T05M06
N205G90G80G40G49
N210G54G00X-2.5Y-2.0S458M03
N215G43Z1.0H05M08
N220G82G98Z-.75R.1F3.6
N225M98P0070
N230G80Z1.0M09

N235G91G28Z0M19
N240M01
N245T06M06
N250G90G80G40G49
N255G54G00X0Y0S343M03
N260G43Z1.0H06M08
N265G01Z-1.25F50.0
N270Y.22F2.7
N275G02J-2.7
N280G01Y0F50.0
N285Z-.83
N290Y.470F2.7
N295G02J-.470
N300G01Y0F50.0
N305G00Z1.0M09
N310G91G28Z0M19
N315M01
N320T07M06
N325G90G80G40G49
N330G54G00X0Y0S343M03
N335G43Z1.0H07M08
N340G01Z-.85F50.0
N345Y.5F2.7
N350G03J-.5
N355G01Y0
N360Z-1.250
N365Y.250
N370G03J-.250
N375G01Y0
N380G00Z5.0M09
N385G91G28Z0M19
N390M01
N395T08M06
N400G90G80G40G49
N405G54G00X0Y-1.5S114M03
N415G43Z1.0H08M08
N420G84G98Z-1.3F8.7R.2
N425M98P0071
N430G80Z1.0M09
N435G91G28Z0M19
N440G28X0Y0
N445M30

Subprograms for Milling Example Program O0024

O0070	O0071
(SUBPROGRAM FOR COUNTERBORES)	(SUBPROGRAM FOR BOLT CIRCLE)
N1X-2.5Y-2.0	N1X0Y-1.5
N2Y2.0	N2X-1.0607Y-1.0607
N3X2.5	N3X-1.5Y0
N4Y-2.0	N4X-1.0607Y1.0607
N5M99	N5X0Y1.5
	N6X1.0607Y1.0607
	N7X1.5Y0
	N8X1.0607Y-1.0607
	N9M99

EXAMPLE OFFSETTING OF TOOL PATH TO COMPENSATE FOR TOOL RADIUS

Machine: Machining Center	Program Number: O0025

Workpiece Zero

G54 X = Lower Left Corner Y = Lower Left Corner Z = Top Part Surface

Setup Description: The material used is free cutting brass .100 thick.

SFM = 300

S = 6100 RPM

Feed per one flute = .0015

$F = 4 \times .0015 \times 6100 = 36$

Tool #	Description	Offset	Comments
T01	.1875 Diameter 4-flute HSS End-mill		

   Note: The example given here is limited to a finishing pass only. The part is clamped to a fixture plate by using the 4 holes machined in a prior operation. In order to write a program, follow the necessary mathematical calculations shown below:

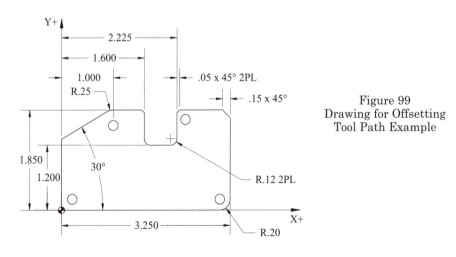

Figure 99
Drawing for Offsetting
Tool Path Example

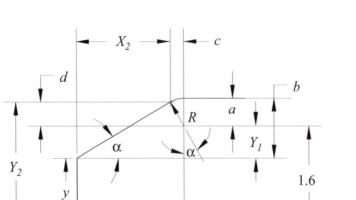

Figure 100
Diagram for Mathematical
Calculations for Offsetting
Tool Path Example 25

$$Y = 1.6 - Y_1 \quad Y_1 = b - a$$

$$\tan \alpha = \frac{b}{1.0} \quad b = 1.0 \times \tan \alpha = 1.0 \times \tan 30° = .5773$$

$$\cos \alpha = \frac{R}{a} \quad a = \frac{R}{\cos \alpha} = \frac{.25}{\cos 30°} = .2886$$

$$Y_1 = b - a = .5773 - .2887 = .2887$$

$$Y = 1.6 - Y_1 = 1.6 - .2887 = \underline{1.3113}$$

$$Y_2 = 1.6 + d$$

$$\cos \alpha = \frac{d}{R} \quad d = R \times \cos \alpha = .2165$$

$$Y_2 = 1.6 + d = 1.6 + .2165 = \underline{1.8165}$$

$$X_2 = 1.0 - c$$

$$\sin \alpha = \frac{c}{R} \quad c = R \times \sin \alpha = .125$$

$$X_2 = 1.0 - c = 1.0 - .125 = \underline{.875}$$

Example of Offsetting Tool Path, Program 25

Program
O0025
N10G90G80G20G40G49
N15G54G00X-.2Y-.2S6100M03
N20G43Z1.0H01M08
N25G00Z-.1
N30G01G41X0D31F36.0
N35Y1.3113
N40X.875Y1.8165
N45G02X1.0Y1.85I.185J-.2165
N50G01X1.55
N55X1.6Y1.8

N60Y1.3
N65G03X1.7Y1.2I.0
N70G01X2.125
N75G03X2.225Y1.3J.1
N80G01Y1.8
N85X2.275Y1.85
N90X3.1
N95X3.25Y1.7
N100Y.2
N105G02X3.05Y0I-.2
N110G01X-.02
N115G40Y-.2
N120G00Z1.M09
N125G91G28Z0M05
N130G28X0Y0
N135M30

EXAMPLE OF DRILLING 1000 HOLES USING ONLY SIX BLOCKS OF CODE

Machine: Machining Center	Program Number: O0026

Workpiece Zero

G54 X = Lower Left Corner Y = Lower Left Corner Z = Top Part Surface

Setup Description: The material used is 4140 steel.

$$S = \frac{12 \times V}{\pi \times D} = 3307$$
$$F = .002 \times 3307 = 6.614$$

Tool #	Description	Offset	Comments
T01	.052 Diameter HSS Drill	SFM = 45	

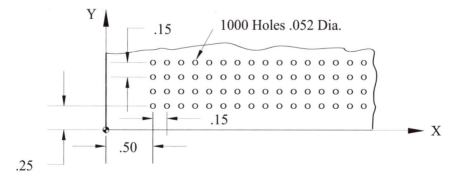

Figure 101 Drawing for Drilling 1000 Holes Example

In the illustration above the grid of holes shown represent only a portion of those on the part. The actual part contains 4 rows of 250 holes each.

Example of Drilling 1000 Holes, Program 26

> **O0026**
> **N001G90G00G80G40G49G54X.5Y.25S3307M03**
> **N002G43M98P0080L4Z1.0H01**
> **N003G91G28X0Y0Z0M30**
>
> **O0080**
> **N001G91G81G98X.15Z-.05L250R.05F6.6M08**
> **N002G80G00X-36.15Y.15**
> **N003M99**

EXAMPLE ILLUSTRATING APPLICATION OF MATHEMATICAL FORMULAS

The cutter used is 1/2 diameter.

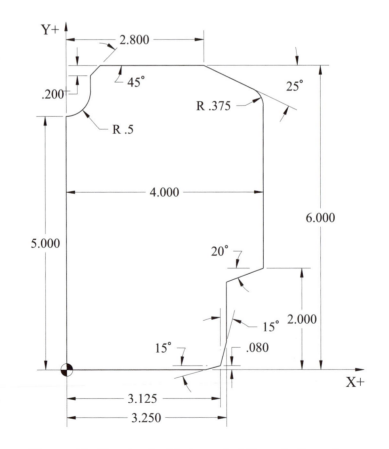

Figure 102   Drawing for Mathematical Formula Example

Example of Mathematical Formulas, Program 27

O0027
N10G90G80G20G40G49
N15G54G00X-.5Y-.5S600M03
N20G43Z1.0H01M08
N25G01Z-.5F5.0
N30X-.25F3.6
N35Y5.25
N40X0
N45G03X.25Y5.5J.25F2.0
N50G01Y5.9036F3.6
N55X.5964Y6.25F2.5
N60X2.8554F3.6
N65X3.8891Y5.768F2.5
N70G02X4.25Y5.2015I-.2641J-.5665F2.0
N75Y1.8249F3.6
N80X3.5Y1.552
N85Y.4203F3.6
N90X3.3541Y-.1241F2.5
N95X2.8844Y-.25
N100X-.25F3.6
N105G00Z1.M09
N110G28Z1.0
N115G28X-.2500Y-.2500M05
N120M30

Notes: Reference Position Return to Machine Zero is performed in the absolute system. The following is an explanation of the parts of the program that require mathematical calculations.

N50G01Y5.9036F3.6

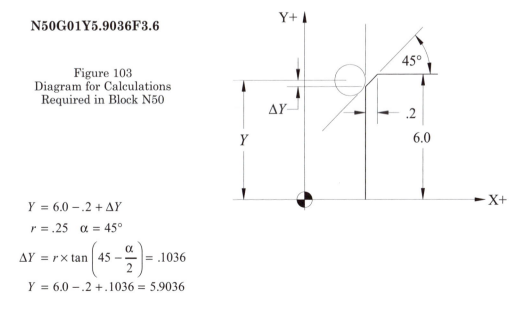

Figure 103
Diagram for Calculations
Required in Block N50

$$Y = 6.0 - .2 + \Delta Y$$

$$r = .25 \quad \alpha = 45°$$

$$\Delta Y = r \times \tan\left(45 - \frac{\alpha}{2}\right) = .1036$$

$$Y = 6.0 - .2 + .1036 = 5.9036$$

N55X.5964Y6.25F2.5

Figure 104
Diagram for Calculations
Required in Block N55

$X = .5 + .2 - \Delta X$

$r = .25 \quad \alpha = 45°$

$\Delta X = r \times \tan\dfrac{\alpha}{2} = .1036$

$X = .5 + .2 - .1036 = .5964$

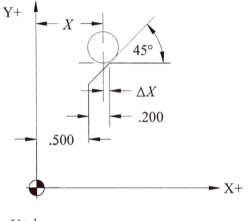

N60X2.8554F3.6

Figure 105
Diagram for Calculations
Required in Block N60

$X = 2.8 + \Delta X$

$r = .25 \quad \alpha = 25°$

$\Delta X = r \times \tan\dfrac{\alpha}{2} = .0554$

$X = 2.8 + .0554 = 2.8554$

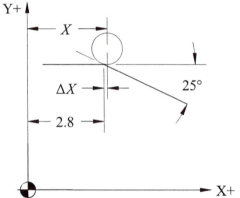

N65X3.8891Y5.7680F2.5

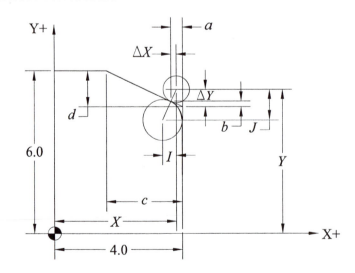

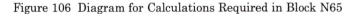

Figure 106 Diagram for Calculations Required in Block N65

$$X = 4.0 - a + \Delta X$$
$$R = .375 \quad \alpha = 25°$$
$$a = R \times \tan\left(45° - \frac{\alpha}{2}\right) \times \cos\alpha$$
$$a = .2165 \quad r = .250$$
$$\Delta X = r \times \sin\alpha = .1056$$
$$X = 4.0 - .2165 + .1056 = \underline{3.8891}$$
$$Y = 6.0 - d + b + \Delta Y$$
$$c = 4 - 2.8 = 1.2$$
$$\tan\alpha = \frac{d}{c} \quad d = c \times \tan\alpha = .5595$$
$$b = R \times \tan\left(45° - \frac{\alpha}{2}\right) \times \sin\alpha = .1009$$
$$\Delta Y = r \times \cos\alpha = .2266$$
$$Y = 6.0 - .5595 + .1009 + .2266 = \underline{5.7680}$$

N70G02X4.25Y5.2015I-.2641J-.5665F2.0

$$Y6.0 - d - I \quad d = .5595$$
$$I = R \times \tan\left(45 - \frac{\alpha}{2}\right) = .2390$$
$$Y = 6.0 - .5595 - .2390 = 5.2015$$

$$I = (R + r) \times \sin\alpha = .2641$$
$$J = (R + r) \times \cos\alpha = .5665$$

Figure 107
Diagram for Calculations
Required in Block N70

N75G01Y1.8249F3.6

$$Y = 2.0 - \Delta Y$$
$$r = .25 \quad \alpha = 20°$$
$$\Delta Y = r \times \tan\left(45° - \frac{\alpha}{2}\right) = .1751$$
$$Y = 2.0 - .1751 = 1.8249$$

Figure 108
Diagram for Calculations
Required in Block N75

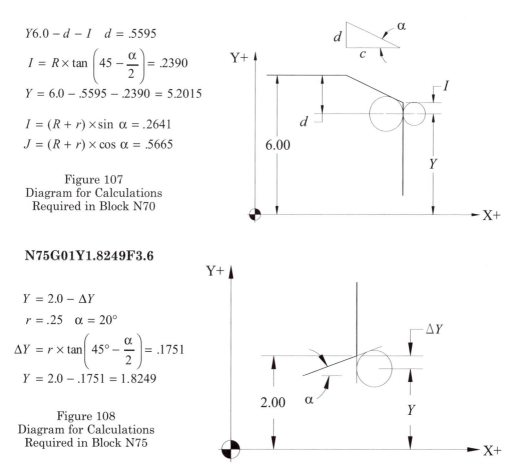

N80X3.5Y1.5520

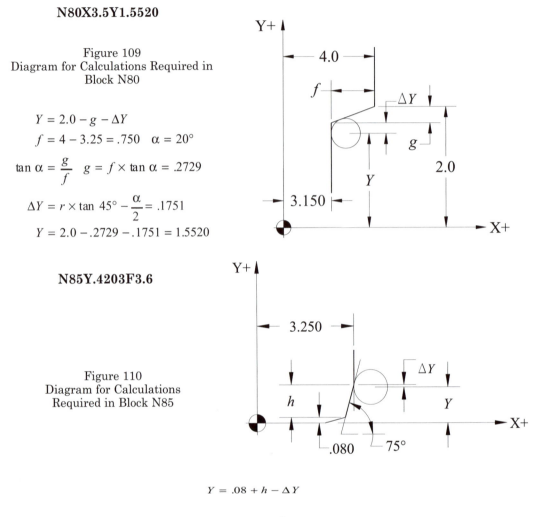

Figure 109
Diagram for Calculations Required in
Block N80

$Y = 2.0 - g - \Delta Y$

$f = 4 - 3.25 = .750 \quad \alpha = 20°$

$\tan \alpha = \dfrac{g}{f} \quad g = f \times \tan \alpha = .2729$

$\Delta Y = r \times \tan\ 45° - \dfrac{\alpha}{2} = .1751$

$Y = 2.0 - .2729 - .1751 = 1.5520$

N85Y.4203F3.6

Figure 110
Diagram for Calculations
Required in Block N85

$Y = .08 + h - \Delta Y$

Figure 111
Diagram for more Calculations Required in Block N85

$\alpha = 75° \quad \tan \alpha = \dfrac{h}{K}$

$h = K \times \tan \alpha = .3732$

$\Delta Y = r \times \tan\left(45° - \dfrac{\alpha}{2} \right) = .0329$

$Y = .08 + h - \Delta Y = .4203$

N90X3.3541Y-.1241F2.5

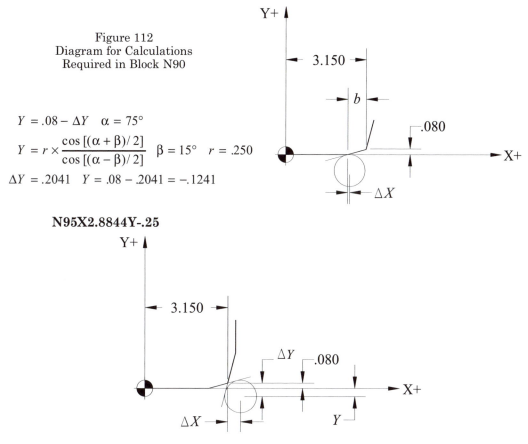

Figure 112
Diagram for Calculations
Required in Block N90

$$Y = .08 - \Delta Y \quad \alpha = 75°$$

$$Y = r \times \frac{\cos [(\alpha + \beta)/2]}{\cos [(\alpha - \beta)/2]} \quad \beta = 15° \quad r = .250$$

$$\Delta Y = .2041 \quad Y = .08 - .2041 = -.1241$$

N95X2.8844Y-.25

Figure 113 Diagram for Calculations Required in Block N95

$$X = 3.15 - b + \Delta X$$

Figure 114 Diagram for
more Calculations
Required in Block N95

$$\tan \alpha = \frac{.08}{b} \quad \alpha = 15°$$

$$b = \frac{.08}{\tan \alpha} = .2985 \quad r = .250$$

$$\Delta X = r \times \tan \frac{\alpha}{2} = .0329$$

$$X = 3.15 - .2985 + .0329 = 2.8844$$

321

EXAMPLE ILLUSTRATING APPLICATION OF COORDINATE SYSTEM ROTATION

Figure 115
Drawing for Coordinate System
Rotation Example

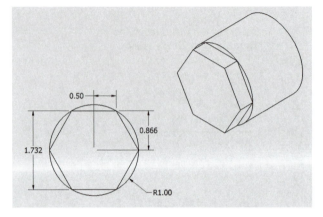

Example of Coordinate System Rotation, Program 28

Machine: Machining Center	Program Number: O0028

Workpiece Zero:

G54 X = Part Center Y = Part Center Z = Top Part Surface

Setup Description: Use a V-Block to hold the part in a vise

The material is low carbon steel.

Tool #	Description	Offset	Comments
T07	1/2 Diameter 4-flute End-Mill	H7	D7 = Offset #

Program
O0028
(1/2 DIA HSS 4FL END MILL)
N10G90G20G80G40G49
N15T7M6
N20G54G0X.75Y1.366S720M3
N25G0G43Z1.H07M08
N30G0Z1.
N35G68X0Y0R0.0
N40M98P291
N45G68X0Y0R60.
N50M98P291
N55G68X0Y0R120.0
N60M98P291
N65G68X0Y0R180.

N70M98P291
N75G68X0Y0R240.
N80M98P291
N85G68X0Y0R300.
N90M98P291
N95G69
N100G00Z1.
N105G40G80Z5.M09
N110G91G28Z0
N115G91G28X0Y0
N120M30

Subprograms for Coordinate System Rotation Example Program O0028

O0291
(Rotation Subprogram)
N1Z.1
N2G1Z-.4F20.
N3G2X.5Y1.116R.25F7.0
N4G1X-.5
N5G2X-.75Y1.366R.25
N6G0Z.1
N7M99

17. When using canned drilling cycles, which of the following codes are used to return the drill to the reference plane?

        a. G99

        b. G98

        c. G90

        d. G92

18. When using canned drilling cycles, what other letter address is necessary to identify the reference plane position?

        a. P

        b. Q

        c. R

        d. S

19. When using cutter diameter compensation in a program, what letter address is used to identify the location of the value of the offset?

        a. A

        b. R

        c. H

        d. D

20. What is the G-Code used to cancel Cutter Diameter Compensation?

        a. G40

        b. G41

        c. G42

        d. G43

21. Match the following definitions with the proper M-Code.

Program Stop	M30 ___
Optional Stop	M06 ___
End of Program	M03 ___
Spindle on Clockwise	M00 ___
Spindle on Counterclockwise	M01 ___
Spindle Off	M19 ___
Flood Coolant On	M05 ___
Flood Coolant Off	M98 ___
Spindle Orientation	M08 ___
Subprogram Call	M04 ___
Subprogram End	M09 ___
Tool Change	M99 ___

22. Match the following definitions with the proper G-Code.

Linear Interpolation	G90 ___
Circular Interpolation Clockwise	G00 ___

Rapid Traverse   G81 ___
Dwell   G40 ___
Absolute Programming   G28 ___
Incremental Programming   G41 ___
Canned Cycle Cancellation   G91 ___
Peck Drilling Cycle   G42 ___
Drilling Cycle   G43 ___
Cutter Diameter Compensation Left   G01 ___
Cutter Diameter Compensation Right   G83 ___
Cutter Diameter Compensation Cancellation G84 ___
Zero Return Command   G91 ___
Inch Programming   G54 ___
Metric Programming   G92 ___
Tapping Cycle   G21 ___
Absolute Program Zero Setting   G02 ___
Fixture Offset Command   G80 ___
Positive Tool Length Compensation   G20 ___

23. Match the following definitions with the proper letter address.

Program Number   A ___
Sequence Number   B ___
Preparatory Function   C ___
Miscellaneous Function   D ___
X Coordinate   F ___
Y Coordinate   G ___
Z Coordinate   H ___
Reference Plane Designation   I ___
Feed rate   J ___
Spindle Function   K ___
Subprogram Repeats   L ___
Tool Function   M ___
Tool Length Compensation Register   N ___
Diameter Compensation Register   O ___
Rotational Axis about the X   P ___
Rotational Axis about the Y   Q ___
Rotational Axis about the Z   R ___
Dwell in Seconds   S ___
Peck Amount in Canned Drilling   T ___
Arc Center X Axis   X ___
Arc Center Y Axis   Y ___
Arc Center Z Axis   Z ___

24. If a linear move is programmed, G01X1.5Y1.5, what is the angle of the resulting cut?
        a. 30°
        b. 180°
        c. 45°
        d. 90°

25. If a rapid traverse positioning move is programmed along the X and Y axes and the distances are unequal, the shortest distance will be achieved first.
        T or F

26. When using fixture offset programming G54-G59, it is possible to have multiple offsets in one program.
        T or F

PART 5

COMPUTER AIDED DESIGN & COMPUTER AIDED MANUFACTURING (CAD/CAM)

OBJECTIVES:

1. Become familiar with common CAD/CAM capabilities.

2. Become familiar with CAD/CAM software Graphical User Interface.

3. Learn terminology specific to CAD/CAM.

4. Learn Geometry Creation techniques common to CAD/CAM.

5. Learn how to use the CAD/CAM software to create tool path and CNC program code.

WHAT IS CAD/CAM?

Computer Aided Design and Computer Aided Manufacturing (CAD/CAM) utilizes computers to design drawings of part feature boundaries in order to develop cutting tool path and CNC machine code (a part program). By using CAM, the cutting tools and all data specific to them are defined and then tool paths are identified by selecting drawing geometry that determine how they are going to be used for cutting. Drawing in CAD is simply constructing a drawing using lines, arcs, circles and points, and positioning them relative to each other on the screen. One of the major benefits of CAD/CAM is the time saved. It is much more efficient than writing CNC code line-by-line.

CAD/CAM is now the conventional method of creating mechanical drawings and Computer Numerical Control (CNC) programs for machine tools. CAD is the standard throughout the world for generating engineering drawings. The personal computer has become a powerful tool used by manufacturing for these and many other purposes. Engineers seldom use the drafting board to design their projects, they now use computers extensively. Designers can create the drawings needed in other CAD programs and share them electronically with the manufacturing department. Drawings are converted to a common file format, such as the Initial Graphics Exchange Specification (IGES) or Drawing Exchange Format (DXF). There have been huge advancements in the design field and Solid Models have become prevalent over two-dimensional drawings. Many CAM systems will import the Solid Model files directly for tool path creation. Then, the manufacturing engineer can create tool path and assign cutting tool information relative to the desired results. CAD is limited, in nature, to the generation of engineering drawings, while CAD/CAM combines both design and manufacturing capabilities. When using CAD/CAM, the drawing may be created from scratch or imported from a CAD program. It is not necessary to have the drawing dimensioned for this operation, but the full scale of the part is required. The CAD/CAM operator assigns the tools and their order of usage while creating the tool path. There are many CAD/CAM programs on the market today. The most popular ones are easy to use, have a solid background and reliability. To make good use of this computing power, it is important to fully understand the machining processes to be carried out. Just as CNC doesn't change the actual machining, the same is true of CAD/CAM for programming. Remember, the overall objective of CAD/CAM is to generate tool path for a CNC machine in the form of a CNC program. It is imperative to have a full understanding of the rectangular and polar coordinate systems. It is also necessary to have a complete understanding of cutting tool selection, speeds and feeds. Nearly all CAD/CAM programs will automatically develop speed and feeds data based on the tool and work material selection, however, adjustments are frequently necessary. This data base of information can and should be updated to match the requirements for your shop for the best results.

When constructing the part geometry, consider the type of machining operation. For instance, if the desired result is to drill a hole using a standard drill, construction of only the point that represents the center coordinate location of the hole is necessary.

In this book, the most recent versions of Mastercam and EdgeCAM are featured as CAD/CAM software examples and some of the programs created in this text have been verified using the simulation capabilities inherent to these software. This chapter is only intended as an introduction to CAD/CAM. Many other CAD/CAM programs use similar techniques to accomplish the same result.

The following is a short description of the process of using Mastercam to create geometry, tool path and program code for CNC machines:

PERSONAL COMPUTER

The computer needed to run this type of software has important minimum requirements. Normally, a large screen is desirable for ease of viewing the geometry created. CAD/CAM programs require a lot of hard disk space so a large hard drive is also recommended. Because CAD/CAM is used to create complex drawings and perform graphic simulations, the computer has the following basic needs: The memory the computer uses to access files while working on them is called RAM, (Random Access Memory). For CAD/CAM, a large amount of RAM is highly recommended (individual software manufacturers have recommended minimums). The computer's processing speed is listed in MHZ (Megahertz), again the higher the number, the better. The computer's graphics card and monitor controls the screen resolution. A powerful graphics card is strongly advised. Remember to consult the specific minimum requirements for the CAD/CAM program you chose to work with.

WINDOWS

To be successful using CAD/CAM, it is necessary to understand the use of a Personal Computer operating system and software programs. Microsoft Windows is the most widely used operating system on personal computers. In this section, we will be referring to the Windows XP operating system. With CAD/CAM, the operator must understand the operating system and have basic skills for mouse usage, including: double-click, right mouse button and the mouse pointer. Just like most computer programs in use today, CAM programs use a Graphical User Interface (GUI) for ease of input. CAM programs use icons, toolbars and menu systems to guide the user through program use. Many functions can be accessed via short-cut keystrokes and mouse button clicks. Some functions require a single click with the left mouse button to activate a command.

CONVENTIONS

In the example that follows, when the Menu selection method is used, the Menu item used will be **bolded** in the instructions given and the short-cut keystroke Underscored to match the Mastercam menu bar. When Icon selection method is used, the Icon graphic and directions will be included in the instructions. ***Specific instructions will be given in bold italics.***

MASTERCAM X2, PROGRAM STARTUP

From the Windows main screen, look at the desktop to see if there is a shortcut icon and ***double-click the left mouse button on the Mastercam X2 icon*** (Figure 1). If there is not a desktop shortcut icon, ***press the start button in the lower left corner with the left mouse button. Slide the mouse pointer up to the right and you will see a list of all available programs. Slide the mouse pointer to find Mastercam X^2 and another list will appear to the right. Again, slide the mouse pointer through the list to find the desired program, Mastercam X^2.*** Once the program is started, by default, Mastercam is in the Design mode. We are going to use Milling for this example, so change to the Mill system by selecting **Machine Type** from the Menu Bar, then **Mill** and lastly **Default**. To use the mouse to activate the Mill system, ***single click the left mouse button on Machine Type, then select Mill, and then select Default for the machine configuration.*** The Mastercam main interface screen will be displayed as shown below in Figure 2 but without the item descriptions.

Figure 1
Mastercam
X^2
Shortcut Icon

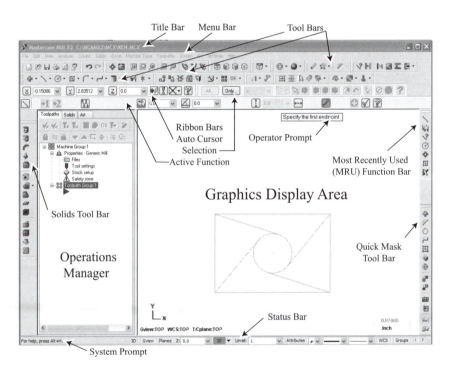

Figure 2 Mastercam X2 Graphical User Interface

MASTERCAM X^2 USER INTERFACE

Getting familiar with the user interface is imperative for efficient use. Brief descriptions of the main components of the interface are given here. The more you use the functions, the more natural it will become. Don't forget that there is extensive help available within the program by selecting Help from the Menu Bar or the question mark (?) icon for context sensitive questions.

Title Bar

The title bar lists the software name, version and current mode; Design, Mill, Lathe, Wire or Router and the file name and location. For example:

Mastercam Mill X2 C:\MCAMX2\MCX\KEN.MCX

Menu Bar

Many of the program functions are accessible by clicking the left mouse button while over the icon on the tool bars. A list of the same functions is available on the Menu Bar. The functions from the Menu Bar can be activated by using the mouse or by using short-cut keystrokes. Press the alt key to activate access to these keystrokes. They are generally the first letter of the word or, if not, they are identified by an underscore of the letter needed to activate the command.

Graphics Display Area

This is where the part geometry and tool path are displayed during creation and verification. The Graphics Display Area is blue (color #9) by default and can be changed to another color, if necessary, by selecting **Settings** from the Menu Bar, then **Configuration**, then **Colors** from the Topics section of the System Configuration dialog, then select Graphics background color. Use caution when changing the background colors, because some of the other color combinations will not allow visibility of selection, chaining and tool path, etc. White has been used in this text for printing reasons only.

Tool Bars

Tool bars contain icons that act as short cut buttons to all types of functions and are accessible via mouse clicks. When the mouse selection arrow is hovered over one of the tool bar icons, a fly out descriptor appears. The user has the ability to activate/deactivate the on screen tool bars by accessing the Toolbar States dialog box by selecting Settings from the Menu Bar and then Toolbar States.

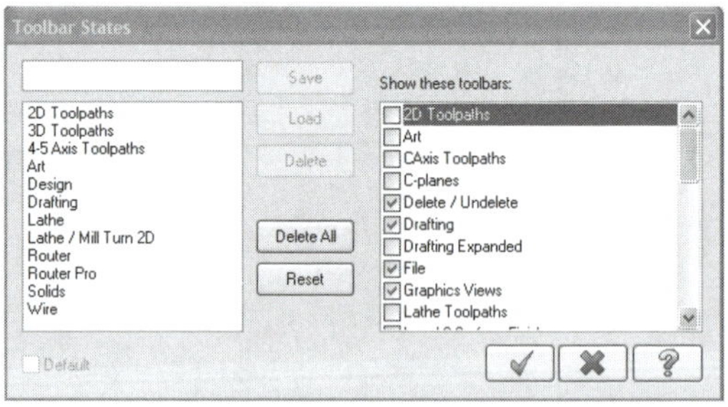

Figure 3 Toolbar State Dialog Box

Auto Cursor Ribbon Bar

Figure 4 Auto Cursor Ribbon Bar

You can use this tool bar (Figure 4) to set coordinate values for X, Y or Z axis input. This toolbar can also be used to set Auto Cursor selection types, activate Fast Point mode and change the configuration for Auto Cursor.

General Selection Ribbon Bar

Use this tool bar to set general selection masking functions for chaining and deleting of entities (See Figure 5).

Figure 5 General Selection Ribbon Bar

Active Function Ribbon Bar

Once a drawing command is active, the Ribbon Bar shown in Figure 6 becomes available for input that is specific to the type of entity being created. Some examples in the line creation mode are: Multi-line, Line length, Polar Angle, Horizontal or Vertical, and Tangent.

Figure 6 Active Function Ribbon Bar

Status Bar

The Status Bar (Figure 7) is placed at the bottom of the screen and gives the user feedback about the current drawing status with information about: 2D/3D, Geometry views (Gview), Construction and Tool Planes (CPlane and TPlane), Colors, Levels, Attributes, Point Styles, Line Styles, Line Weights, World Coordinate Systems (WCS) and Groups. The user can set any of the aforementioned data as well by left clicking on the item in the Status Bar.

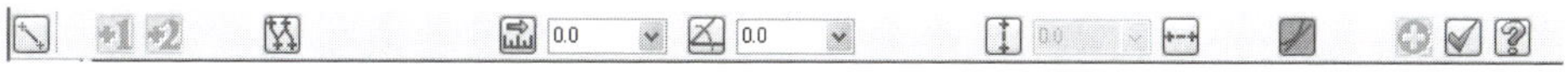

Figure 7 Status Bar

Most Recently Used (MRU) Tool Bar

The Most Recently Used (MRU) Tool bar (Figure 8) is usually docked vertically on the right side of the User Interface. It is a temporary tool bar which contains the commands most recently used. It is made handy for quick access to recently used commands.

Figure 8
MRU
Tool Bar

Quick Masks Tool Bar

The Quick Masks tool bar (Figure 9) allows selection control by left clicking of the type of mask desired. By default, the types of masks available on the Quick Mask tool bar are: Points, Lines, Arcs, Splines, Surfaces, Solids, Wireframe, Result, Group, Color, Level and Drafting. By selecting a function buttton on the Quick Masks tool bar, masking is set to select all entities of the type chosen. If the right mouse button is pressed while hovering over the

Figure 9
Quick
Masks Tool
Bar

335

desired Quick Mask type button, masking is set to single selection.

If you are working on a complex drawing and need to select Points only, for example, you can left-click on the Quick Masks Points button, only Points will be selected. By right-clicking the button, points can be selected one at a time until selection is complete.

Operations Manager

The Operations Manager Pane (Figure 10) contains tabs with information about Toolpaths, Solids and Art. Under each tab there is an information tree. The Toolpath tab describes the Machine Group and its Properties and each Toolpath Group, its properties and parameters. At the top of this pane are the icons specific to Toolpath Selection, Backplotting, Verification, Post processing and Editing.

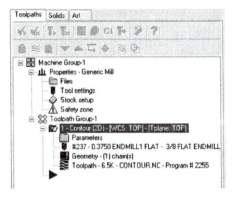

Figure 10 Operations Manager Pane

Note: When the escape key (esc) is pressed, the software accepts the current command and steps back to the main level in the menu, essentially clearing the active command. If the Enter key is pressed after creation of a feature of the geometry, the current command is accepted and the mode of geometry entry is retained.

MACHINE GROUP SETUP AND GEOMETRY CREATION

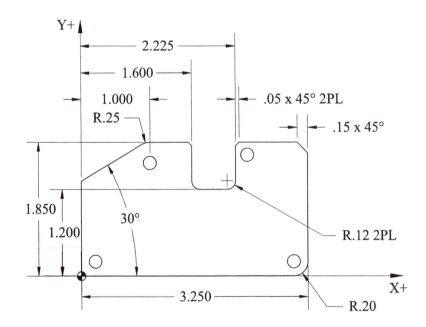

Figure 11 Drawing for Mastercam Example

In many cases, the machinist/programmer is not provided with an electronic file for the part geometry they must create. CAD/CAM programs provide the ability to recreate the drawing for the purpose of making the tool path program. For this example, all of the geometry shown in Figure 11 must be recreated. The first consideration when recreating geometry is where to set the Workpiece Zero or origin. A zero location has been indicated on the print by the symbol in the lower left hand corner. It makes good sense to use this same location to start drawing. The part will be clamped with cap screws to a holding fixture that is held in a vise. In this example, only the contour will be machined with a 3/16 inch 2-Flute end mill. The part blanks provided are 1/4 inch thick by 2 inch wide 2024 Aluminum that is pre-machined to 3.35 inches in length.

MACHINE TYPE

When you begin the process planning steps to write any CNC program, you must select the type of machine required to perform the job. The software configuration purchased dictates the types of machines available to select from, including: Mill, Lathe, Wire, Router, or Design. The information entered here establishes the data needed for the program such as: stock size tools and part origin.

Note: When Mastercam is started, the Design mode is active. You can create the design and then activate the Machine Type before tool path creation can begin but, it is a matter of preference.

• Start a new Mastercam X2 file by selecting **File, New**. Respond NO to the dialog stating, The Mastercam X file has changed, save it? Alternatively, **Press the New icon on the File toolbar. Reopening Mastercam X^2 will also work.**

• From the Menu Bar, single click the left mouse button on **Machine Type.** With the mouse

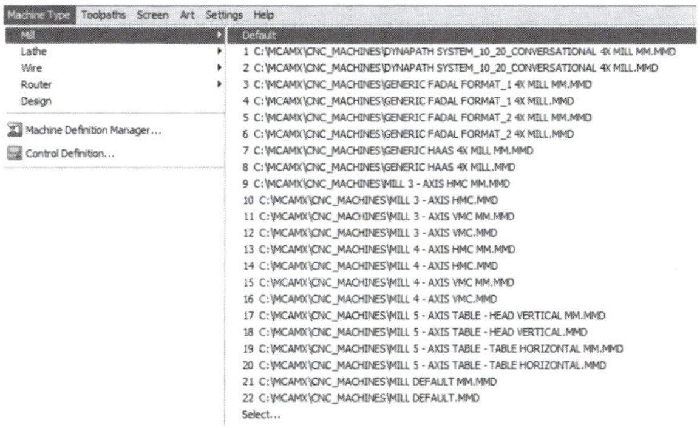

Figure 12 Machine Type Selection

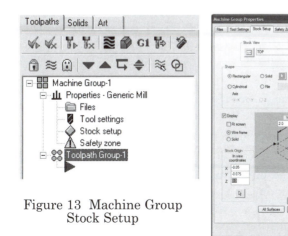

Figure 13 Machine Group
Stock Setup

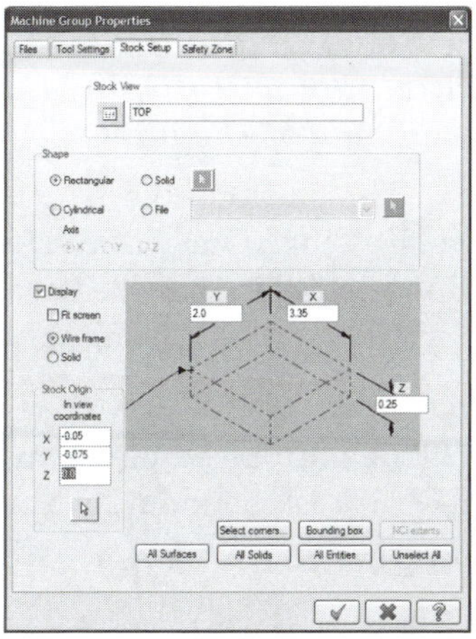

Figure 14 Stock
Setup Dialog

pointer select **Mill** and then **Default** from the drop down list for this example (Figure 12). Machine Group 1 will appear in the Operations Manager Pane.

- To access the Stock setup dialog, *left click on the plus sign next to Properties – Generic Mill in the Operations Manager Pane* as shown in Figure 13. Then *left-click on Stock setup to access* the dialog displayed in Figure 14.

Explanations are given below for each item in the Machine Group Properties dialog in (Figure 14) that is used in this example for Stock Setup.

- <u>Stock View</u> should be set to correspond with the standard geometry view to be used for the tool path needed. *In this case, the Top view is correct.*

- <u>Shape</u> should be set to match the raw material shape, or it can be set to an existing Solid by selecting the model, or any file, by identifying the files location. To set the stock size, key in the dimensions for the *X, Y* and *Z*, including raw material (the *Z* value input here must be positive). *In this case, the rectangular part blank information is input to the fields for X (3.35), Y (2.0) and Z (.25)*

- <u>Display</u> establishes the way the material is displayed in the graphics window. Whether the stock is included when Fit to the Screen function is used, the part file is displayed as a Wire Frame or as a Solid. *Put a check mark on Display and choose the Wire frame radio button.*

- <u>Stock Origin</u> identifies the coordinate locations of the raw material zero values. The origin for the part may be moved to a desired corner location by left clicking once on the black arrow and then left clicking over the new location. *For our example, use the left mouse button and left click on the arrow that is in the center of the part, then, click the mouse button over the lower left hand corner of the diagram.* The arrow indicator of the location of the part origin will move to the corner selected. The exact numerical locations may also be input into the *X Y* and *Z* to accomplish this if they are known. The *Y* and *X* coordinates define the outer boundary of the raw stock of the part. When the arrow button below the Stock Origin section is pressed, the display will revert to the drawing file and allow the manual selection of the stock origin by mouse, directly from the drawing. In this example, there is one hundred-thousandths excess material in the length and one hundred fifty-thousandths excess material in the width. *By inputting half the difference in the length and width into the Stock Origin fields for X (-.05), Y (-.075) and Z (0) the graphical verification will display the appropriate amount of metal removal for each side of the part.*

- <u>Select Corners</u> . . . This button allows for manual selection of the stock corners from the active drawing. Once the selection is made, the values are input automatically into the fields that identify the boundaries of the part.

- <u>Bounding Box</u> . . . This button allows the user to return to the drawing to manually make a boundary around the drawing to represent the material.

- <u>NCI Extents</u> . . . This button compares and calculates all of the tool paths in the NCI file and creates stock boundaries that include all tool movement extents.

- *Left click on the Tool Settings tab*

Explanations are given below for each item in the Tool Settings tab of the Machine Group Properties dialog shown in (Figure 15) that will be used in this example.

- ***Program # Enter any desired # in the first field, i.e. 2255.***

    - Feed Calculation may be based from the tool, material, system default, or be defined by the user. In our case, ***we will select From tool.***

    - Sequence #'s for the program code are set here. Sequence # Start is the starting number for the line sequence "N" numbers in the program. Any starting value can be entered here. Common starting numbers are 1, 2, 5, 10 or 100. Increment, is the sequence number increments for the line sequence numbers. Any value may be entered here. Common increments are 1, 2, 5 or 10. ***We will accept the default values of 100 and 2.***

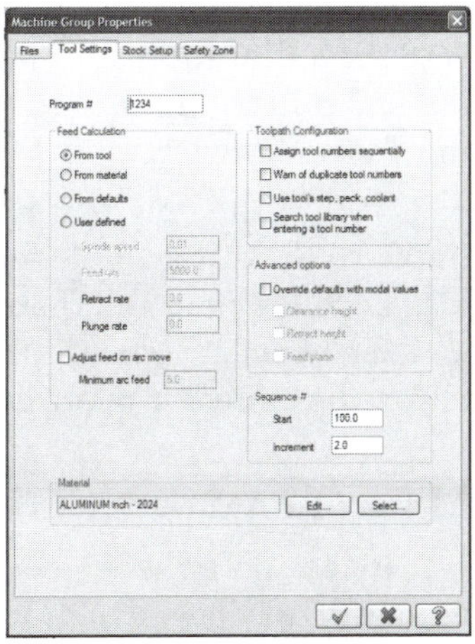

Figure 15
Tool Setting Dialog

- Material type is assigned by selecting from the materials library after pressing the Select . . . button. The user can also change feeds and speeds settings by pressing the Edit . . . button to access the Material Definition Manager. **We will accept the default material, Aluminum inch 2024. When all of the settings are completed, press the check mark button to accept them.** The stock will be displayed in the graphics display area.

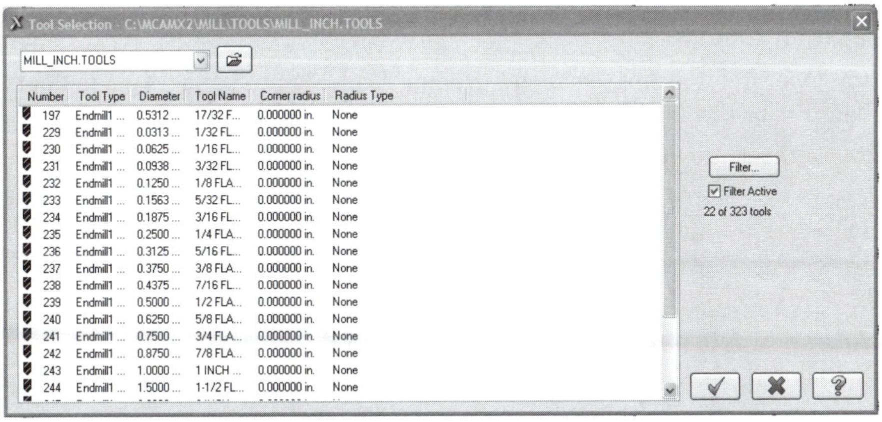

Figure 16   Tool Manager Dialog

From the Menu Bar, select **Toolpaths** and then **Tool Manager** . . . from the drop down menu. The dialog in Figure 16 will then be displayed.

Place a check in the Filter Active check box and then press the Filter . . . button to activate the filters. If this checkbox is not checked, then all 323 tools in the standard database will be displayed in the list. By checking this box to activate this filter the number and type of tools displayed can be refined.

From the Tool List filter dialog, **press the button labeled as None. Then press the "Endmill 1 flat" button, so that end mills only will be displayed and press the check mark to accept.** A list of only end mill tools will be displayed. **Select the 3/16 inch diameter end mill from the list (#234) by double clicking on it.** The chosen end mill will now be moved into the upper portion of the tool display area window of the Tools Manager.

This is the only tool needed for the program in our example so **press the check mark to accept and close the dialog.** If additional tools are needed, the same process may be followed to add them. Now, all the Machine Group setup information is complete and the geometry creation may begin. Any of the items that have been set so far can always be edited, if required.

MASTERCAM X2 GEOMETRY CREATION STEP-BY-STEP

Please note the steps listed here, are by no means, the only method by which this geometry may be created. Individual preferences and speed will ultimately be the determining factor in how drawings are created. The Menu Bar method will be listed as our method of geometry creation.

To create the drawing in Figure 11, follow these steps:

Note: to activate the keystroke shortcuts press the alt key first.

1. From the Main Menu Bar, position the mouse pointer over <u>C</u>reate and press the left button, or *press the letter "C" to activate the create menu.*

Note: Even though the short-cut keystroke letter is in uppercase, it is unnecessary to key it that way. In other words, a lower case "c" will accomplish the same thing.

2. Use the mouse to *select Line, to activate the line menu.* Select, **Create Line Endpoint . .**

In the Operator Prompt area, the prompt reads: "Specify the first endpoint." (The system defaults to the sketch method of point entry.)

3. *Press the spacebar to activate the Fast Point entry mode.*

4. *Key in the coordinates for the start point of the line: 0, 0 and press Enter* (where 0, 0 = X value Y value. The Z value is zero unless otherwise specified via the Status Bar or keyed here.)

5. *Press the letter H on the keypad, or the Horizontal Icon on the Active Function Ribbon Bar* to lock the line into the horizontal line mode.

In the Operator Prompt area, the prompt reads: "Specify the second endpoint." The first point is drawn and a horizontal line is attached to it. As the mouse is moved the line stretches like a rubber band to the point. The default mode is Sketch so whenever the mouse is moved, the line length changes correspondingly in the Ribbon Bar readouts.

6. The specific line <u>L</u>ength (value of the line length) *may be entered by pressing L on the keyboard or the Length Icon on the Active Function Ribbon Bar to lock the line into the line length mode. To enter the signed value of the length, key in 3.25.* To accept the input, left click the mouse button and then press Enter again to accept the Y value for the line. To adjust the value of

the Y axis, press the Enter key and then enter the new Y value into the Active Function Ribbon Bar to set the Y location and press Enter, to accept the input.

Note: Alternatively, if the second endpoint coordinate values are known, press the spacebar to activate the Fast Point mode and begin typing in the values. The line will be created when Enter is pressed a second time. If the coordinate value for a point stays the same, no entry is necessary for that axis. For example: Key in: 3.250 for the second end point and press Enter, Enter.

The Line creation mode is still active, except now we need to change to a vertical line type.

7. ***Press the letter V to activate the Vertical Line mode, or left click on the Vertical icon in the Active Function Ribbon Bar to lock the line into the vertical line mode.***

In the Operator Prompt area, the prompt reads: "Specify first endpoint."

8. ***Use the mouse to select the endpoint of the line created last.*** The Auto Cursor Configuration allows easy selection of the line endpoint. ***Left click any where near the line endpoint to select it.***

9. ***Press the letter L to set the specific Length and then key in 1.850. Left click the mouse button to accept the length. Press Enter to accept the X coordinate of 3.25.***

The Line creation mode is still active, now we need to change back to a horizontal line type.

10. ***Use the mouse to again select the endpoint of the line created last.***

11. ***Press the spacebar and key in 2.225, 1.85, press Enter, Enter.***

12. ***Use the mouse to again select the endpoint of the line created last.***

13. ***Press the letter H to exit the Horizontal line mode.***

14. ***Press the spacebar and key in (,1.2) and press Enter, Enter.*** Note: Be sure to enter the comma first in order to accept the current location for the X axis value.

15. ***Use the mouse to select the endpoint of the line created last.***

16. ***Press the spacebar and key in 1.6 and press Enter, Enter.***

17. ***Use the mouse to select the endpoint of the line created last.***

18. ***Press the spacebar and key in (,1.85) and press Enter, Enter.***

19. ***Use the mouse to select the endpoint of the line created last.***

20. ***Press the spacebar and key in 1.0 and press Enter, Enter.***

Note: the start of the .25 inch radius arc is known to be that location in X and Y.

Next, a construction line must be created in order to obtain the proper location for the .25 inch radius arcs endpoint.

21. ***Press the spacebar and key in 1.0, 1.6*** (the coordinate value of Y is obtained by subtracting the known radius value of .250 from 1.85 = 1.6)

22. ***Press the letter A on the keyboard to select the Angle line, type***

In the Operator Prompt area, the prompt reads: "Specify the second endpoint."

23. ***Key in 120° and press Enter*** (because the angle of 30° is given on the print and when the 30° is added to 90° it = 120°).

24. ***Press the letter L on the keyboard and enter in .25, Enter, Enter*** to accept the line.

25. From the Menu Bar select <u>C</u>reate, *use the mouse to select Arc, Create Arc Endpoints.*

26. ***Select the upper endpoint of the construction line last created.***

In the Operator Prompt area, the prompt reads: "Enter the second point".

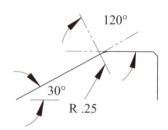

Figure 17
Construction Line for Arc Creation

27. ***Select the endpoint of the horizontal line that was last created and then press Enter.***

28. ***Press the letter R on the keyboard to set the radius of the arc to .25 and press Enter.***

Upon doing this two full circles will be drawn. In the Operator Prompt area, the prompt reads: "Select an Arc".

29. ***Carefully position the mouse selection point over the arc you wish to keep and left click to accept the selection, press Enter to accept.***

Notes: It may be necessary to zoom in on the area of the drawing for easier selection of the correct arc.

If the Auto Cursor is enabled, a small box will be drawn when the cursor is positioned over the endpoints. The Auto Cursor may be Enabled or Disabled by placing a check mark to Enable or removing the check mark for Disable. When this function is ON, it is useful for simple drawings, but may be annoying on complex drawings.

30. From the Menu Bar select <u>C</u>reate, <u>L</u>ine, Create Line Endpoint . . .

31. ***Press the tangent function button, in the Active Function Ribbon Bar to activate it.*** It may once again be useful to zoom in on the area of the drawing for easier selection of the line arc tangency point.

32. ***Right click the mouse while anywhere within the graphics display area and select Auto Cursor to Disable All, then place a check mark to Enable only the Tangent setting and press the check mark button to accept the changes.***

33. ***Use the mouse to position the selection point over the .25 radius arc and left click to pick the arc, press Enter to accept the selection.***

34. ***Press the letter A on the keyboard to activate the Angle mode.***

35. ***Key in an angle of 210° and then press Enter*** (because 180° + 30° = 210°).

36. **Press the letter L on the keyboard to activate the Length mode and enter a value of 2.0 inches.** This length will be sufficient to intersect with the vertical line rising from the X, Y zero location we will draw next. A line 2 inch es long will be drawn in both directions from the tangency point.

37. ***Select the line you wish to keep (left of the tangent arc) by left clicking on it and then pressing Enter, Enter.***

38. *Right click the mouse anywhere within the graphics display area and select Auto Cursor again to adjust the settings to include Endpoints.*

39. *While the line command is still active, select the endpoint of the line at the X, Y zero location.*

40. *Press the letter V on the keyboard to activate the vertical line function.*

41. *Press the spacebar and then key in Y1.5, Enter, Enter, Enter to accept the entry.*

If the drawing exceeds the display area of the screen, press the F2 function key to Unzoom the current display by .5.

Now we will take the steps necessary to trim and delete the excess lines, chamfer and fillet the corners, as specified on the print.

42. From the Menu Bar select <u>C</u>reate, *and then use the mouse to select Fillet, and then Fillet Entities.*

In the Operator Prompt area, the prompt reads: "Fillet select an entity". (The default fillet size is a .250 radius). Press the letter R on the keyboard and enter the desired value.

43. *Press the letter R on the keyboard and key in .2 for the radius and press Enter.*

Note: The Style should be set to Normal on the Active Function ribbon Bar.

44. The user prompt states "Fillet: select an entity". *Use the mouse to select the entities in the lower right hand corner of the part. Choose line 1 and then choose line 2 as shown in Figure 18.*

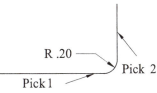

Figure 18 Selection Points for the .2 Radius Fillet

45. While still in fillet mode, *change the radius to .12 for the 2 other fillets by pressing the letter R on the keyboard and entering .12, Enter.*

46. *Pick each vertical and horizontal line as shown in Figure 19, and press Enter after all four picks to accept the fillets.*

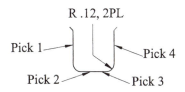

Figure 19 Selection Points for the .12 Radius Fillets

All fillets are done, now we need to trim the excess lines off from the left angular and vertical lines.

47. *From the Menu Bar select Edit, Trim/Break, Trim/Break/Extend.*

48. *On the Active Function Ribbon Bar press in the "Trim 2 entity" button.*

In the Operator Prompt area, the prompt reads "Select the entity to trim/extend".

49. *Choose the angular line, then the last vertical line as shown in Figure 20.*

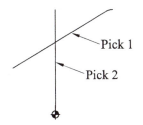

Figure 20 Selection Points for Trimming

The excess lines will be trimmed. Now the chamfers must be created.

50. *From the Menu Bar select $\underline{C}$reate, Chamfer, Chamfer Entities . . .*

51. *Set the value for the chamfer to .05 by pressing the number 1 on the keyboard and entering .05 and pressing Enter.* (The default value for distances is .25, change as needed.)

52. *Pick the necessary lines as shown in Figure 21 and press Enter to accept.*

53. *Press the number 1 on the keyboard again to change the chamfer distance setting to .15 for the last chamfer.*

54. *Pick the necessary lines as shown in Figure 22 and press Enter to accept.*

The construction line that was used for the creation of the .25 arc, tangent to the angular line, should be deleted now.

55. *From the Menu Bar, select $\underline{E}$dit, Delete, Delete $\underline{e}$ntities . . .*

56. *Use the mouse to position the cursor over the construction line, press the left mouse button to select it and then press Enter.* The line will be deleted. The drawing is complete and the file should be saved.

57. *From the Menu Bar select $\underline{F}$ile, Save $\underline{A}$s*

The windows dialogue box titled "Save As" will appear.

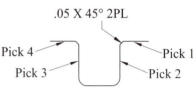

.05 X 45° 2PL

Pick 4

Pick 3

Pick 1

Pick 2

Figure 21
Selection Points for Chamfering

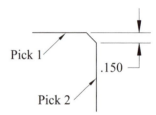

Pick 1

.150

Pick 2

Figure 22
Selection Points for Chamfering
the .15 Distance

58. *Key in the File Name desired and Press the Save button. For this example, we have chosen the File Name of Contour.*

The completed drawing should be displayed in the graphics area as shown in Figure 23.

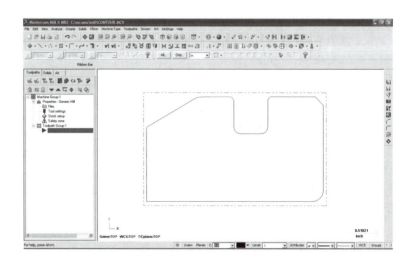

Figure 23
Finished Geometry

TOOL PATH

Once the part geometry has been created, then tool paths can be generated from it. Choose the type path that applies to the geometry. For this example we will use contour.

1. ***From the Menu Bar   select Toolpaths, Contour Toolpaths ...*** At this point, a dialog box will appear. Enter new NC name. The name CONTOUR will be present since we named the file earlier when we saved it. You may leave it as is or change it to whatever is desired. ***Press the checkmark to accept the entry.***

2. In the Operator Prompt area, the prompt reads "Select Contour chain 1".

Note: The default setting is Chain, so the contour may be picked without first activating the chaining mode. There are many options available for setting chaining options via the Chaining dialog. Take a moment to review each setting and refer to the online help for detailed descriptions.

3. ***Left click on the vertical line geometry just above the Workpiece Zero location.***

This selection determines where the actual tool path will begin, so take care to select an appropriate point. An arrow on the drawing contour will be displayed indicating the direction of the chain. The chain direction determines the tool travel direction. Tool offset direction is always dependant on chaining. For instance, if the arrow is pointing up, as shown in Figure 24, the tool will travel in the *Y* positive direction on the part geometry and to the left of the line.

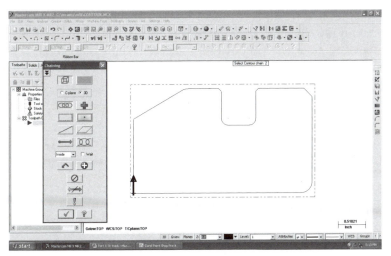

Figure 24 Contour Toolpath Chain

If the arrow direction is not correct, press **Reverse** on the Chaining dialog to change the direction. The offset direction from the contour determines climb or conventional cutting. Press **Reverse** again and the direction toggles. If the starting location is not the desired location, press the Expand/contract dialog button (down arrow in the upper left corner of the dialog), to expose the Start buttons (for "Move start of chain forward or Move start of chain back") until the desired starting position for the tool start is found.

4. *Press the green check mark to accept the toolpath chain.*

The dialog box titled, "Contour (2D) will appear as shown in below in Figure 25.

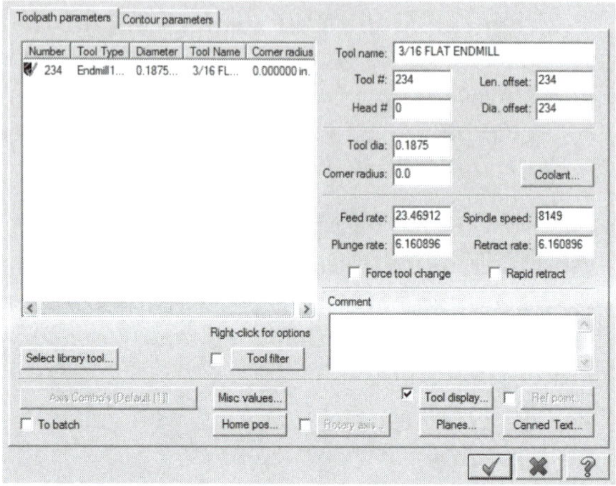

Figure 25
Contour (2D) Toolpath Dialog

Notice that the .187 diameter flat end mill that was identified earlier from the Tool Manager is present. If additional tools are required, they can be added from this dialog by following these directions: Press the "Select library tool . . ." button, double-click on the desired tool to select it and add it to the active list. Filters may be used to sort for specific tools by using the Filter button and Tool List Filter. Once a tool is added to the Toolpath parameters list, it can be edited by right clicking on it and selecting "Edit tool . . ." from the menu list. By right clicking in the open area of the tool list, you can select "Create new tool . . ." to create an entirely new tool. You can also adjust the feeds and speeds parameters for a selected tool by right clicking in the open area of the tool list and selecting the "Feed speed calculator . . ."

CONTOUR (2D)

The contour (2D) window has two tabs: Tool parameters and Contour parameters. The default tab is Tool parameters.

Tool Parameters

The Tool Parameters tab has several other items that can be adjusted by the programmer.

The Tool #: defaults to the number that matches the library tool number (234) but it can be changed to any number desired.

Head #: refers to the magazine where the tool is mounted in the case of multiple tool magazines.

Length offset: defaults to correspond with the number selected from the tool library on the first selected tool and sequentially after that. The length-offset number will normally correspond with the tool number but any number may be used.

Dia. Offset: defaults to match the tool selected from the tool library but it also can be changed to any number, dependant upon the available pockets in the tool changer of the machine being programmed. For instance, on some older machines, if there are 30 pockets for tools the logical diameter offset number for tool number 1 would be 31. With the common use of geometry offsets today, the tool diameter offset will have the same number as the tool and the value of the radius will be entered in the appropriate column of the offset register.

Tool dia: defaults to the diameter of the selected tool.

Corner radius: allows the input of the corner radius of a bull or ball nose style cutter.

Coolant allows for the use of flood coolant, mist coolant (if available), through the tool coolant or setting it to OFF.

Feed rate: is the linear feedrate at which the tool will travel and is determined by the tool and workpiece material selected. This feed can be modified in this field.

Plunge rate: is the linear Z axis feedrate at which the tool will travel determined by the tool and workpiece material that is selected. This feed can be modified in this field.

Spindle speed: allows setting of the r/min desired for each tool.

Retract rate: is the speed at which the tool will retract to the reference plane at completion of the cutting cycle. This feed can be modified in this field.

Comment: this section allows for the input of program comments that are intended to aid the operator and will be inserted within the program for operator guidance.

Misc values . . . can be used to set up to ten integer values. The most commonly adjusted integers are; Work coordinates, (G92 or G54's), Absolute/Incremental (0=ABS) and Reference return, (G28 or G30).

Home pos . . . allows for setting of a specific tool change position.

Rotary axis . . . when selected, these parameters are used to create toolpath motion where a rotating axis is used for cylindrical parts.

Tool display . . . these parameters are used to manipulate how the toolpath appears in the graphics window during backplot

Planes . . . opens the Work Coordinate System, Tool Plane and Comp/Construction Plane dialog box. By using this box, the planes for construction and machining, origins, and work offset for the toolpath can be set.

Ref point . . . when selected is used to adjust Approach and Retract tool path intermediate points.

Canned Text . . . allows addition of program comments from a list of some commonly used text statements.

Contour Parameters

Press the Contour parameters tab to open the dialog shown in Figure 26.

Input the appropriate values for Clearance . . ., Retract . . ., Feed plane . . ., Top-of-stock . . . and Depth . . . as required. In this example, *Clearance is set to 1-inch, Feed plane is set at .1 inch in Absolute, Top-of-stock is set to 0 in Absolute and Depth is set to -.25 in Absolute.*

When checked, the first button Multi Passes . . . enables roughing and finishing

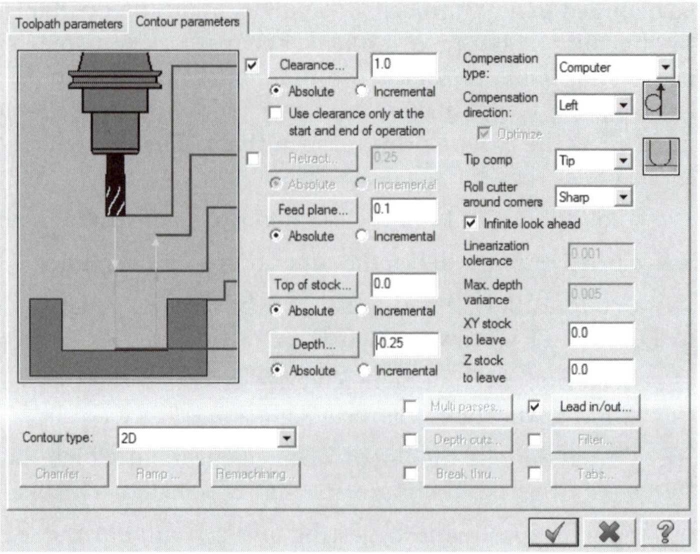

Figure 26
Contour Parameters Dialog

passes in the XY direction with control of finish passes at final depth or all depths.

The button labeled, Depth cuts . . . enables multiple passes for the Z step amount. A maximum roughing step can be established, the number of finish cuts, the finish step and the amount of stock to leave for finishing cut can be set.

The Lead in/out button enables toolpath entry and exit lines and arcs into the cut relative to the direction of the chain mentioned earlier.

The selection box Compensation type allows selection of Computer, Control, Wear or Reverse Wear. If Control is selected, the post processor will develop a diameter compensation number specific to the tool and insert function G41 or G42 and a D# into the code of the program.

The selection box Compensation direction determines the offset direction of the cutter in the computer (be sure the correct direction is indicated, for this example, it should be left).

- *For our example select Computer.*
- *Press the check mark to accept the settings.*

The resulting tool path will be displayed in the graphics window resembling Fig. 27.

- *From the Operations Manager pane, select the backplot button* (at the middle of the button bar), as shown in Figure 28. Be sure that the Contour (2D) folder has a green check mark on it before proceeding. If it does not have a green check mark on the folder, choose "Select all operations" from the button bar and then select "Regenerate all selected operations"

- *Press Back Plot, and then press Play (R) from the player buttons,* to observe a simulation of the cutters actual path. There are multiple setting options available within the Backplot dialog for the display of the type of tool path, the tool holder and more (See Figure 29). It is also helpful to set the view to isometric for better visualization.

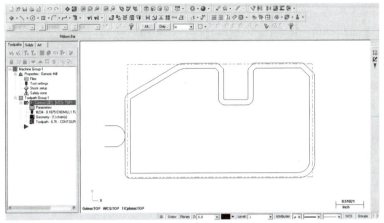

Figure 27 Tool Path

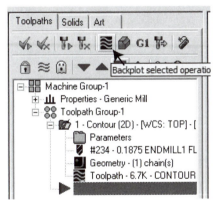

Figure 28 Backplot

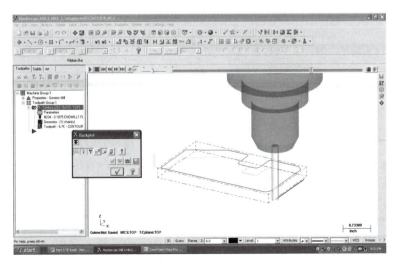

Figure 29 Completed Backplot

VERIFICATION

Now it is possible to display a solid model style representation of the finished tool path.

- *Press the "Verify selected operations" button from the button bar* for a solid model simulation of the part being cut.

- *Press the Machine button* (arrow pointing right) at the top of the Verify dialog to play the simulation. The speed with which the simulation runs can be controlled by adjusting the slider or pressing the Machine slowly or Machine quickly buttons. The resultant toolpath is simulated and should look similar to Figure 30.

Figure 30 Verify Toolpath Simulation

After viewing the Backplot and Verify simulations, you can still make any needed adjustments to the toolpath by selecting the Parameters folder for the Toolpath Group-1 and making the changes. Be sure to select and regenerate any operations you have adjusted before Backplotting or Verifying again. At this point, if all looks well, *Press the x to close the floating Verify dialog.*

POST PROCESSING

In order to actually machine the toolpath contour we have generated, we must first post process it into machine usable CNC code. This step converts the tool path just created into CNC code that the machine tool can read in order to machine the part.

- *Press the "Post selected operations" button (labeled G1).*

- *Press the green check mark Post Processing dialog to post the file.*

- When the Save As dialog appears, *identify the program number to save the file as.* The default file name is the same as the tool path file name, if a different file name is desired it should be changed now. The machine tool requires a program number, so now is a good time to establish the program number. *Key in 2255 Enter.*

• **Press Save,** and in a matter of seconds, the program will be post processed and will open the program file editor as shown below in Figure 31.

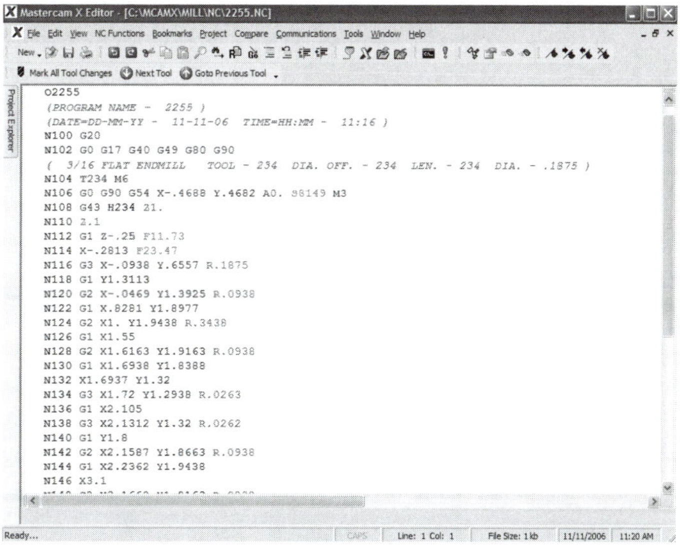

Figure 31
Post Processed File

There post processors specifically designed for each different type of machine controller. The default controller type is a Fanuc style G-code and this is what we have used here.

• *Close this screen by pressing the close window button or choosing file, then Close, then press the close window button (x).*

• *To close the Mastercam X^2, select File from the Main menu bar, then Save.*

• Once the file is post processed and saved, it is ready to be sent to a specific machine controller. Check the manufacturer's manual for specific directions for this procedure.

After the completion of all the items mentioned above, the programmer can print out a set-up sheet by right clicking within the Toolpath manager and selecting Setup sheet. The style of this setup sheet can be changed by selecting options through the Settings, Configuration, Setup sheet program drop down menu. This is a form of automated process planning and this document can be used by CNC setup persons to aid in the machine setup.

ASSOCIATIVITY

The concept of associativity is inherent to most modern CAD/CAM programs. The information input to the program regarding the tool path, tool, material, and parameters specific to each, are linked to the geometry. This means that if any of the parameters for the parts mentioned above are changed, the other related data can be regenerated, to take these changes into account without recreating the entire operation.

CAD/CAM is the tool of choice for creating CNC programs and the power it has now will only be magnified in the future. The basic concepts that were demonstrated here are merely a taste of the capabilities CAD/CAM has to offer.

PART 5 STUDY QUESTIONS

1. The acronym CAD/CAM stands for Computer Aided Design and Computer Aided Manufacturing. T or F.

2. Toolbars can be docked, undocked, docked vertically, docked horizontally or be turned off completely. T or F.

3. The Machine Group is displayed in the
>   a. Graphic display area.
>   b. Status Bar.
>   c. Verify Dialog.
>   d. Operations Manager.

4. What is a fillet?

5. How is the fillet command accessed within Mastercam?

6. What is a chamfer?

7. How is the chamfer command accessed within Mastercam?

8. Geometry creation consists of
>   a. CAD.
>   b. Lines, Arcs, Points, etc.
>   c. Both a. and b.
>   d. Toolpath creation and Post Processing.

9. Where do you establish the stock size for your part drawing?
>   a. From the Machine Type menu item.
>   b. Toolpath.
>   c. Machine Group Properties – Stock Setup.
>   d. Operations Manager.

Part 5 CAD/CAM

10. Parameters specific to the toolpath can be adjusted by accessing the

      a. Contour (2D) dialog.   c. Tool Parameters.

      b. Operations Manager.   d. XForm dialog.

11. The general steps to creating a Mastercam program for a CNC Machine are: selection of the Machine Type, Stock Setup, Design, Backplot, Verification and Posting.
      T or F.

12. To create a line by inputting the coordinates for its endpoints, how do you enter the "Fast Point" mode?

      a. Press the space bar.

      b. Mouse click in the coordinate fields in the Auto Cursor Ribbon bar.

      c. Press the letter X, Y or Z and enter the values for each separated by a comma.

      d. All of the above are methods.

13. What is the function of the AutoCursor?

14. Describe the meaning of Compensation in computer when referring to cutter compensation?

15. As part geometry is created, lines sometimes cross over each other forming an intersection and extend beyond. How do you remove these line extensions to make a clean corner?

      a. Fillet set to zero radius.   c. Delete.

      b. Chamfer.   d. Trim/Break

16. When chaining part geometry, the direction the chaining arrow points determines whether the tool will travel on the left or right of the selected line.
      T or F.

17. A line can be created from polar coordinates if the _____ function button is selected.

      a. <u>A</u>ngle   c. <u>T</u>angent

      b. Sketcher   d. <u>M</u>ulti-line

18. The Most Recently Used (MRU) toolbar stores the last several commands that you have accessed. You can use these buttons to reactivate any command on the MRU button bar.
      T or F.

PART 6

COMPUTER AIDED MANUFACTURING FROM SOLID MODELS

OBJECTIVES:

1. Become familiar with EdgeCAM CAD/CAM capabilities.

2. Become familiar with the EdgeCAM User Interface.

3. Learn terminology specific to Solid Modeling for CAD/CAM.

4. Learn how to use Solid Model files to create tool path and CNC program code.

5. Create CNC Machine Setup instructions from a CAD/CAM program.

SOLID MODELING BASICS

Solid Modeling of machined parts is becoming the method of choice for design to manufacture. It has great power and flexibility to aid the implementation of the model to metal concept. The designer has freedom to modify designs with ease and all related dimensional characteristics will automatically be updated to include the changes. This carries over to the CAM program as well. Once the adjusted file is opened again, after any changes, the program recognizes that changes have been made to the original and asks the user if the existing file should be updated to reflect these changes. This speeds the process when changes are necessary and that is nearly always. The software for Solid Modeling is very user friendly and machinists, programmers and design engineers learn to use its power fairly quickly.

In this section, we will examine a popular CAM program. With EdgeCAM Solid Machinist users can insert, open or drag-and-drop native files from other Solid Modeling software such as: Solid Works, Autodesk Inventor, etc. without the need for translators. Seamless and reliable integration means less chance for errors, making it an excellent choice for yielding quick productivity gains.

Solid Models are Feature Based Parametric designs. A feature is any entity of a part that has volume or area. Some basic features are: holes, bosses, pockets, profiles and faces. Parametric means that the entity has parameters that define its shape and size. For instance, a prismatic part may have a length of 3.0 inches, a width of 2.0 inches and a height of 1.0 inches. These values are parameters that can be adjusted resulting in a similar prismatic part.

ASSOCIATIVITY

As mentioned in the prior chapter, associative capabilities are imperative to effective CAD/CAM programs. The information input to the program regarding the part geometry, tool path, tool, material, and parameters specific to each, are linked to the geometry. This means that if any of the parameters for the parts mentioned above are changed, the other related data can be regenerated, to take these changes into account, without recreating the entire operation including tool paths. When changes are made, the software detects them and notifies the user that the model has been updated. The user is then asked to update the file to have the software reload the file in order to accept the changes.

Notes: Please note that CPL's are not associative. If changes are made to a model, the CPL will not update and will have to be reestablished.

Part 6 Computer Aided Manufacturing From Solid Models

In order to include threaded-hole data, EdgeCAM must be launched from the design package using the EdgeCAM CAD Links Option.

It is not possible to open Solid Model files in EdgeCAM without having the Solid Machinist module.

THE EDGECAM USER INTERFACE

When using any software, the first thing users have to get accustomed to is the User Interface (UI). This is the driver's seat from which you will create your product designs, manipulate the Solid Model files, create toolpath and post process CNC code for manufacture of the part. The more comfortable you are behind the wheel and the more you know where all of the controls are, the better you will drive.

Note: The EdgeCAM software is available on demonstration discs through your local reseller and can be used in the Student Edition mode to practice. Everything can be completed just as is shown here except Post Processing.

Some brief descriptions are given below for the Design and Manufacture mode User Interfaces.

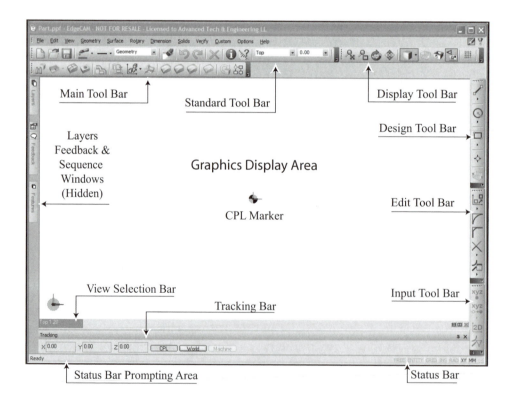

Figure 1 EdgeCAM User Interface

Design Mode

Menu

File   Edit   View   Geometry   Surface   Rotary   Dimension   Solids   Verify   Custom   Options   Help

Figure 2 EdgeCAM Menu Bar

Just below the title bar is the menu bar. It contains access to complete function of the software, if desired. It contains File, Edit, View, Geometry, Surface, Rotary, Dimensions, Solids, Verify, Custom, Options and Help. The menu items can be accessed by selecting the desired item with a left mouse click or by using a keystroke by first pressing the alt key on the keypad, and then pressing the Underscored letter of the menu item. It is a good idea to become familiar with any shortcut keys that aid in speeding the programming process. Select Help from the menu and key in Shortcut keys and press Enter for an index search of the EdgeCAM User Guide contents. Click on fixed assignments for a list of all of the shortcut keys. In the left hand corner of the Menu bar the Design and Manufacture icons are given. To access Manufacture mode, press the icon of the cutting tool to switch to the Manufacture mode.

Graphics Display Area

The graphics display area is the large open area at the center of the interface where the part model is displayed when created, imported or opened. The Coordinate Position Locator (CPL) is located in the center of the window. By default, the color is blue with a gradient, but it can be changed to any color desired by selecting Options from the menu and then Colours from the drop down menu to access the Color Configuration dialog box. White is used in our examples.

Figure 3 CPL

Coordinate
Position Locator

Layers, Feedback, Properties and Features Windows

These windows are typically docked on the left hand side of the screen and give the user feedback information that is used during design and manufacturing phases of programming. In Figure 4, the Windows are shown in the Auto Hide mode. The windows open up when the mouse selection is over the desired window and they close when positioned into the graphics area allowing the greatest amount of display area. The windows can be pinned open by pressing the thumb-tack in the upper right hand corner of the window. Additional windows may be accessed by right-clicking while over the title of any one of the windows and choosing Windows.

Figure 4

Windows

Tool Bars

The tool bars shown in Figure 5 are: Standard, Edit, Solids and Display. They can be docked horizontally, as shown, floating or vertically. There are other tool bars, not shown here, that are commonly active and docked vertically. They are: Design and Input. Tool bars can be added by right-clicking anywhere on a tool bar and selecting the desired Tool Bars from the menu.

Figure 5 Tool Bars

Status Bar

| Start point of line | FREE ENTITY | GRID | INS | RAD | XY | MM |

Figure 6 Status Bar

The Status bar is located at the bottom of the display area. On the left hand side of the bar the user is given interactive prompts during an active command. On the right hand side information is given about the machining environment, (XY or ZX) and the measurement system (IN or MM), etc.

Viewport Selection Band

The viewport selection Band is located in the lower left hand corner of the display just below the World Coordinate System (WCS) indicator and above the Tracking bar. Right click on the bar to select one of the seven standard views, make a user defined view or adjust view Properties.

Tracking Toolba

| Tracking | | | | | | 中 × |
| X -67.08 | Y -107.29 | Z 0.00 | CPL | World | Machine |

Figure 7 Tracking Toolbar

The Tracking tool bar gives live coordinate feedback regarding the cursor location while in geometry creation mode. The toolbar can be pinned open (as shown in Figure 7) or set to the Auto Hide mode. By pressing the CPL, World or Machine buttons the coordinate feedback will be relative to the system selected.

MANUFACTURE MODE

Menu

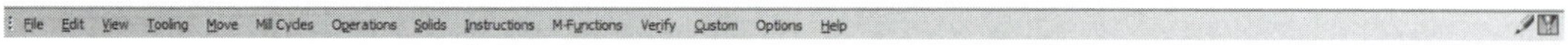

Figure 8 Manufacture Mode Menu

The Manufacture Mode menu is practically identical to the Design Mode except that the emphasis here is on manufacture. It contains File, Edit, Tooling, Mill Cycles, Operations, Solids, Instructions, M-Functions, Verify, Custom, Options and Help. The keystroke shortcuts described before are effective here, as well.

Tool Bars

Figure 9 Manufacture Mode Toolbars

The tool bars for the Standard and Display Tool bars from the Design mode are the same for the Manufacture mode. The other Tool bars that are specific to the Manufacture mode are: Main, Operations, Rotary, 5-Axis Simultaneous (shown in Figure 9). They can be manipulated in the display area in the same fashion as described before. There are other tool bars, not shown here, that are commonly active and docked vertically. They are: Mill Cycles and Input.

Layers, Features, Sequence, Machine Tree, Fixtures, Properties, Feedback and Preview Windows

Once again, these windows are typically docked on the left hand side of the screen and give the user feedback information that is used during design and manufacturing modes of programming. They are manipulated in the same fashion as described before in the Design mode. Additionally, Sequence, Machine Tree, Fixtures, and Preview are windows specific to the Manufacture mode.

Simulation Bar

Figure 10 Simulation Bar

The Simulation bar allows for the cutting Toolpath cycles to be graphically checked from within the software by pressing the play button in the upper right-hand corner of the bar. The speed of the simulation can be controlled by moving the bottom slider along the horizontal bar. The user can run the simulation forward and backward as needed to pinpoint any possible issues by moving the upper slider from right to left or left to right respectively.

The following are the general steps involved in creating CNC code by using a solid Model in EdgeCAM: Starting the EdgeCAM program; Establish the Environment; Identify Component Material; Open a Solid Model File; Establishing a CPL; Create Stock; Feature Finder; Identify ToolSore Database or Toolkit; Execute Machining Simulation; and, Generating the CNC Code.

CONVENTIONS

In the example that follows, when the Menu selection method is used, the Menu item used will be **Bolded** in the instructions given and the short-cut keystroke Underscored to match the EdgeCAM menu bar. When Icon selection method is used, the Icon graphic and directions will be included in the instructions. ***Specific instructions will in bold italics*** and steps will be numbered.

EDGECAM PROGRAM STARTUP

From the Windows main screen, look at the desktop to see if there is a shortcut icon and ***double click the left mouse button on the EdgeCAM icon*** (Figure 11). If there is not a desktop shortcut icon, ***press the Windows Start button in the lower left corner with the left mouse button. Slide the mouse pointer up to the right and you will see a list of all available programs. Slide the mouse pointer to find EdgeCAM and another list will appear to the right. Again, slide the mouse pointer through the list to find the EdgeCAM program.***

Figure 11
EdgeCAM
Shortcut Icon

Establish the Environment

By default, EdgeCAM starts in the Design mode. Confirm that the environment is in the XY mode for milling by observing the lower right-hand corner of the Status Bar. If it is not in the X̲Y mode follow these directions: *Select* **Options** *from the menu and then* **XY Environment** *from the drop down menu. Also confirm that the measurement system is set to the inch mode by observing the lower right hand corner of the status bar. If it is not in the inch (IN) mode follow these directions:* **Select** **Options** *from the menu and then* P̲**references** *from the drop down menu. From the General tab of the Preferences dialog, select the Inch radio button and press OK.*

OPENING A SOLID MODEL FILE IN EDGECAM

When a new design is used, one of the first things that must be done is similar to that in machining setups. After a file is imported or opened, the orientation of the part needs to be established in relation to the machining required in order to begin the machining operations.

Coordinate Position Locator (CPL)

The Coordinate Position Locator is the same thing as the Workpiece Zero location for machining of the part. When a file is created in Solid Modeling program, there is no guarantee that the designer used a location system that will relate to the machining operations so it must be established now.

Establishing a CPL

Using "Align Body for Milling" is the simplest method of establishing the correct orientation of the CPL. Follow these basic steps to accomplish the alignment:

1. *Pick Align Body for Milling from the Solids Menu and follow the status bar command prompts.*

2. Select face to define XY plane (or return). If the Z axis marker is pointed in the wrong direction, pick the same face again and it will be reversed.

3. *Right click to accept the selection and follow the status bar command prompts.*

4. Select linear edge to define CPL plane Axis (or return). *Using the mouse select the edge that represents the X axis in relation to how you intend to machine the part.*

5. Select linear edge to define CPL plane Axis (or return). *Using the mouse again, select the edge that represents the Y axis in relation to how you intend to machine the part. When the X and Y orientation is correct, right click to accept.*

6. Select point to translate to origin (or return). *Use the mouse to left click on the desired origin location.* In our case, it is the center of the 2.0 diameter pocket. This will position the CPL in alignment with the Workpiece Zero location for machining. Finally, right click with the mouse to accept the CPL location. For our model, the display should look like Figure 12 on the following page :

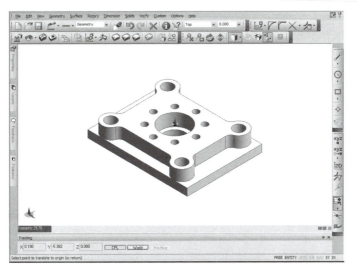

Figure 12 Align Body for Milling

Automatic Stock Creation

7. ***From the Menu bar, select Geometry and then select Stock/Fixture from the menu choices. The stock dialog box will be displayed. Put a check mark in the Automatic Stock check box and input zero for the remaining entries and press O.K.*** A wire frame will be drawn around the model that represents the stock. You can press the Toggle Stock button in the Display tool bar to get a better visualization of the stock.

USING THE FEATURE FINDER

The Automatic Feature Recognition (AFR) function has the ability to recognize features that are common to milling operations, such as: Pockets, Open Pockets, Bosses, Holes, Faces and Profiles.

Activate the Feature Finder by pressing the tool bar button as shown in Figure 13, or by selecting **Solids** from the main menu and then **Feature Finder** from the drop down menu. The General tab of the Feature Dialog will be displayed. Complete the entries as follows to match Figure 14:

Figure 13
Feature
Finder

Feature Finder dialog descriptions:

Highest Wall Level: By checking this item the pocket walls will be extended to the highest point. This is especially useful if there are multiple pockets within the same surface.

Automatically Name: By checking this item, the features that are found are automatically named and displayed in the Features Window. For instance in our example they are named: Through Hole; 2D Boss; 2D Pocket and Open Pocket.

Holes, Find: When this item is checked, any hole-type features are recognized.

Figure 14
Feature Finder Dialog

363

Holes, Maximum Diameter: The value entered into this field controls whether the feature is considered a hole to be drilled or a pocket. Those equal to or below are considered to be a hole and those above are considered to be a pocket, thereby determining the machining strategy.

Holes, Threading Information: If this item is checked, and the threading information identified in the model is passed through to the Manufacture mode. ***Note: This function is only available via the CAD Link utility offered through EdgeCAM.***

Holes, Group Similar Holes: When this item is checked, all holes with similar attributes are selected together as a group. In our example, the bolt-circle of holes qualify as one group and the rectangular pattern as another group.

Mill, 2D Pocket: When this item is checked, the Feature Finder will recognize two-dimensional pockets.

Mill, Open Pockets: If this item is checked, pockets that are not fully closed are found. This allows the toolpath to approach through the existing opening.

Mill, 2D Boss: When this item is checked, all two-dimensional Bosses will be recognized.

Nesting, Single: If this item is check marked, to Single, the pocket cannot have a boss within it.

Note: The descriptions given are only those related to the example part. For more details, press the Help button within the dialog.

Immediately after the function is completed the Feedback Window will be displayed as shown in Figure 15. Open the Features Window, if it is not displayed, and notice all of the Features that were found. As you move the mouse selection over each item in the list the Feature on the drawing is highlighted in the graphics display area.

Figure 15
Feature Finder Feedback Window

Note: Item 3 has been deleted in our example because it is not needed for the desired toolpath.

This concludes the preparation of the model for machining and the next step is to enter the Manufacturing Mode.

MANUFACTURE MODE

Enter the Manufacture Mode by pressing the icon in the upper right-hand corner of the main menu bar. The Machining Sequence dialog will be displayed.

Set the items in the General tab as follows:

Sequence Name: Operation 1

Kit Name: EdgeCAM Example

Choose a Code Generator: Discipline, Mill; Machine Tool, fanuc3x-in.mcp

Figure 16
Feature Window

    Figure 17 Manufacture Mode Icon

Confirm that the Initial CPL is set to TOP and that the Datum Type is set to Absolute and press O.K.

Identify Component Material

Based on the selection here, the part material affects the accurate calculation of feeds and speeds and the software will even make suggestions for tool selection based on the chosen geometry to be machined. There are 6 tabs for filtering material selection and new materials can be input for use.

In order to have the software calculate the correct feeds and speeds, the component material must be identified. The tool material will be established later from within the ToolStore section. *Select Options from the menu and then Model from the drop down menu. Choose the Browse . . . button from the Model dialog* to display the Component Material library as shown in Figure 18. *Choose Aluminum-Wrought alloy aged and press the Select button.*

Figure 18
Materials Dialog

ToolStore

Tools can be selected from the standard cutting tool database or users can create, modify or import tooling data while in the manufacture mode. Toolkits can be assembled that are specific to Jobs, Materials or Machines. The database is editable and allows access to third-party tooling vendor product data giving it even more power. The Technology Assistant is another programming tool that uses the cutting tool data and the part material data to automatically calculate feeds and speeds. It can even be set up to identify which cutting inserts are appropriate for what materials. All of the ToolStore and Toolkit information is linked to the Technology Assistant. The idea here is to incorporate the maximum amount of technology into the building of the CNC program saving time at the machine and cutting quality parts on the first run. The next step is communicating this information to the machinist on the shop floor. With

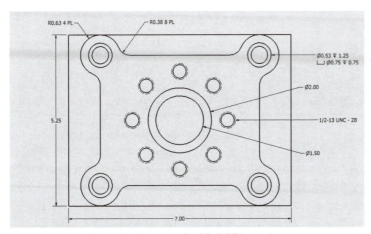

Figure 19 Drawing for Model Dimensions

intranet-based Job Reports, the tooling and specific setup information, including images, can be viewed on any computer with an internet browser leading to the ultimate goal of a "paperless" factory. This concept will be demonstrated later in this section.

We will build a Toolkit specifically for the example given here and label it: "Edgecam Example" job. The tools used in our example are as follows:

Tool 1 = 1/2 inch diameter, 2-flute, Center Cut End Mill – KC720. This tool will be used for the 1.5 and 2.0 inch diameter pockets and rough and finish of the outer contour.

Tool 2 = 1/2 inch diameter, 90° Spot Drill – KC720. For all of the holes.

Tool 3 = 3/4 inch diameter, 2-flute, Center Cut End Mill – KC720. This tool will be used to finish the Counterbore.

Tool 4 = 17/32 inch diameter HSS Jobber Drill. For the counterbored through hole.

Tool 5 = 27/64 inch diameter HSS Jobber Drill. For the 1/2-13 UNC threaded holes.

Tool 6 = 1/2-13 UNC Tap. For the threaded holes.

Notes: The depth of the contour step is 3/4 inch.

The Toolkit could be developed prior to and independent of the EdgeCAM session. Many shops create Tooling Libraries and Toolkits specific to their application. EdgeCAM comes with a standard Tool Library and Tool Kits that are a starting point.

To build the Toolkit for our job, follow these directions:

1. ***Open the Toolkit Assistant by pressing the Windows start button, select All Programs, EdgeCAM and then Toolkit Assistant.***

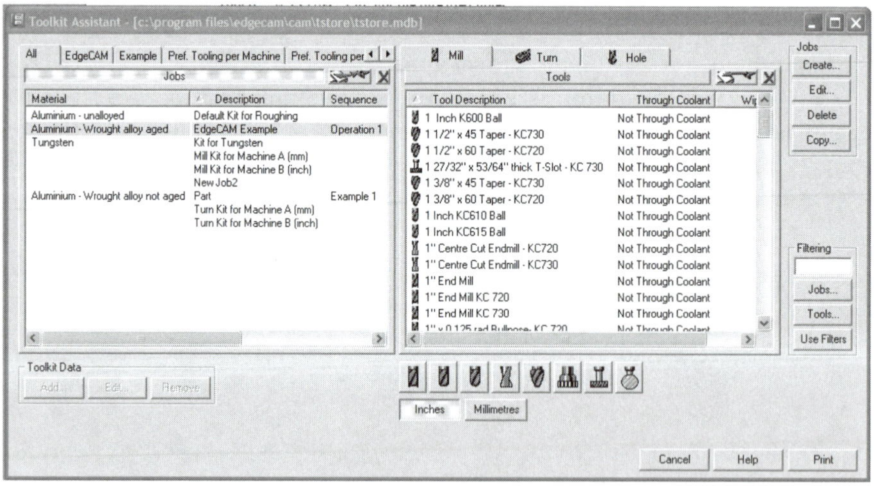

Figure 20 Toolkit Assistant

2. ***Press the Create button in the Jobs section of the Toolkit Assistant dia log (upper right hand corner Figure 20).***

A dialog will appear similar to Figure 21, with a default Description of New Job 1, etc. Enter the following specific data: ***Description: EdgeCAM Example; Family: EdgeCAM; Sequence: Operation 1; Programmer: Your Name; Machine Tool: Press the browse button and select*** *the fanuc3x-in.mcp; Material: Press the Browse button and select Aluminum – Wrought alloy aged;*

Note: Links to the CAD File, CAM File and the NC File can be identified by pressing the browse button to set the locations.

3. When the information has been completed, ***Press OK button,*** and the new job will be listed under the Jobs section of the Toolkit Assistant dialog (Figure 20).

4. ***Left click on newly created EdgeCAM Example job in the Jobs list to highlight it.***

5. ***Select the 1/2 inch diameter, Center Cut End Mill – KC720 from the Tool Descriptions list of the Mill tab and press the Toolkit Data Add button.***

6. ***Complete the following data in the resulting dialog: Turret Position 1; Tool Offset 1 and Radius Offset 1, and press OK.*** This will add the tool to the tooling associated with the new EdgeCAM Example Job you have created.

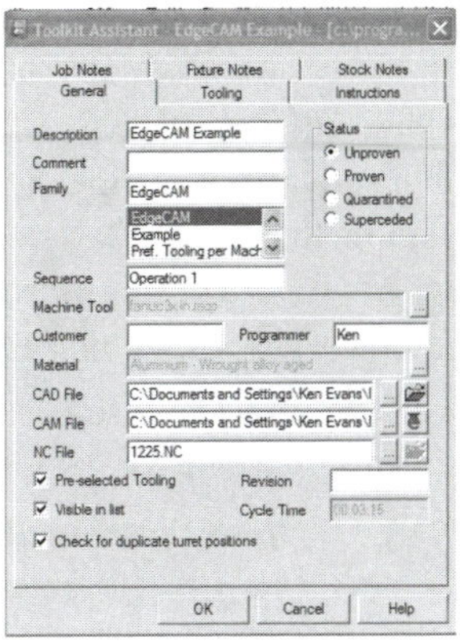

Figure 21
Toolkit Assistant Dialog

Repeat steps 4 through 6 for each of the four remaining tools in our list after changing to the Hole tab of the Tool Description list.

7. ***Select the 1/2 inch diameter, 90° Spot Drill – KC720 and set it to Tool Offset 2.***

8. ***Select the 1/2-13 UNC Tap and set it to Tool Offset 6.***

It will now be necessary to create the two remaining drills that are not in the existing list, as new Hole tools. Follow these steps:

9. ***Ensure that the Hole tab is active.***

10. ***Left click on the existing 3/4 inch Jobber Drill in the Tool Description list.***

11. ***Press the Copy button from the Tools section of the Toolkit Assistant dialog.***

12. ***Change the Tool Description from the General tab of the Toolkit Assistant- New Tool dialog to: 17/32 inch HSS Jobber Drill, Turret Position 4; Tool Offset 4 and press the OK button.***

13. ***Repeat steps 10-12 to enter the remaining 27/64 inch HSS Jobber Drill and set it to: Turret Position 5 and Tool Offset 5. Press OK to close the New Tool dialog.***

This concludes the Toolkit creation for our EdgeCAM Example job. It is now ready for access in the CAM session we will now continue. We identified the Toolkit Name when we first entered the Manufacture Mode.

GENERATING MACHINING TOOL PATHS
Operations vs. Cycles

There are two ways to input the information required for completing the CAM session to setting up machining actions. For beginners, it is easiest to start with the semi-automated method of using the Operations tool icon. By doing this, the user is prompted via descriptive dialog windows

Part 6   Computer Aided Manufacturing From Solid Models

and an Audio Video Interleave (AVI) for tool and cutting data that is required, dependant upon the type of operation selected. There is a limited level of control to the Operations parameters when entering Operations Preferences via the dialog, after accessing it through the menu. Or, for more advanced parameter input and complete control, Cycles may be accessed via the menu or the Mill toolbar. In the later case, the user can control every detail of the machining process manually. Additionally, the user can control individual commands i.e.: tool calls, tool changes and individual cycles by using the buttons in the Main and Mill toolbars.

We are going to use Operations for this example. The following steps are one approach to machining the part:

1. ***Open the Features Window and pin it in the open condition by left clicking on the pin.***

2. ***Pick Operations, Roughing from the Menu Bar or use the Operations Toolbar x and select Roughing Operation Icon.*** *The Status Bar prompt will read "Digitise Geometry to Machine".*

3. ***Use the mouse to select the first 2D Pocket in the Features Window and double click on it*** (Feature 5 in the Features Window).

4. The Status Bar prompt will read *"Select boundary entities (Finish for none)".* ***Right click to finish.***

The Roughing Operation dialog as in Figure 22 will appear. Complete the entries as follows:

5. For the General tab: ***Mill Type: Climb; % Stepover: 75; Offset: 0; Z Offset: No entry; Tolerance: .001; Leave the Rest Rough and Digitise Stock boxes unchecked. When all settings are entered Press OK.***

6. For the Tooling tab: ***Change the Position setting to 1.*** This is our 1/2 inch 2-Flute Center Cut End Mill-KC720.

7. For the Depth tab: ***Leave the settings as they are and press OK.***

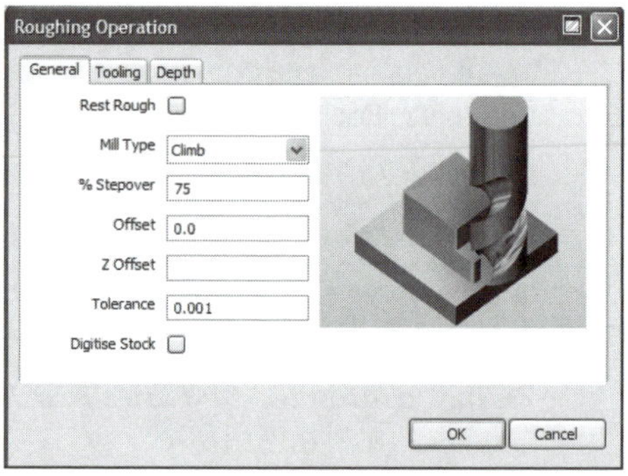

Figure 21 Rooughing Operation Dialog

The roughing toolpath will be generated for the 2.0 inch diameter pocket.

Note: We could set the Offset in the dialog above, to a value other than zero, to leave material for a Profile Operation finishing pass, if required.

8. ***Repeat steps 2-6 for the 1.5 inch diameter pocket*** (Feature 6 in the Features Window).

9. ***Use the mouse to select each of the four Open Pockets on the Geometry.*** (Features 7-10 in the Features Window).

It may be necessary to rotate the model to enable picking of the geometry around the part.

10. The Status Bar prompt will read *"Select boundary entities (Finish for none)."* *Right click to finish.*

11. *Set the Roughing Dialog entries the same as in step 5 above, except set the following value for the General tab: Offset: .015. When all settings are entered, Press OK.*

The roughing toolpath will be displayed for the four open pockets and it will leave .015 stock allowances for the finishing Profile pass that we will enter next.

12. *Pick Operations, Profiling from the Menu Bar or use the Operations Toolbar and select Profiling Operation Icon. The Status Bar prompt will read "Digitise Geometry to Machine".*

13. *Use the mouse to select the 2D Boss in the Features Window and double click on it* (Feature 4 in the Features Window).

14. *Set the Profiling Dialog entries the same as in step 5 above except for the fields on the General tab: Set the Offset back to 0; set the Compensation to Geometry (G41) and the Lead Radius to .3.*

The Profiling toolpath will be displayed for the entire outside contour of the part. All that remain are the hole making operations.

15. *Pick Operations, Holes from the Menu Bar or use the Operations Toolbar and select the Holes Operation Icon.* The Status Bar prompt will read *"Select Points".*

16. *Use the mouse to select any one of the four counterbored holes (left click) and right click to accept the selection.* The Hole Operation dialog will be displayed as in Figure 23.

17. Because the Solid Model and Toolkit are associated with each other, the information needed in the Hole Operation dialog is as follows:

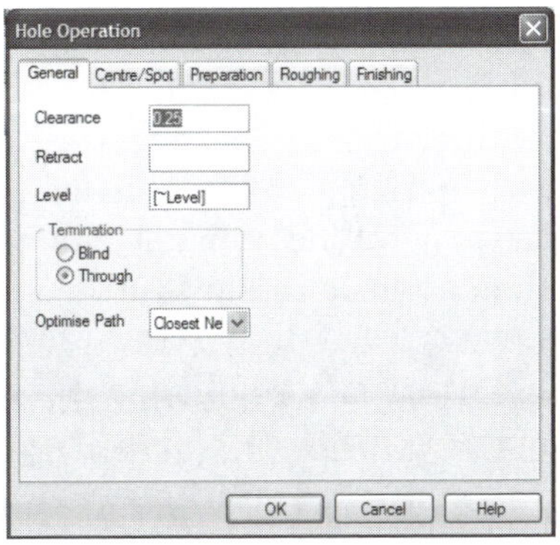

For Centre/Spot tab: *Set the Tooling Position to 2.*

For Preparation tab: *Set the Tooling Position to 4.*

For Roughing tab: *Set the Tooling Position to 3. Set the Roughing Strategy to Peck Drill and set the Cut Increment to .43.*

No information is required for the Finishing tab.

Figure 23 Hole Operation Dialog

18. *Press OK*

The counterbored Hole making toolpath will be displayed.

19. *Repeat steps 15 and 16 and select the points for the Bolt Circle of holes.*

For the Centre/Spot tab: *Set the Tooling Position to 2.*

For the Preparation tab: No information is required.

For the Roughing tab: *Set the Tooling Position to 5. Set the Roughing Strategy*

369

to Peck Drill and set the Cut Increment to .43.

No information is required for the Finishing tab.

20. *Press OK,* and the Hole making toolpath will be displayed

Note: The threaded hole information is only associative to the model if it has been opened through the EdgeCAM CadLink utility.

Follow these steps to complete the session to finish the Holes by tapping them:

21. *Repeat steps 15 and 16 and select the points for the Bolt Circle of holes.*

For the Centre/Spot tab: ***Set the Strategy to None.***

For the Preparation tab: ***Set the Strategy to No Preparation.***

For the Roughing tab: ***Set the Strategy to None***

For the Finishing tab: ***Set the Strategy to Tapping and Set the Tooling Position to 6.***

22. *Press OK,* and the Hole making toolpath will be displayed.

The toolpath is complete, and now it can be checked through simulation to ensure proper operation.

SIMULATION

Tool path simulation is the closest visual representation of the actual machining that must be completed prior to actual machining. The software will display a solid model of the part and the tooling is displayed to help the user visualize machining. Parameters can be set that warn the user if collisions have occurred allowing the corrections to be made before metal is cut saving time and expense. The user has the ability to zoom, pan, rotate and even cutout a section for viewing lathe work with internal features that are hard to see otherwise. During the simulation process, the user can capture an image of the finished part and save it as a Sterolithography (STL) file for later use in setup planning documents. The simulation mode even allows for recording of the simulation to be saved into an AVI file creating a very useful addition to aid machine setup or even in the sales process.

1. ***Select View and then Simulate Machining from the Menu or Press the Simulate Machining Icon on the main Tool bar.***

2. ***Press the Start button on the Main Toolbar to view the Simulation.***

The toolpath will be simulated and look similar to Figure 24. Throughout the entire process of completing any EdgeCAM project you can take screen shots that are useful for communicating setup and tooling information. These images will be included in the Job Report when accessed via an Intranet browser. To save the view of the finished Simulation:

3. ***Pick File, Save Job Images from the Menu Bar.***

4. ***Enter a Job Name in the Job Image dialog and Press OK.***

To save an AVI of the entire simulation, follow these steps:

5. *Open the Simulator and select the desired view from the Display Toolbar.*

6. *Press the "Capture frames to an AVI animation" button on the Main simulator Toolbar.*

7. *Press the Start button and allow the simulation to complete.*

The resultant AVI will be stored within the Job Manager and a link to the animation will be inserted into the Job Report for viewing from an Intranet browser.

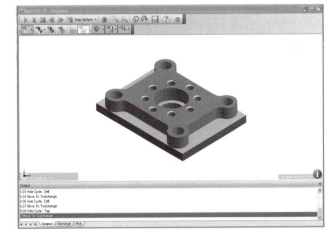

Figure 24   Simulation

OUTPUTTING CNC PROGRAM CODE

Once the Operations and Cycles have been determined and their parameters are set, the next step is Simulate machining. When all corrections have been completed and the simulation looks effective, then it is time to output the CNC code (POST Processing) for the Machine Tool to use. The software comes with a standard set of POST Processors that represent a cross section of the common machine and control types to choose from. Alternatively, users can use the Code Wizard to generate their own POST Processor that is specific to their machine application. See the documentation supplied with the software for more details on using the Code Wizard. To Post the file, follow these steps:

8. *Use the Return to EdgeCAM from the Simulator Main toolbar.*

9. *Press the Generate Code button on the Main toolbar.*

10. *Give the file a CNC Name in the Generate CNC Code dialog. In our case, 2255.*

11. *Job Name: EdgeCAM Example.*

12. *Operation Names and Open Editor checked.*

13. *Press OK.*

Figure 25
Generate
Code Icon

The posted CNC file is displayed and is ready to send to the CNC Machine Tool for setup and dry run. A hyperlink to the posted file is also entered into the Job Report and can be viewed by clicking on the link.

CREATING AN ELECTRONIC JOB SETUP SHEET

Now that the CAM session is complete, the next step is to view the Job Report that will be used to aid in the setup process. To access the Job Report, follow these steps:

1. *Open the Job Manager by using the Windows start button, select All Programs, EdgeCAM and then Job Manager.* The Job Manager dialog will be displayed similar to Figure 26.

2. *Right click anywhere in the Title bar of the dialog and select Job Reports, and then Purge. Allow the Purge to complete.*

3. *Right click again anywhere in the Title bar of the dialog and select Job Reports, and then select View.*

The first display will be the Welcome screen. It lists all of the Jobs in the Job Manager and has links to each.

4. *Press the EdgeCAM link and then select the EdgeCAM Example we have created.*

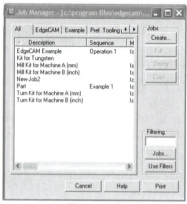

Figure 26
Job Manager Dialog

The information displayed will be what we have entered into the EdgeCAM Example process thus far. In Figure 27, you can see the links to the CAD, CAM and NC files. In Figure 28, you can see a partial tool list and some Job Image graphics we saved along the way, as well as the link to the AVI file we made during the simulation. This information is undoubtedly useful during the setup of the job.

Job Manager Reports
Wednesday, December 27, 2006

Welcome   EdgeCAM   **EdgeCAM Example**   *(Unproven)*

Comment	
Sequence	Operation 1
Cycle Time	00:02:19
Machine Tool	fanuc3x
Customer	
Programmer	Ken
Material	Aluminium - Wrought alloy not aged
Status	Unproven
Revision	
CAM File	Part 6 EC TP
CAD File	Part 6 EC .ipt
NC File	2255.NC
Pre-selected Tooling	Yes

- 3/4" Centre Cut Endmill - KC720
- 1/2-13 UNC
- 1/2" x 90 Spot Drill - KC720
- 1/2" Centre Cut Endmill - KC720

Finished Simulation   **Model**

Job Notes
AVI
Part%206%20EC%20TP_211206_18_38.avi

Fixture & Gauge Notes
Fixture

Figure 27
Job Report Page 1

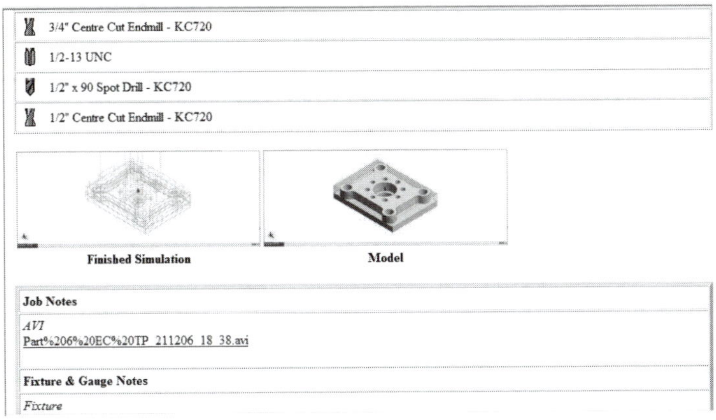

☒ 3/4" Centre Cut Endmill - KC720	
◍ 1/2-13 UNC	
▯ 1/2" x 90 Spot Drill - KC720	
☒ 1/2" Centre Cut Endmill - KC720	

Finished Simulation	**Model**

Job Notes
AVI
Part%206%20EC%20TP_211206_18_38.avi

Fixture & Gauge Notes
Fixture

Figure 28
Job Report Page 2

In our journey to learn about Programming of CNC Machines, we started with the basics of entering the cutting tool, feeds & speeds and setup information manually in Process Planning documents. The next step was to learn about CNC Operation procedures and specific Setup and Operation techniques. We then learned to manually program using G-Codes. Now, all of the programming steps are completed using CAD/CAM, the tool of choice for creating CNC programs. It is efficient and fairly easy to learn and the power it has now will only be magnified in the future. The basic concepts that were demonstrated here are merely a taste of the capabilities that CAD/CAM has to offer.

Happy Programming of CNC Machines!

PART 6 STUDY QUESTIONS

1. When changes have been made to a Solid Model design, they are recognized by EdgeCAM when the file is reopened.

>   T or F

2. Using the Auto Hide function for EdgeCAM Windows allows more display area in the design and manufacture modes.

>   T or F

3. The acronym CPL stands for:

>   a. Coordinate Point Location
>
>   b. Coordinate Position Locator
>
>   c. Coordinate Parameter Location
>
>   d. Configure Preferences Location

4. Which mode is used when developing cutting tool paths and CNC code?

>   a. Design Mode
>
>   b. Sketch Mode
>
>   c. Tool Path Mode
>
>   d. Manufacture Mode

5. What environment is required for a milling project?

>   a. XZ
>
>   b. XYZ
>
>   c. XY
>
>   d. All of the above

6. The acronym AFR refers to:

>   a. Automatic Feed Rates
>
>   b. Automatic Feature Recognition
>
>   c. Feature Finder
>
>   d. Both b and c

7. If the component material is identified incorrectly the feeds and speeds for machining will be correct based on the tool selection anyway.

>   T or F

8. Choosing the correct code generator is dependant upon the specific machine tool to be used.

        T or F

9. During simulation a button can be pressed to capture the movement in the simulation for later use as an aid to setup. What is this type of file called?

    a. AFR, Automatic Feature Recognition

    b. STL, Sterolithography

    c. DXF, Drawing Exchange Format

    d. AVI, Audio Video Interleave

10. Which tool path mode offers complete control of cutting parameters?

    a. Cycles

    b. Operations

    c. Manufacture

    d. Job Report

PART 7

MAZATROL CONVERSATIONAL PROGRAMMING

Figure 1 Mazatrol Control Panel

Courtesy Innovative Precision LLC

OBJECTIVES:

1. Learn about the advantages of conversational programming.
2. Become familiar with the Mazatrol conversational programming language.
3. Learn terminology specific to Mazatrol conversational programming language.
4. Become familiar with typical MAZAK controller screens and menus.
5. Demonstrate how to create a CNC Turning Center program using Mazatrol conversational programming language.
6. Demonstrate how to create a CNC Machining Center program using Mazatrol conversational programming language.
7. Become familiar with offline programming software used to create Mazatrol programs.

WHAT IS CONVERSATIONAL PROGRAMMING

For many years, the concept of programming the CNC machine tool at the controller was thought of as inefficient and tedious. When orders of a small lot size were to be produced, the choice was almost always manual machines. Today, this is not the case, largely because of the advances in conversational programming.

Conversational programming is becoming more widely used throughout the industry and is available as standard on many machine tool controllers. Its major advantage is that it gives the machinist the ability to write programs at the machine quickly and easily. Typically, the process includes a sequence of questions the machinist/programmer must answer, sometimes called "question answer format" or "prompting". As these questions are answered, the program is constructed. Most controls with this capability also allow the machinist/programmer to graphically check the tool path to verify the program. If the program has flaws or missing information, the controller will not execute the tool path and the programmer must remedy the problem. When program errors do occur, an alarm number will show on the screen indicating what the problem is and where, in the program, it occurred. This is obviously a better method of finding errors than actually cutting a part. Another capability of conversational programming is its feature to perform calculations when programming data are missing from the engineering drawing. The programmer constructs intersection points or tangency points points, and with this information, the controller computes the desired geometry. On some controllers, Feeds and Speeds can be calculated automatically based on the workpiece material and cutting tool material. The data necessary to do this is stored in the controller memory in the cutting condition parameters.

Conversational "shop floor" programming uses the concept of operator prompting combined with a Graphical User Interface (GUI). Questions throughout the programming process prompt the user for information necessary to complete the part program. Icons accessed through the function buttons on the controller identify machining operations, i.e. Point Machining (shown in Figure 2), Line Machining and Face Machining, etc.

Part 7 Mazatrol Conversational Programming

POINT MACH-ING	LINE MACH-ING	FACE MACH-ING	MANUAL PROGRAM	OTHER	WPC	OFFSET	END	SHAPE CHECK

Figure 2 MAZATROL Function Buttons

Much like CAD/CAM, the programming process resembles recreation of the part geometry by constructing the shapes using lines, arcs, and points combined with other features.

This information is used in combination with Tool Identification Parameters and Cutting Condition Parameters to generate the tool path code needed to control the machine. The programmer has the added functions of the controller's ability to calculate for unknown coordinate values and have them automatically inserted into the program where they are needed. In most cases, no calculator is needed for trigonometric calculations. These constructions resemble the formerly popular Automatic Programmed Tool (**APT**) method of programming. The actual program the machine executes is still a G-Code format but the operator may never see the actual code on the display.

This type of programming combined with the ability to call G-Code sub programs has tremendous power. Mazatrol, a conversational programming system offered on all MAZAK machine tools allows "G-Code similar" programming, within its conversational language, in what is called a Manual Programming Process. The acronym used to call this type of program process for Mazatrol is **MNP**, where MN stands for Manual and P for Programming.

There are many different conversational languages available for programming available, and one of the main complaints has been is the lack of standardization between the various machine tool builders. MAZAK's Mazatrol language has been at the forefront of the industry in conversational programming for decades and has a proven track record of success has always been MAZAK with their Mazatrol language. The focus of this chapter will be centered on this language only. Other languages contain similar techniques that accomplish nearly the same result.

Just as with any programming endeavor, you must be well prepared. The sequence of events followed by the programmer programmer in order to create a Mazatrol program are very similar to those used in manual programming. Careful examination Study of the technical part drawing for, work holding considerations and tool selection must take place prior to preparation of the part program. With that in mind, it is most efficient to establish a Tool File within the Mazak controller prior to programming. This file should contain a representation of the tools that are available to choose from in your shop. From this information, one of the more powerful aspects of Mazatrol conversational programming can be used to automatically develop tools used in the program. This tool definition, when properly defined, can also be used to automatically calculate the proper feeds and speeds used for machining. Once this file is setup, programming may begin.

When the program is completed, the tool path must be verified by graphical simulation. At this point, if all checks well, the operator takes over for the measuring of tool and work offsets and one final program test by dry run. And, finally, the first part of CNC machining begins.

CONVENTIONS

For this section, the following text format convention is used. For the MENU selection, the letters will be in CAPITALS while the user prompts will be in capital *ITALICS*. Mazatrol acronyms are given in capital letters and **BOLD TYPE.**

TURNING CENTER PROGRAM CREATION

During the programming process, many unique abbreviations and acronyms are used to simplify prompting and input. The following are some of the acronyms encountered in the sequence of creating Mazatrol conversational Turning Center programs:

TURNING CENTER ABBREVIATIONS AND ACRONYMS

WKNO = Workpiece Number

MAT = Material

FC = Ferrous Cast Iron

FCD = Ferrous Cast Ductile Iron

S45C = Low Carbon Steel

SCM = Alloy Steel

SUS = Stainless Steel

AL = Aluminum

CU = Copper

CB ST = Carbon Steel

ALOY = Alloy Steel

CASIR = Cast Iron

9310 = 9310 Alloy Steel

BRASS = Brass

A2 = Tool Steel

MAX = Maximum

MIN = Minimum

OD = Outside Diameter

ID = Inside Diameter

RPM = Revolutions per Minute

FIN-X = Finish Allowance - X axis

FIN-Z = Finish Allowance - Z axis

BAR = Bar Machining e.g. solid Barstock

CPY = Copy Machining i.e. net shape material, casting, etc. Uniform material all around, all surfaces

CNR = Corner Machining e.g. re-machining of corners where the tool cannot reach, due to tool geometry + more

EDG = Edge Machining

FCE = Face

BAK = Back

THR = Threading Inside Diameter (I.D.) or Outside Diameter (O.D.)

GRV = Grooving, I.D., O.D., Face or Back

MTR = Workpiece Shape, is a user defined arbitrary shape that is other than bar or net shape and requiring non-uniform material removal

DRL = Drill

MNP = Manual Program Unit

M-CODE = Miscellaneous codes e.g. coolant M8

FCE = Face e.g. Edge FCE or BAR FCE

CPT-X = Cutting Point - X axis

CPT-Z = Cutting Point - Z axis

RV = Surface Speed for Rough Cut (V = Velocity)

FV = Surface Speed for Finish Cut (V = Velocity)

V ROUGHNESS = Surface Roughness determined by in/rev setting

R-FEED = Roughing Feed rate in/rev or mm/rev

R-DEP = Roughing Maximum Depth of Cut

R-TOOL = Rough Tool No.

F-TOOL = Finish Tool No.

ID CODE = Tool Identification Code for Spare Tool Usage

LIN = Linear Feed Move

TPR = Tapered Feed Move

S-CNR = Start <CNR-C> or <CNR-R> This means Start Corner -C = Chamfer - R = radius

SPT-X = Geometry Starting Point - X axis

SPT-2 = Geometry Starting Point - Z axis

FPT-X = Geometry Final Point - X axis

FPT-2 = Geometry Final Point – Z axis

F-CNR = Final <CNR–C> <CNR-R>/Necking Final corner chamfer or radius or necking

CTR = Center Point for Radius Programming

BAK = Back Machining

CHAMF = Chamfer for thread ending

ANG = Angle of thread

HGT = Thread Height

V = Velocity Cutting Speed Threading

END = End Unit

SHIFT = Second/third part, etc., shift amount

TPC = Temporary Parameter Change/Toolpath Control

Following, are brief descriptions of the general programming process for Turning Centers:

The control must be in the program-editing mode and a work number (program number) must be identified in order to begin.

• Press the soft key labeled "Work No." and key in the desired program number, and press Input.

Note: There is no need for the letter address O to precede the program number with Mazatrol programs.

Before any programming can take place, the programmer must determine the type of program needed EIA/ISO or Mazatrol. All MAZAK machines use Mazatrol as their standard program type, with EIA/ISO (G-Code) on some older generation machines as an optional feature. Turning Center programs are made up of these four basic parts; a Common Data Process, Machining Process, Sequence Data and an End Process.

COMMON DATA PROCESS

The information at the head or top line of the program applies to the entire program. The programmer is prompted to answer the following questions for this common data.

WORKPIECE MATERIAL <Menu>

The controller memory is preset with standard materials of Carbon Steel, Alloy Steel, Cast Iron, Aluminum and Stainless Steel to select from in the menu. This choice affects the automatic calculation of cutting feeds and speeds throughout the program. It is possible to add user defined materials to the cutting condition parameters if the material needed is not available.

MAX. OUTER DIA. of WORKPIECE

This value input here is dependant upon the diameter geometry of the raw workpiece.

Note: If the programmer inputs a value that exceeds this diameter, an alarm will result on the controller display which will prevent execution of tool path verification and automatic operation.

MIN. INNER DIA. of WORKPIECE

If the workpiece geometry is of solid bar stock, this value may be set to zero. If an inner diameter exists, such as with tubing, the programmer must input this value. Doing this prevents the generation of tool path where material is nonexistent.

WORKPIECE LENGTH

The overall length of the workpiece along the Z axis, including the clamping amount, should be entered for this value. It must be at least the maximum machined length. If a programmed value exceeds this length, the controller will set off an alarm display preventing execution of tool path verification and automatic operation. This value is not meant to represent the extension value for the setup of the part in the chuck jaws.

MAX. SPINDLE RPM LIMIT (rpm)

This enables the programmer to limit the spindle RPM to a predetermined amount (G50 in G-Code programming). If no value is input into this data field, the controller will execute the maximum spindle RPM when at the centerline in the X axis. This maximum RPM may be undesirable in some cases.

FINISH ALLOWANCE-X

The amount of material to be left for a finishing pass in the X axis is input at this time. This value is input in consideration of the diameter of the workpiece. For example: if a value of .040 inch is input, the amount taken off the diameter is = to .080 inch for the finishing pass.

FINISH ALLOWANCE-Z

The amount of material to be left for a finishing pass in the Z axis is input at this time.

STOCK REMOVAL of WORKFACE

It is common to machine material from the face of the workpiece in order to attain a finished surface that establishes the Z axis Workpiece Zero for the part. This amount is dependant upon the condition of the material and programmer preference.

MACHINING PROCESS

In this section, of the program, the individual machining process data are identified in order to complete the workpiece definition. In other words, what the type of machining that is is to be done. In the Figure 3 below, note the choices are **BAR, CPY, CNR, EDG, THR, GRV, WORKPIECE SHAPE and END.**

BAR	CPY	CNR	EDG	THR	GRV	WORKPIECE SHAPE	END		>>>

Figure 3 Machining Process Menu

Bar (**BAR**) machining is used for outside diameter (O.D.) or, inside diameter (I.D.) turning and boring. Copy (**CPY**) machining is used for O.D. or I.D. machining of existing geometries like castings or forgings, where a uniform amount of material is to be removed and is other than solid bar stock. Corner (**CNR**) machining is used when additional cutting tools are needed to finish corners that cannot be cut because of tool geometry limitations. Edge (**EDG**) machining is used to perform machining on the face of the workpiece. Thread (**THR**) is for machining of external and internal screw threads.

Grooving (**GRV**) is for machining of external and internal grooves. Workpiece Shape machining is similar to **CPY** except that the material removal shape does not need to be uniform. **END** is used to end the program. The arrow keys at the right offer some additional options of Drilling, Tapping and Manual Programming (as mentioned above i.e. G-Code within the Mazatrol program).

Once a selection is made for the type of machining operation, then more information is needed to identify how to apply it. In Figure 4 below, **BAR** machining has been selected and a new set of menu choices are displayed. Those items that are bold in the graphic are captured-type of cuts. The first on the left, **OUT**, is used to perform general O.D. machining and the third from the left, **IN**, is used for general I.D. machining like boring.

Figure 4 BAR Machining Menu

Once the type of machining is selected (**BAR**), the necessary related information is as follows: Feeds and Speeds, are automatically calculated by pressing a function soft key (AUTOSET) and are based on parameter information directly associated to the selected cutting tool and workpiece material identified in the Common Data Process; tool selection for roughing and finishing cycles; the Starting Point in X (**SPT-X**); the Starting Point in Z (**SPT-Z**); the Finish Point in X (**FPT-X**) and, the Finish Point in Z (**FPT-Z**).

Sequence Data

The finished workpiece shape is identified by the input of point data until the desired geometry exists using lines (**LIN**), tapers (**TPR**), arcs, chamfers and fillets, limited only by the tool geometry configuration. The same type data are necessary for internal bar machining. In Figure 5 below, the two types of arcs shown represent convex and concave shapes, respectively, and the CENTER menu selection command is needed to identify the arc center point.

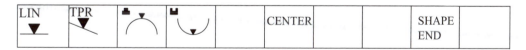

Figure 5 Sequence Data Menu

When all the geometric data are entered and the shape is defined properly, then the SHAPE END menu key is pressed to end the process. The remainder of the program is constructed in the same manner until the workpiece geometry is complete.

END

This process ends the program (similar to M30 in G-Code programming) and offers the opportunity to the programmer to: set a counter for the number of workpieces to be machined; set the return position of the turret after machining ends; identify a next program number to machine, whether to continue the same program repeatedly or not, and how many repetitions; and, also, the shift amount for the coordinates system can be set.

Shape Check

Once the program data are entered, the programmer may perform a Shape Check of the geometry to verify its accuracy. The controller will not allow the shape check if there are serious problems with the geometry definition and an alarm will be displayed on the screen identifying the program number, process number and sequence number of the mistake. Performing a shape check will draw the finished workpiece geometry on the screen in two-dimensional form.

Tool Path Verification

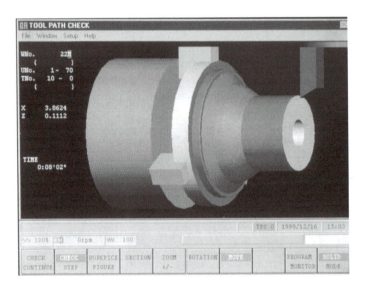

Figure 6 Tool Path Check
Courtesy MAZAK Corp

Another step completed before running the part is Tool Path Verification and it may be used to check the the part geometry and tool path. The newest controllers are equipped with solid model rendering of the workpiece raw stock configuration. The programmer can simulate on-screen, the actual machining of the workpiece in real-time by pressing CHECK CONTINUE from the menu.

If an area of the graphic is hard to see because of its size, the programmer can ZOOM into that area and magnify it for better viewing. Tool shapes are graphically simulated as well, allowing an excellent visual aid for correcting any program problems.

Often, a workpiece has internal features that are difficult to see even with solid modeling. The newest controls have the graphical capability to section the workpiece, allowing a full visual representation that offers even more assistance to the programmer for verifying programs.

Once these verification steps are complete, the machinist may begin the first article of CNC automatic operation.

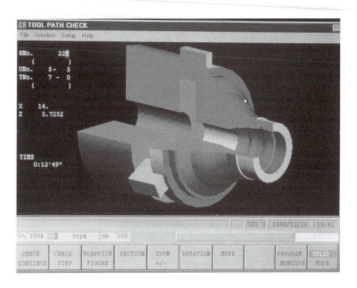

Figure 7 Tool Path Section
View Check

MACHINING CENTER PROGRAM CREATION

During the programming process, many unique abbreviations and acronyms are
used to simplify prompting and input. The following are some of the acronyms encountered in the sequence of creating Mazatrol conversational Machining Center programs.

Figure 8 Mazatrol Program Display
Courtesy MAZAK Corp

387

Machining Center Abbreviations and Acronyms

CST IRN = Cast Iron

DUCT IRN = Ductile Iron

CBN STEEL = Carbon Steel

ALY STEEL = Alloy Steel

STNLESS = Stainless Steel

AL = Aluminum

COPPER = Copper

WPC = Workpiece Coordinate

ATC = Automatic Tool Change

UNo. = Unit Number

ARBITRY = Arbitrary

FACE MIL = Face Mill

TOP EMIL = Top End Mill

PCKT MT = Pocket Mountain

PCKT VLY = Pocket Valley

SRV-Z = the amount of stock removal in Z

SRV-R = the amount of radial stock removal in X-Y

BTM = Bottom

RGH = Surface roughness

WAL = Wall roughness (Same number system applies)

WID-R = Width of Cut (radial)

CHMF = Chamfer amount

FIN-Z = Finish allowance for the Z axis

FIN-R = Finish allowance for the X-Y axis

NOM Ø = Nominal diameter of the tool

PRI–NO. = Priority number for the tool

APPRCH-X = Approach point for the tool in the X axis

APPRCH-Y = Approach point for the tool in the Y axis

CW CUT = Clockwise cutting direction

CCW CUT = Counterclockwise cutting direction

ZFD = Z Feed, Move to depth of cut at linear feedrate in the Z axis. Rapid may be selected here but the default setting is G01.

Caution should be used here because a rapid movement into solid material will break the tool and possibly damage the machine.

DEP-Z = Depth of Cut

C-SP = Cutting Speed

FR = Feedrate

SNo. 1 = Sequence Number

X BI-DIR = BI-Directional cutting both directions along the X axis

Y UNI-DIR = UNI-Directional (Single) cutting along the Y axis

CN1 = Corner 1

LNE = Line

CHMF = Chamfer

CTR-DR = Center Drill

PT = Point

INIT = Initial

R = Reference or Radius

Following are brief descriptions of the programming process for Machining Centers: The control must be in the program-editing mode and a work number (program number) must be identified. *Note: There is no need for the letter address O to precede the program number with Mazatrol programs.* Before any programming can take place, the type of program to create either EIA/ISO or Mazatrol must be determined. All MAZAK machines use Mazatrol as their standard with EIA/ISO (G-Code) on some older generation machines as an optional feature. The construction of a Mazatrol Machining Center program contains these basic parts: a Common Data Unit, identification of a Coordinate System, the Machining Units and their Sequence Data and, an End unit.

Common Data Unit

The information at the head of the program applies to the entire program. The programmer is prompted to answer the following questions for this common data.

MATERIAL <Menu>?

The controller is preset with standard materials of Cast Iron, Ductile Cast Iron, Carbon Steel, Alloy Steel, Stainless Steel, Aluminum and Copper Alloy to choose from. This choice, combined with tool material, affects the automatic calculation of cutting feeds and speeds throughout the program. It is possible to add other materials to the cutting condition parameters, if the material needed is not available.

INITIAL POINT Z (CLEARANCE)?

This value identifies where all of the tools will move to, at rapid traverse, before machining begins. In G-Code programming, this is the same as the Initial Reference Plane. A common reason for setting this at a particular height is to provide for clearance of work holding clamps.

ATC MODE ZERO RETURN <Z.X+Y:0, X+Y+Z:1>?

The choice of Zero Return method establishes how the tool is returned to the Automatic Tool Change (ATC) position for a tool change. For example, a selection of 0 returns the Z axis to the tool change position and then, the X and Y, simultaneously. This is the safest choice, in most cases; however, if no clearance issues are evident, then simultaneous movement of X, Y and Z may be chosen by selection of 1 here. *Note: that this movement is always at rapid traverse.*

MULTI MODE <Menu>?

There are three choices for establishing a work coordinate systems: MULTI OFF, MULTI 5 x 2 and OFFSET TYPE.

When MULTI OFF is selected, the next unit is started and this unit is commonly used for the Workpiece Coordinate system or WPC. Values are set for the location of the origin of the workpiece coordinate system, just as with G-Code programs, additionally, offsets G54–59 may be used.

MULTI 5 X 2

By selecting MULTI 5 x 2, the machining of multiple repetitions of the same program for several workpieces can be used. A MULTI FLAG is required, in conjunction with this call, in order to identify how many repetitions and where. This technique is limited to each of the repetitions having corresponding distances from part to part. For example, the distance between all repetitions in the X axis must be equal and the Y axis distances must be the same. Up to ten duplications can be set.

OFFSET TYPE

Using this type of coordinate system arrangement allows the arbitrary location of the Workpiece Zero or origin for multiple workpieces within the working envelope of the machine. The locations may be random and polar rotation of the coordinate system is allowed. Up to 10 individual offsets are allowed.

COORDINATE SYSTEM

In this unit, the actual physical locations for the coordinate system axis zero points are entered. As mentioned earlier, this can be in the form of a WPC or work offset using G54-G59 as is common in G-Code programs. The operator measures the distances in X, Y and Z in relation to the Machine Zero and enters these values into the program in this unit by using the WPC MEASURE function.

MACHINING UNIT

Here, the individual machining units are identified in order to complete the workpiece. In Figure 9 below, note that the choices are: POINT MACHINING, LINE MACHINING, FACE MACHINING, MANUAL PROGRAM, OTHER, **WPC**, OFFSET, **END** and SHAPE CHECK. Additional Machining Units may be entered until the final part geometry is completed as needed. Some examples are: "POINT MACHINING" used for Drilling Tapping, Reaming, etc.; "LINE MACHINING" used for Line Center, Line Left, Chamfer Left, etc.; or, "FACE MACH-ING" used for Face Mill, Top End Mill, Pocket, Step, Slot, etc. The following descriptions will be limited to Point machining and Line machining.

POINT MACH-ING	LINE MACH-ING	FACE MACH-ING	MANUAL PROGRAM	OTHER	WPC	OFFSET	END	SHAPE CHECK	

Figure 9 Machining Process Menu

POINT MACHINING

Point machining constitutes a large percentage of Machining Center work. In Figure 10, the different choices for types of point machining are shown:

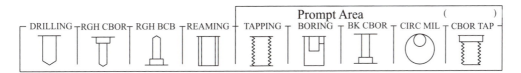

Figure 10 Point Machining Menu

When a selection is made from one of these choices, unit data are required and automatic tool development is completed, based on this information. The required information for Drilling is: the diameter, the depth and whether the hole is to be chamfered. The basic required tooling is developed based on this information. For example, a center drill or spot drill, a drill of the size stated and a chamfering cutter will be developed. This tooling information is taken from a predetermined tool file in the control. The tool file should be constructed (for all other types of machining units) by the machinist/programmer prior to programming but can be done as the program is completed.

SEQUENCE DATA

Tool Sequence Data

Each individual tool has specific sequence data that are required as follows:

* definition of the actual size of the tool
* the priority in which this tool is to be used
* the diameter of the hole
* the hole depth
* pre-existing hole diameter
* pre-existing hole depth
* the desired surface finish
* the type of drilling cycle (i.e. drilling, pecking, etc.)
* the cutting speeds and feeds
* and, the use of any M-Codes.

Shape Sequence Data

Finally, the shape sequence data is set for the machining unit. In other words, the actual figure pattern or shape is identified. In the case of drilling, the choices are: POINT, **LNE,** SQUARE, GRID, CIRCLE, ARC and CHORD, shown in the Figure 11. As soon as the pattern is completed, SHAPE END is pressed to end the unit.

Just as with all units, CHECK allows the shape to be checked graphically for each individual unit. The remainder of the program is constructed in the same manner until all geometry shapes are complete.

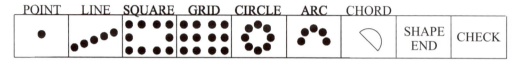

Figure 11 Point Sequence Data Menu

LINE MACHINING

Line machining (linear contouring) is another very common activity performed by Machining Centers. In Figure 12, choices for types of line machining are shown.

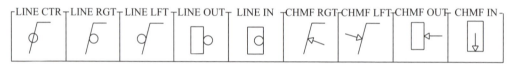

Figure 12 Line Machining Menu

When a selection is made, unit data are required and automatic tool development is completed, based on this information. The required information for line machining is: the depth of cut; the amount of stock removal in Z (**SRV-Z**); the amount of radial stock removal in X-Y (**SRV-R**); the desired finished surface roughness; chamfer width, if required; the allowance for the finish Z depth cut; and, the allowance for the radial finish width cut. The values input to these items determine the automatic tool development.

Tool Sequence Data

Each individual tool developed has specific sequence data that are required, as follows: the nominal diameter of the tool; the priority in which the tool is to be used; the approach point along the X axis, ; the approach point along the Y axis, ; the cutting direction of either CW, or CCW, the plunge cutting feedrate along the Z axis; the depth of cut, the cutting speed, and the cutting feedrate; and, any M-Codes as required.

Shape Sequence Data

Finally, the shape sequence data is created in the machining unit where the actual figure pattern or shape is identified. In the case of line machining, the choices are: SQUARE, CIRCLE, and **ARBITRY**, shown in Figure 13.

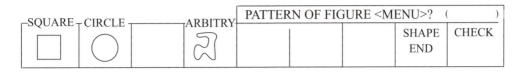

Figure 13 Shape Sequence Data Menu

The geometric shape is constructed by input of point data that describe each feature of square, circular or arbitrary shape. As soon as the pattern is completed, SHAPE END is pressed to end the unit.

Just as with all similar units, the CHECK menu button allows the shape to be checked graphically for each individual unit. The remainder of the program is constructed in the same manner until all geometry shapes are complete.

End Unit

This unit ends the program in the same manner as M30 does in a G-Code program. The programmer has the option in this unit of continuing the program for a number of repetitions and to control the Automatic Tool Change positioning.

Shape Check

When the entire program is written, it is beneficial to verify its accuracy by performing first a SHAPE CHECK and then TOOL PATH CHECK. The SHAPE CHECK verifies the geometry and the TOOL PATH CHECK verifies the actual relationship between the geometry and the tools actual cutting path. The machinist/programmer has the ability to change from two-dimensional to three-dimensional views or split the display screen to show the X-Y and X-Z, and zoom-in on features that are hard to see. Once these checks are completed, without errors, the program is ready for set-up of the tool and work offsets. The machinist may then begin automatic operation.

MAZATROL TURNING CENTER PROGRAM EXAMPLE

For this example, we will use an offline program called MazaCAM to create the Mazatrol program. It very closely emulates the programming process that happens when programming at the controller and the output is essentially the same. Some of the figures are screen shots from the MazaCAM program.

Start the Geopath software by double clicking the GeoPath Icon (Figure 14) or by selecting Start, All Programs, SolutionWare CAD-CAM, GeoPath from the Windows start menu system.

From the SolutionWare window, select File-Utilities from the Menu and then select MazaCAM CAD/CAM.

From the MazaCAM Utilities screen, select Make Mazatrol from the menu and then select Mazatrol Editor from the drop down list. The Select Program File dialog will come up. Input the file name of

GeoPath

Figure 14
GeoPath Icon

Courtesy
Solution Ware Corp.

the program you wish to create and press Open. Answer yes to create the file and select the controller type you wish to program. In this case, use the Fusion 640T controller type.

To write a program in Mazatrol at the machine controller, you must use the soft keys located under the CRT monitor. There are 14 soft keys. The twelve in the middle are considered Menu selection keys while the left most key is considered the Display Key and the right most key is considered the Page key. Titles are given above each of the menu keys which aid in the program creation process.

All Mazatrol programs consist of: a Common Data Unit; one or more Workpiece Coordinate Units; one or more Machining Units; and, an End Unit.

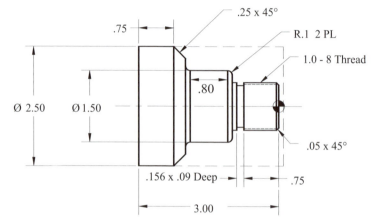

Figure 15
Drawing for Mazatrol
Turning Center Program

Your Company
CNC Setup Sheet

Date: Today	Prepared By: You
Part Name: Turning Center Project	Part Number: 4321
Machine: CNC Turning Center	Program Number:

Workpiece Zero: X = Centerline Y = NA Z = Finished Face

Setup Description:

Clamp the part in a 3-jaw chuck with 3.25" extended out of the chuck

Tool #	Tool Description	Offset #	Comments
1	Rough Turning Tool .031 TNR	1	SFPM 600
2	Finish Turning Tool .015 TNR	2	SFPM 850
5	O.D. Grooving Tool	5	.156 W ide .005 TNR
7	O.D. Threading Tool	7	SFPM 150

Steps to Create a Mazatrol Turning Program

Turn on the machine and Home all of the axes.

- Press the left most Display key.
- Select PROGRAM from the Menu keys.
- Press the WORK NO. Menu key.

Note: During programming, there is a user prompt area in the lower right corner of the CRT monitor that gives a brief description that leads you through the programming process. Simply answer the questions.

The following message will be displayed in the prompt area:

WORKPIECE NO. ?   (   ).

- Key in the desired work number (program number) i.e. 2525 and press Input. It is a good idea to check the Program File prior to this step in order to identify program numbers that have been used.

At this point, there is a Menu option to program in either EIA/ISO (G-Code or MAZATROL).

Note: On some older generation controls (M2 - M32 – M+), the EIA/ISO selection was an option and may not be available.

- Select MAZATROL from the Menu keys

You are now ready to create the Common Data Unit which will be listed on the screen as Process #0. The first prompting question encountered is:

WORKPIECE MATERIAL <MENU>?   (   )

From the Menu options you should select the material that most closely matches the material you are using. This selection, along with the tool file information, affects the automatic calculation of feeds and speeds later in the programming process.

- Select **AL** from the Menu keys and Press Input.

The material choices are given above the menu keys on the control and abbreviations for materials are: **FC** = Cast Iron; **FCD** = Cast Ductile Iron; **S45C** = Low Carbon Steel; **SCM** = Alloy Steel, ALY STEEL = Alloy Steel; **SUS** = Stainless Steel; **AL** = Aluminum and, **CU** = Copper. The last Menu key on the right has arrows pointing to the right (→→→→) access a second page of additional material selections. These can be used for user defined materials.

Note: All Cutting Condition Parameters are based from Carbon Steel (e.g. Plain Carbon Steel 1018) and all other materials are a percentage factor of this. You may develop new materials specific to your application with this criteria in mind. See the section in your Mazak User Manual provided with the machine for details on assigning Cutting Condition Parameters.

MAXIMUM OUTSIDE DIAMETER <MENU>?   (   )

- Input 2.5 inches

MIN INSIDE DIAMETER <MENU>?   (   )

- Input 0

WORKPIECE LENGTH?   (   )

- Input 5.0 inches

This amount allows for 2 inches of the material to be clamped, in the chuck.

MAX SPINDLE RPM LIMIT (rpm)?   (   )

- Input 3000

FINISH ALLOWANCE X?   (   )

- Input .01

FINISH ALLLOWANCE Z?   (   )

- Input .005

STOCK REMOVAL OF WORKFACE?   (   )

- Input .1

MODE MENU?

From the Menu keys, notice the arrow keys at the left end (F9). When pressed, a next menu screen is displayed.

- Choose M-Code (F5)
- Input 08 for Flood Coolant (F6)
- Use cursor down key to start a new machining process

MODE MENU?   (   )

- Choose EDGE

SECTION TO BE MACHINED ? <MENU>

- Choose FCE from the Menu keys

ROUGHING PERIPHERIAL SPEED?   (   )

- Press AUTO SET (F1)

FINISH PERIPHERIAL SPEED?   (   )

- Press AUTO SET (F1)

ROUGHING FEEDRATE?   (   )

- Input .016

ROUGHING DEPTH OF CUT PER PASS   (   )

- Input .08

ROUGHING TOOL TYPE?   (   )

Choose from tools that are identified in the tool file and the turret (at the machine). In our case, we will use tool number 2 for roughing and tool number 4 for finishing, these have been predefined in the software.

- Input 2 for rough turning tool
- Input tool 4 for the finish turning tool

STARTING POINT X?   (   )

• Input 2.5

STARTING POINT Z?   (   )

• Input .1

FINIAL POINT X?   (   )

• Input 0

FINAL POINT Z?   (   )

• Input 0

SURFACE ROUGHNESS or FINISHING FEEDRATE? <MENU>(   )

• Select Roughness (F1) and then choose 6 (F6)

Once this data is entered a new process begins.

MODE MENU?   (   )

• Choose BAR from the Menu keys (F1)

• Choose BAR OUT from the Menu keys (F1)

CUTTING PATTERN MENU?   (   )

• Chose #1 for the cut style (F2)

The difference between #0 and #1 is the way the tool moves at the end of the cut sequence. For BAR OUT 1, the tool feeds up the back wall, out to the O.D., at the end of the cut on every pass; whereas, on BAR OUT 2, the tool rapids away from the cut at 45° to a clearance point set by parameter and then consecutive passes are completed and the back wall is finished on the last pass only.

CUTTING POINT X?   (   )

• Input 2.5

CUTTING POINT Z?   (   )

• Input 0

SURFACE SPEED FOR ROUGH CUT?   (   )

• Select AUTO SET from the Menu keys (F1)

FINISHING PERIPHERIAL SPEED?

• Select AUTO SET from the Menu keys (F1)

ROUGHING FEEDRATE?

• Input .016

ROUGHING DEPTH OF CUT PER PASS?

• Input .080

ROUGHING TOOL TYPE?

• Input 2, for the rough turning tool

FINISHING TOOL TYPE?

• Input 4, for the finish turning tool

SHAPE PATTERN?

• Choose **LIN** (F1)

CHAMFERING (C) vs. ROUNDING R AT START POINT

• Input .05 to create the .05 x 45° chamfer (C)

Note: if a Radius is required you must press the CORNER R Menu key (F1), prior to entry of the amount.

X END POINT?

• Input 1.0 to create the 1.0 diameter

Z END POINT?

• Input 1.0 to create the length of the 1.0 diameter

CHAMFERING vs. ROUNDING AT END POINT?

• Press Input to omit this setting

RADIUS OF ARC OR TAPER ANGLE?

• Press Input to omit this setting

FINISHING FEEDRATE FOR SURFACE ROUGHNESS?

• Input 6 (F6)

SHAPE PATTERN?

• Choose **LIN** (F1)

CHAMFERING (C) vs. ROUNDING R AT START POINT?

• Press CORNER R (F1)

• Input .1 to create the .1 radius (R)

X END POINT?

• Input 1.5 to create the 1.5 diameter

Z END POINT?

• Input 2.0 to create the length of the1.5 diameter

CHAMFERING vs. ROUNDING AT END POINT?

• Press CORNER R (F1)

• Input .1 to create the .1 radius (R)

RADIUS OF ARC OR TAPER ANGLE?

• Press Input to omit this setting

FINISHING FEEDRATE FOR SURFACE ROUGHNESS?

• Input 6 (F6)

SHAPE PATTERN?

• Choose **LIN** (F1)

CHAMFERING (C) vs. ROUNDING R AT START POINT

• Input .25 to create the .25 x 45° chamfer (C)

X END POINT?

• Input 2.5 to create the 2.5 diameter

Z END POINT?

• Input 3.0 to create the length of the 2.5 diameter

CHAMFERING vs. ROUNDING AT END POINT?

• Press Input to omit this setting

RADIUS OF ARC OR TAPER ANGLE?

• Press Input to omit this setting

FINISHING FEEDRATE FOR SURFACE ROUGHNESS?

• Input 6 (F6)

Select SHAPE END from the Menu keys (F9)

A new Process Number is started. Select the Machining Unit type next.

• Select **GRV** for Groove from the Menu keys (F6)

SECTION TO BE MACHINED?

• Choose OUT (F1)

MACHINING PATTERN?

• Choose type #1 (F2)

NUMBER OF GROOVES?

• Input 1

SPACING AMOUNT OF MULTIPLE GROOVES?

• Press Input to omit this setting

GROOVE WIDTH?

• Input .156 for the groove width

FINISH REMOVAL ALLOWANCE?

• Input .01 for the groove finish allowance

ROUGHING PERIPHERIAL SPEED?

• Press the AUTO SET Menu key (F1)

FINISHING PERIPHERIAL SPEED?

• Press the AUTO SET Menu key (F1)

ROUGHING FEEDRATE?

• Input .010

ROUGHING DEPTH OF CUT PER PASS?

• Input .04

ROUGHING TOOL TYPE?

• Input 8 for the rough grooving tool

FINISHING TOOL TYPE?

• Input 8 for the finish grooving tool

CHAMFERING (C) vs. ROUNDING R AT START POINT?

• Input .04 for chamfering at the start of the groove

X START POINT?

• Input 1.0

Z START POINT?

• Input .75

X END POINT?

• Input .9

Z END POINT?

• Input .75

CHAMFERING vs. ROUNDING AT END POINT?

• Press Input to omit this setting

FINISHING FEEDRATE FOR SURFACE ROUGHNESS?

• Input 6 (F6)

A new Process Number is started. Select the next Machining Unit type.

• Select **THR** for Threading from the Menu keys (F6)

SECTION TO BE MACHINED?

• Choose OUT (F1)

MACHINING PATTERN?

• Choose #0 STANDARD (F1)

CHAMFER ANGLE <0: NO CHAMFER, 1:45 DEGREES, 2:60 DEGREES>?

• Choose #0 for no chamfer

THREADING LEAD?

• Input .125 for the thread lead

This amount is determined by dividing one by the number of threads per inch. In this case, 1/8 = .125.

THREADING ANGLE?

• Input 59 for the threading angle

NUMBER OF THREADS?

• Input 1 for the number of thread starts

THREADING HEIGHT?

• Input .0801 for the thread height

NUMBER OF TIMES THREADING?

• Input AUTO SET (F1)

SPINDLE PERIPHERIAL SPEED?

• Input AUTO SET (F1)

FIRST THREADING AMOUNT?

• Input .010

TOOL TYPE?

• Input 6 for the threading tool number.

X START POINT?

• Input 1.0

Z START POINT?

• Input 0

X END POINT?

• Input 1.0

Z END POINT?

• Input .85

• Select SHAPE END from the Menu keys

A new Process Number is started. Select the next Machining Unit type.

Select END from the Menu keys (F8)

COUNT NUMBER OF MACHINED WORKPIECES <YES = 1, NO = 0>?

• Input 0

TOOL RETURN POSITION <0 = CHANGE POSITION, 1 = HOME,
 2 = FIXED POINT>?

• Input 1

Inputting this value will send the turret to the Home position at the end of the program.

WORK NUMBER OF FOLLOWING PROGRAM?

• Press Input to omit this setting

EXECUTE PERPETUALLY = 1; EXECUTE NUMBER OF TIMES IN NUM = (0)

• Press Input to omit this setting

NUMBER OF TIMES TO REPEAT PROGRAM?

• Press Input to omit this setting.

Z SHIFT AMOUNT OF PROGRAM ORIGIN?

• Press Input to omit this setting.

This concludes programming of the part. A sample of the program shape plot is displayed in Figure 16 and the program output is displayed in Figure 17.

Figure 16 Program Shape Plot
Courtesy Solution Ware Corp.

```
C:\SLN\Lathe.BIN (Fusion640T)                                    _ □ X

 File  Program  Edit  Find/Search  Tool-File  Calculate  Settings  Help

PNo. MAT     OD-MAX  ID-MIN  LENGTH  RPM  FIN-X  FIN-Z  WORK-FACE MPX FIN-LENGTH
  0 AL       2.5     0.      5.      3000 0.01   0.005     0.1       0

PNo. MODE   #1  #2  #3  #4   #5  #6  #7  #8    #9 #10 #11 #12
  1 M        8

PNo. MODE                        RV   FV  R-FEED R-DEP    R-TOOL   F-TOOL
  2EDG FCE                       ?    ?   0.016  0.08       2        4
SEQ           SPT-x      SPT-z      FPT-x      FPT-z               ROUGH
  1           2.5        0.1        0.         0.                    6

PNo. MODE  #  CPT-X    CPT-Z      RV   FV   R-FEED R-DEP    R-TOOL  F-TOOL
  3BAR OUT 1  2.5      0.         ?    ?    0.016  0.08       2       4
SEQ SHP S-CNR   SPT-X      SPT-Z     FPT-X      FPT-Z   F-CNR/$ RADIUS/A ROUGH
  1 LIN C0.05   <>         <>        1.         1.                        6
  2 LIN R0.1    <>         <>        1.5        2.      R0.1               6
  3 LIN C0.25   <>         <>        2.5        3.                         6

PNo. MODE  # No. PITCH   WIDTH  FINISH RV   FV   FEED   DEP    R-TOOL  F-TOOL
  4GRV OUT 1 1          0.156  0.01   ?    ?    0.01   0.04     8       8
SEQ      S-CNR    SPT-X      SPT-Z     FPT-X      FPT-Z    F-CNR  ANGLE   ROUGH
  1       C0.04   1.         0.75      0.9        0.75                      6

PNo. MODE  # CHAMF LEAD   ANG  MULTI HGT  NUMBER  V    DEPTH      TOOL
  5THR OUT 0  0    0.125  59   1    0.046   ?     ?    0.01        6
SEQ           SPT-x      SPT-z      FPT-x      FPT-z
  1           1.         0.         1.         0.85

PNo. MODE  COUNTER  RETURN  WK.No.   CONTinue  NUM.    SHIFT
  6END       0        1
```

Figure 17 Finished Turning Program
Courtesy SolutionWare Corp

MAZATROL MACHINING CENTER PROGRAM EXAMPLE

For this example, we will give the steps to input the program as you would at the controller. The graphics for some of the Menu keys, Part Shape and Program output are taken from the offline program called MazaCAM that can also be used to create the Mazatrol program. It very closely emulates the programming process that happens when programming at the controller and the output is essentially the same. Some of the figures used are screen shots from the program and are labeled as such.

To write a program in Mazatrol at the machine controller, you must use the soft keys located under the CRT monitor, (there are 14 soft keys). The twelve in the middle are considered Menu selection keys while the left most key is considered the Display Key and the right most key is considered the Page key. Titles are given above each of the menu keys which aid in the program creation process.

All Mazatrol programs consist of a: Common Data Unit; one or more Workpiece Coordinate Units; one or more Machining Units; and, an End Unit.

STEPS TO CREATE A MAZATROL MILLING PROGRAM:

• Turn on the machine and Home all of the axes.

- Press the left most Display key.
- Select PROGRAM from the Menu keys.
- Press the WORK NO. Menu key.

Note: During programming, there is a user prompt area in the lower right corner of the CRT monitor that gives a brief description that leads you through the programming process. Just answer the questions.

The following message will be displayed in the prompt area:

Your Company
CNC Setup Sheet

Date: Today	Prepared By: You
Part Name: Machining Center Project	Part Number: 5432
Machine: Machining Center	Program Number:

Workpiece Zero: X = Centerline Y =Centerline Z Top Most Finished Surface

Setup Description:

Clamp the part in a vise on parallels.
A minimum of 3/16" must be above the vise jaws.

Tool #	Tool Description	Offset #	Comments
1	3.0" Ø 5 Tooth Face Mill	1	SFPM 1800, in/rev = .012
2	.750" Ø 2 Flute End Mill	2	SFPM 200, in/rev = .002
3	.50" Ø x 90° Spotting Drill	3	SFPM 200 in/rev = .002
4	.4219" Ø 27/64 Drill	4	SFPM 200, in/rev = .002
5	.4375" Ø Reamer	5	r/min = 1/3 of lowest drill speed
			feed = 1/2 of lowest drill feed

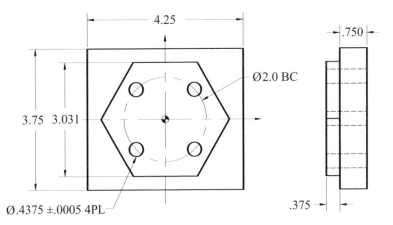

Figure 18 Drawing for Mazatrol CNC Machining Center Program
CNC Machining Center Setup Sheet

WORK NO. (NAME SEARCH <?INP>   (   )

• Key in the desired work number (program number) i.e., 2525 and press Input. It is a good idea to check the Program File prior to this step in order to identify program numbers that have been used.

At this point, there is a Menu option to program in either EIA/ISO (G-Code or MAZATROL). Note: *On some older model controls (M2 – M32 – M+) the EIA/ISO selection was an option and may not be available.*

• Select MAZATROL from the Menu keys

You are now ready to create the Common Data Unit which will be listed on the screen as Unit #0. The first prompting question encountered is:

MATERIAL <MENU>?   (   )

From the Menu options, you should select the material that most closely matches the material you are using. This selection, along with the tool file information, affects the automatic calculation of feeds and speeds later in the programming process.

• Select **AL** from the Menu keys and Press Input

The material choices are given above the menu keys on the control and abbreviations for materials are; : **CST IRN** = Cast Iron; **DUCT IRN** = Ductile Iron; **CBN STEEL** = Carbon Steel; **ALY STEEL** = Alloy Steel; **STNLESS** = Stainless Steel; **AL** = Aluminum; **COPPER** = Copper. The last Menu key on the right has arrows pointing to the right ($\rightarrow \rightarrow \rightarrow \rightarrow$) for a second page of additional material selections. These can be used for user defined materials.
Note: All Cutting Condition Parameters are based from Carbon Steel (e.g. Plain Carbon Steel 1018). All other materials are a percentage factor of this. You may develop new materials specific to your application with this criteria in mind. See the section in your Mazak User Manual provided with the machine for details on assigning Cutting Condition Parameters.

INITIAL POINT- Z (CLEARANCE)?   (   )

For specific details, see the preceding text in this section for a description of Initial Z Clearance question to establish the Z Plane.

• Key in 1.0 and press Input

ATC MODE ZERO RETURN <Z.X+Y:0, X+Y+Z:1>?   (   )

This sets the path the spindle is to take for the Automatic Tool Change Mode.

See existing text for a more detailed description of ATC MODE.

• Key in 0 to select the Z first then XY movement and Press Input

MULTI MODE <Menu>?

See the existing text in this section for details regarding whether there are multiple parts on the table, or not.

• Select MULTI OFF from the Menu keys

MACHINING UNIT <MENU>?   (   )

The following choices are available to choose from: POINT; LINE; FACE; MANU-

AL PROGRAM; OTHER; WPC; OFFSET; END and, SHAPE CHECK. (See Figure 8).

- Choose WPC from the Menu select keys

This is UNo.1 WPC = Workpiece Coordinate. Use this unit to identify the location for the part zero of the geometry from machine home. These values are usually measured using an edge finding device or probe system and input within the program or the offset registers. You may also set this item to use G54-G59 and additional offsets of A – K. Consult the Operation Manual of your machine for specific instructions on setting these values. A number can be assigned, in order to include multiple WPC's within the same program.

- Press Input or use the right pointing cursor key to accept a value of 0 for our WPC number
- Press Input again to pass over the Additional Offsets

At this point, you will be prompted as follows:

WORKPIECE COORDINATE, WPC - X? (   )

WORKPIECE COORDINATE, WPC - Y? (   )

WORKPIECE COORDINATE, WPC - θ? (   )

WORKPIECE COORDINATE, WPC - Z? (   )

WORKPIECE COORDINATE, WPC - 4? (   )

- Input 0 in each case for now

Setup Notes: The exact coordinate values for each of these offsets will be measured during the setup process and entered via WPC-MEASURE, WPC SEARCH. Position an edge finding device in order to find the workpiece edge along the X axis. Press the TEACH Menu key and key in a value (0) that represents the location of the spindle, in relation to the Workpicece Zero, and Press Input. Don't forget to compensate for the radius of the edge finder. Repeat these steps for the Y axis. Input 0 for the workpiece rotation angel of Theta. All cutting tools used in the program should be installed in the magazine and measured to the tool sensor prior to workpiece coordinate setting. For the WPC – Z, position any tool that is used in the program and touch off the top most surface of the part. Remember, sometimes this top-most surface is above the finished surface zero. This is the case in our example, so once the tool is touched-off along the Z axis, Press TEACH, key in -.05 and Press Input. Input 0 for the 4th axis workpiece coordinate.

When all these data are entered, a new unit (UNo.2) is started.

MACHINING UNIT <MENU>?   (   )

Often, it is necessary to machine the work surface to establish a Z -0 work face. It is commonly at the beginning of the program for this reason. The following choices are available to choose from for Face Machining: **FACE MIL; TOP EMIL; PCKT MT; PCKT VLY**. (See Figure 8).

- Select **FACE MIL** from the Menu keys

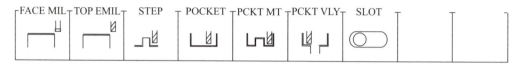

Figure 19 Face Machining Menu Keys

Once the Face Machining menu key is pressed, there are columns with the following headings displayed on the screen: DEPTH, **SRV-Z, BTM and FIN-Z.**

DIST: WPC Z0 TO FIN SURFACE?   (   )

- Input 0 here because this represents our finished surface for Z

Z AXIS STOCK REMOVAL?   (   )

- Input .05 here because there is that much excess material to remove

BOTTOM ROUGHNESS <MENU>?   (   )

- Choose from 6 from the Menu keys

This will automatically set the Z depth of cut for the roughing (.0402) and finishing (.0098) passes. A number system of 1-9 is used here that relates to surface finish. Higher number the more fine the finish. This number affects the finish depth of cut.

- Press Input

▼	▼▼	▼▼	▼▼	▼▼▼	▼▼▼	▼▼▼	▼▼▼▼	▼▼▼▼	
1	2	3	4	5	6	7	8	9	

Figure 20 Bottom and Wall Roughness Menu Keys

The following Figure identifies the approximate surface Micro Finish that is created when each of the numbers are selected. For example, selection of Menu key 6 will produce a 32 Micro Finish.

1000	500	250	125	63	32	16	8	4	

Figure 21 Surface Micro Finish Chart

On the left hand side of the screen, the acronym, **SNo.**, will be displayed and the columns will have the headings: **NOM-Ø, No., APRCH-X, APRCH-Y, TYPE, WID-R, C-SP, FR** and **M**. Because we selected Face Machining earlier, the rough and finish tools will be automatically developed. On the next line, R 1 F-Mill will be displayed.

NOMINAL DIAMETER?   (   )

- Input 3.0 for the Face Mill diameter

TOOL FILE CODE?   (   )

• Press Input to omit setting of this special identification code

MACHININING PRIORITY NUMBER? ( )

• Press Input to omit

APPROACH POINT X, AUTO ?<MENU>( )

• Press the AUTO SET menu key

This action will automatically set the approach point a safe distance from the part geometry in the X axis. This value is controlled by parameter setting and is typically 66% of the tool diameter.

Note: When AUTO SET is pressed, question marks are temporarily entered until the part geometry is entered and the values are changed relative to this geometry. The programmer can manually enter these values, if known, or other values are preferred.

APPROACH POINT Y, AUTO ?<MENU>? ( )

• Press the AUTO SET menu key

CUTTING DIRECTION <MENU>? ( )

X BI-DIR	Y BI-DIR	X UNI-DIR	Y UNI-DIR	X BI-DIR SHORT	Y BI-DIR SHORT			
F1	F2	F3	F4	F5	F6	F7	F8	F9

Figure 22 Face Machining Cutting Direction
Courtesy of SolutionWare Corporation

• Select **X BI-DIR** from the Menu keys (F1 from the function keys).

X BI-DIR stands for Bi-Directional cutting along the X axis; whereas, **X** or **Y UNI-DIR** stands for cutting along either axis in a single direction. **X** or **Y BI** or **UNI-DIR SHORT** means that the tool will not position completely off the part but by only 33%.

Note: the subheadings F1-F9 are not present on a MAZAK controller. These are function keys on a standard keyboard and are used within the MazaCAM Editor only.

DEPTH OF CUT? ( )

• Press the AUTO SET Menu key

By selecting the AUTO SET function, the controller calculates the difference between the full depth required and the finish allowance determined earlier in the program, and inputs this amount. In our case, it is .0402, because .0098 was the allowance and .050 is the stock removal amount.

WIDTH OF CUT? ( )

• Press the AUTO SET Menu key

This value is based on 66% of the diameter of the tool selected, which, in our case, is the three-inch face mill, so, by pressing AUTO SET, the control will output approximately 2.7 for the width of cut.

CUTTING SPEED, AUTO?<MENU>?   (   )

• Press the AUTO SET Menu key

Pressing the AUTO SET Menu key uses the tool and material data previously entered into the program to determine the cutting speed (sf/min)

FEEDRATE, AUTO?<MENU>?   (   )

• Press the AUTO SET Menu key

•Pressing the AUTO SET menu key uses the tool and material data previously entered into the program to determine the cutting feed rate (in/min)

M CODE? (   )

Coolant flow is activated here. Two M-Codes may be activated per tool.

01 OPT. STOP	03 SPNDL FWD	04 SPNDL REV	05 SPNDL STOP	07 MIST COOLANT	08 FLOOD COOLANT	09 OFF COOLANT	50 AIR BLAST	->->->
F1	F2	F3	F4	F5	F6	F7	F8	F9

Figure 23 M-Codes Menu Keys
Courtesy SolutionWare Corp.

• Select 08 from the Menu keys (F6 from the function keys) for Flood Coolant

• Press Input to omit the second M-Code selection

On the next line, F 2 F-Mill will be displayed for the Finish Face Mill.

• Repeat the entries in a similar fashion, as above, for the required data.

After identifying what style of machining is to take place, the next information needed is the Figure Pattern. This is the coordinate data that makes up the part geometry. When the entire shape has been input, you must end the unit by pressing the "SHAPE END" menu key. This action closes the Unit and starts a new one. It is also a good idea to complete a SHAPE CHECK, at this point, to ensure the data that has been input is correct. The Shape of the Figure Pattern will be displayed on the screen.

PATTERN OF FIGURE <MENU>?   (   )

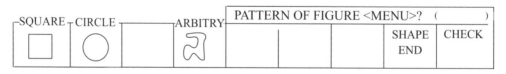

Figure 24 Pattern of Figure Menu Keys

• Choose Square from the Menu keys

On the display there are columns with the headings: P1X/CX, P1Y/CY, P3X/R, P3Y, CN1, CN2, CN3, CN4.

CORNER 1 COODINATE X?   (   )

• Key in -2.125 and press Input

This value represents the coordinate for the lower left hand corner of the part geometry for the X axis.

CORNER 1 COODINATE Y? ( )

• Key in -2.125 and press Input

This value represents the coordinate for the lower left hand corner of the part geometry for the Y axis.

CORNER 3 COODINATE X? ( )

• Key in 2.125 and press Input

This value represents the coordinate for the upper right hand corner of the part geometry for the X axis.

CORNER 3 COODINATE Y? ( )

• Key in 2.125 and press Input

This value represents the coordinate for the upper right hand corner of the part geometry for the Y axis. This entry completes the required geometry for the Face Milling operation.

Note: CN1 through CN4 allow the entry of chamfers or radii at each of the individual corners of the square.

• Press Input 4 times to omit entries for CN1 through CN4

• Press the SHAPE END Menu key

A new Unit is started and is labeled UNo. 3.

MACHINING UNIT <MENU>?

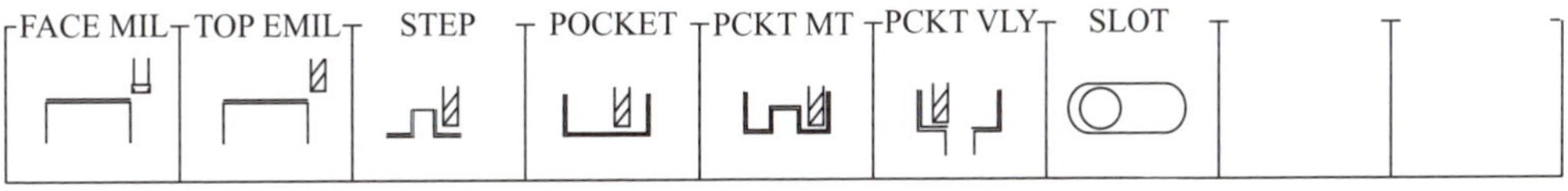

Figure 25 STEP Machining Unit Menu Keys

• Select FACE MACHINING

• Select STEP from the Menu selection keys

The displayed column headings are the same for STEP machining as in facing. The acronyms described at the beginning of this section are the same. Focus your attention on the prompting question in the lower right-hand corner of the display for data entry.

DIST: WPC 0 TO FIN. SURFACE? ( )

• Since our finished Z-zero surface is our reference, we will Input zero (0) here

Z AXIS STOCK REMOVAL? ( )

• Since we know there is .375 of material to remove in our STEP, Input .375. No sign is necessary

BOTTOM FACE ROUGHNESS <MENU>? ( )

• Input 6 here for the surface roughness control for the bottom (**BTM**) of the STEP

WALL ROUGHNESS <MENU>?   (   )

• Input 6 here for the surface roughness control for the walls (**WAL**) of the STEP

FINISH ALLOWANCE Z ?   (   )

This value is automatically calculated based on the bottom (**BTM**)surface roughness number selected. It can be modified, at any time; however.

> • Accept the pre-calculated value (.0098) and press Input

FINISH ALLOWANCE R ?   (   )

This value is automatically calculated based on the wall (**WAL**) surface roughness number selected. It can be modified, at any time; however.

> • Accept the pre-calculated value (.0098) and press Input

The tool data information is input for the unit now. The column headings are titled the same as in the face machining unit for SNo.

WHICH TYPE OF TOOL <MENU>?   (   )

ENDMILL	FACE MILL	CHAMFER CUTTER	BALL ENDMILL		OTHER			
F1	F2	F3	F4	F5	F6	F7	F8	F9

Figure 26 Type of Tool Menu Keys
Courtesy SolutionWare Corp

> • Select ENDMILL from the Menu keys (F1 function key) and press Input

Complete the roughing (R1) tool data as follows:

NOMINAL DIAMETER?   (   )

• Input .750 for the roughing end mill diameter

TOOL FILE CODE?   (   )

This allows further identification of the tool by a letter code. For example, roughing, carbide, high speed steel, etc. The shop you are working in should establish a system to follow.

> • Press Input to omit this setting

MACHINING PRIORITY NO.?   (   )

This allows prioritization of the sequence in which the tool is used. If no value is input, the tool will be used by program priority.

> • Press Input to omit this setting

APPROACH POINT X, AUTO →<MENU>   (   )

• Press the AUTO SET menu key

APPROACH POINT Y, AUTO Z →<MENU>   (   )

• Press the AUTO SET menu key

CUTTING DIRECTION <MENU>?   (   )

The choices here are clockwise (**CW-CUT**) cutting, or counterclockwise (**CCW-**

CUT). This input controls whether the cut is climb or conventional cutting.

- Press the CW-CUT Menu key

FEEDRATE Z, <MENU>/DATA<INPUT>? ( )

The menu choices are: CUT-G01 or a RAPID-G00. This value is set by default for CUT G01.

Caution: If this value is set to rapid and material exists, at the penetration point, a collision will occur.

- Press the CUT-G01 Menu key

DEPTH OF CUT? ( )

- Press AUTO SET

WIDTH OF CUT? ( )

- Press AUTO SET

CUTTING SPEED, AUTO→<MENU>? ( )

- Press AUTO SET

FEEDRATE, AUTO→<MENU>? ( )

- Press AUTO SET

M-CODE? ( )

- Press 08 and press Input

Complete the finishing tool sequence (F2) data in a similar fashion as for the rough tool. A different finishing tool may be selected or the same tool can be used. The depth of the finishing pass depends on the amount remaining from the roughing tool which was identified in the prior information.

PATTERN OF FIGURE <MENU> ( )

Here is where we identify the shape of our STEP geometry, there are two shapes involved in every case. Always identify the outside shape first when performing a STEP geometry.

- Select SQUARE from the Menu keys

CORNER 1 COORDINATE X? ( )

- Key in -2.125 and press Input

CORNER 1 COORDINATE Y? ( )

- Key in -2.125 and press Input

CORNER 3 COORDINATE X? ( )

- Key in 2.125 and press Input

CORNER 3 COORDINATE Y? ( )

- Key in 2.125 and press Input

This concludes the first shape for the STEP machining unit. The second pattern is neither SQUARE or CIRCLE so we must use arbitrary (**ARBITRY**). We have pre-calculated all of the necessary coordinate points for creation of the geometry.

Select **ARBITRY** from the Menu keys

Once **ARBITRY** type geometry has been selected, the following choices become available (See Figure 27 below):

LINE	CW ARC	CCW ARC	SHAPE ROTATE	SHAPE SHIFT	REPEAT END	STATING POINT	SHAPE END	
F1	F2	F3	F4	F5	F6	F7	F8	F9

Figure 27 Arbitrary Shape Definition Menu Keys   Courtesy SolutionWare Corp.

Select LINE from the Menu keys (F1 function key)

COORDINATE X OF FIGURE ? ( )

• Key in .875 and press Input

COORDINATE Y OF FIGURE ? ( )

• Key in -1.5515 and press Input

• Press Input five more times to omit the other entries or use the cursor down button to skip to the next line.

• Select LINE from the Menu keys (F1 function key).

Repeat the same entry technique for the following remaining coordinates:

• *3LINE   X-.875 Y-1.5515*

• *4LINE   X-1.75 Y0*

• *5LINE   X-.875 Y1.5515*

• *6LINE   X.875   Y1.5515*

• *7LINE   X1.75   Y0*

As units are completed, it is wise to perform a SHAPE CHECK to ensure that you have entered everything correctly. If all looks well, proceed to the next unit.

• Press SHAPE END twice to end the unit

MACHINING UNIT <MENU>? ( )

• Choose POINT MACH-INING.

POINT MACH-ING	LINE MACH-ING	FACE MACH-ING	MANUAL PROGRAM	OTHER	WPC	OFFSET	END	SHAPE CHECK	

Figure 28 Machining Unit Menu Keys

MACHINING UNIT <MENU>? ( )

• Choose REAMING.

HOLE DIAMETER? ( )

• Input .4375.

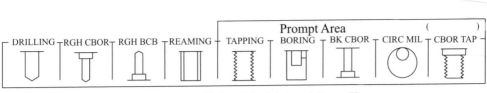

Figure 29 Point Machining Menu Keys

HOLE DEPTH?   (   )

• Input 1.125

Note: When drills are identified in the tool data register, theat drill point compensation is added to assure full drilling depth.

CHAMFER WIDTH?   (   )

• Input 0.

PRE-REAMING , OPERATION <MENU>?   (   )

Choices are: DRILL, BORING BAR or END MILL.

• Select the the DRILL Menu key.

CHIP VAC. CLEANER<Y:1, N:0>?   (   )

• Press Input to omit entry here

At this time, the tools are automatically developed for the reaming operation and listed based on the prior input for the tool sequence, SNo.

Sequence 1 is the **CTR-DR** of .3 diameter.

Sequence 2 is the DRILL of , .43 diameter.

TYPE OF DRILLING CYCLE <MENU>?   (   )

• For Sequence 2 the Drilling Cycle needs to be changed to PECKING CYCLE 2 and the DEPTH OF CUT to .43

Note: This will initiate the PECK DRILLING CYCLE for the deep holes in order to break and clear chips. Each peck will penetrate the material .43 deep until the finished depth is reached.

Sequence 3 is the REAM or .44 diameter.

The only thing that needs completion for the remaining tool information is is the feeds and speeds and coolant.

• Press HSS AUTO.

• Input 08 for the M-Ccode

• Repeat the same entries steps for each tool.

POINT CUTTING PATTERN <MENU>?   (   )

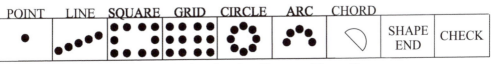

Figure 30 Point Cutting Pattern Menu Keys

Choose CIRCLE from the Menu keys.

Z VALUE OF THE WORK SURFACE?   (   )

• Input zero for the work surface value.

CIRCLE CENTER X?   (   )

• Input 0 for the circle center, coordinate for the *X* axis.

CIRCLE CENTER Y?   (   )

• Input 0 circle center coordinate for the *Y* axis

ANGLE OF START PT FROM X-AXIS?   (   )

By looking at the drawing, we can calculate that the first hole in the pattern is 45° from the polar coordinate reference.

• Input 45°.

CIRCLE RADIUS?   (   )

• Input 1 inch.

NUMBER OF HOLES?   (   )

• Input 4.

RETURN POSITION <,INIT: 0, R:1>?   (   )

• Input 1 for the tool to return to the R-Plane

• Select SHAPE END.

That concludes the creation of our part program geometry.

• Select END from the menu keys

CONTINUE <Y:1,N:0>?   (   )

• Input 0 for No.

PARTS COUNTER <Y: 1, N: 0>?   (   ).

• Input 0 for No

AUTO TOOL CHANGE <Y: 1, N: 0>?   (   )

• Input 0 for No

End coordinates for *X, Y* and *Z* and the fourth axis, may be input in order to position the spindle away a from the part for easy access. These values are in relation to Machine and Zero.

• Press the Menu Select soft key

• Press the PROGRAM COMPLETE Menu key.

• Press the TOOL PATH Menu key.

The choices are:, PATH CONTINUE, PATH, STEP, PART SHAPE, SHAPE ERASE, PATH TRACE, PROGRAM, STORE, PLANE CHANGE, and SCALE CHANGE.

• Choose PART SHAPE menu Menu key.

The part geometry shape will be displayed similar to Figure 31, and if no errors

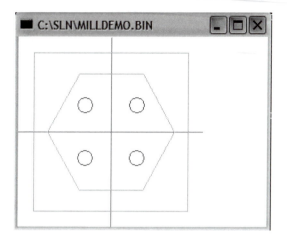

Figure 31 Finished Mill Program
Shape Display
Courtesy SolutionWare Corp.

```
MILLDEMO.BIN                                                    _ □ X
 File  Program  Edit  Find/Search  Tool-File  Calculate  Settings  Help

UNO     MAT    INITIAL-Z  ATC-MODE  MULTI-MODE  MULTI-FLAG  PITCH-X  PITCH-Y
 0      AL         1.         0        OFF          <>         <>       <>

UNO UNIT            X        Y        THETA       Z          4
 1 WPC- 0          0.       0.         0.         0.         0.

UNO  UNIT    DEPTH   SRV-Z   SRV-R    BTM        WAL       FIN-Z     FIN-R
 2 FACE-MIL 0.       0.05     <>       6          <>       0.0402      <>
SNO  TOOL  NOM-D  NO  APRCH-X  APRCH-Y TYPE ZFD DEP-Z  WID-R C-SP  FR   M M
R  F-MILL  3.        ?        ?       X-BI <> 0.0402 2.7  1200  0.006  8
F  F-MILL  3.        ?        ?       X-BI <>  <>    2.7  1800  0.012  8
FIG PTN  P1X/CX  P1Y/CY  P3X/R   P3Y    CN1     CN2      CN3      CN4
 1 SQR   -2.125  -2.125  2.125  2.125

UNO  UNIT    DEPTH   SRV-Z   SRV-R    BTM        WAL       FIN-Z     FIN-R
 3 STEP-M  0.375   0.375    <>       6          6        0.0098    0.0098
SNO  TOOL  NOM-D  NO  APRCH-X  APRCH-Y TYPE ZFD DEP-Z  WID-R C-SP  FR   M M
R  E-MILL  0.75      ?        ?       CW   G01 0.3652 0.5  150  0.003  8
F  E-MILL  0.75      ?        ?       CW   G01  <>    0.5  200  0.002  8
FIG PTN  P1X/CX  P1Y/CY  P3X/R   P3Y    CN1     CN2      CN3      CN4
 1 SQR   -2.125  -2.125  2.125  2.125
 2*LINE        0.875   -1.5515
 3 LINE       -0.875   -1.5515
 4 LINE       -1.75    0.
 5 LINE       -0.875    1.5515
 6 LINE        0.875    1.5515
 7 LINE        1.75     0.

UNO  UNIT    DIA    DEPTH  CHAMFER  PRE-REAM   CHP
 4 REAMING 0.4375 1.125   0.        DRILL       0
SNO  TOOL   NOM-D NO HOLE-D HOLE-DEP PRE-DIA PRE-DEP RGH DEPTH C-SP  FR   M M
 1 CTR-DR 0.5        0.4375 <>        <>      <>     90ø CTR-D 200 0.002  8
 2 DRILL  0.43       0.4219 1.29             100     PCK2 0.855 200 0.002  8
 3 REAMER 0.44       0.4375 1.125    <>       <>      6   G01  67  0.001  8
FIG PTN   Z      X        Y       AN1     AN2     T1      T2     F M N P Q R
 1 CIR   0.      0.       0.       45.     <>     1.      <>    <>4 <> <> <>1

UNO UNIT  CONTINUE  NUMBER  ATC   X        Y        Z      4     ANGLE
 5 END       0        0
```

Figure 32 Finished Mill Program Display
Courtesy SolutionWare Corp.

occur and all looks well, proceed to tool path checking.

• Press PATH CONTINUE.

The tool path will now be displayed and, again, if all looks well, a first piece can be machined. The graphics capabilities allow for zooming, rotating, scaling, etc., so the programmer has lots of capability to see what the tool path will perform.

While the steps given here for Mazatrol conversational programming may seem like a lot of effort, the truth is, that when you get used to this style of data entry, the programming process can be very quick and simple.

Figure 33 MAZAK Integrex
Courtesy Daco Precision
Manufacturers

THE FUTURE OF CNC PROGRAMMING

The ultimate goal of any manufacturing is to increase productivity by improving efficiency, minimizing wasted and idle time and to cut lead times, while maintaining accuracy. Modern technology enables these goals to become reality.

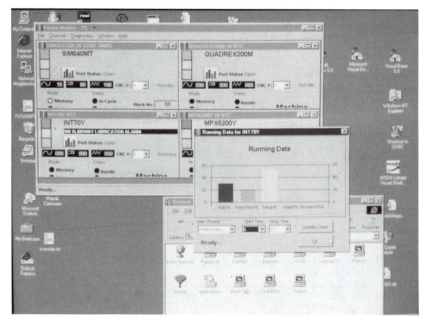

Figure 34 Machine Maintenance Window   Courtesy MAZAK Corp.

Part 7 Mazatrol Conversational Programming

The PC has revolutionized our society and manufacturing has been a benefactor because of the direct effect on controlling CNC machines. Now PC's are used to network machines together, in order to manage workloads. These networks also make it possible to communicate between machines and the office, to manage and download programs, to obtain machine status and operation reports in real time, and monitor other network locations. Machine monitoring includes automatic operation, machine stop and feed hold, set-up and alarm. Spindle load and spindle speed are recorded to provide reports and information on completed workpiece counts. Some machines utilize both a PC and CNC fused into one, providing bi-directional communication between the PC and CNC. This makes an intelligent CNC control system. Intelligent CNC responds to questions, makes suggestions and provides detailed reports on machine operation and production status.

The most advanced RISC CPU (Reduced Instruction Set Computer, Central Processing Unit) technology is used to provide faster processing speeds which, in turn, help to achieve reduced set-up and cycle times.

Knowledge-based navigation functions allow the determination of optimum metal cutting conditions prior to actual cutting, dependant upon stored data. Based on the part program, tool data and workpiece material, the navigation functions suggest the optimum cutting conditions and shows where improvements in cycle time can be achieved through changes in spindle speed, feed rate and tools.

Advancements in the graphical cutting simulation allow the 3-dimensional solid part model to be displayed. This feature can be used to show part sections for checking inside diameters and deep holes, and, views can be rotated. These added capabilities aid in program verification of tool paths.

To keep machine utilization high, modern controls allow self-diagnostic functions for service and maintenance. Self-diagnostic menus quickly trouble shoot the cause of an alarm and suggest possible solutions for trouble-shooting. Alarm displays indicate when

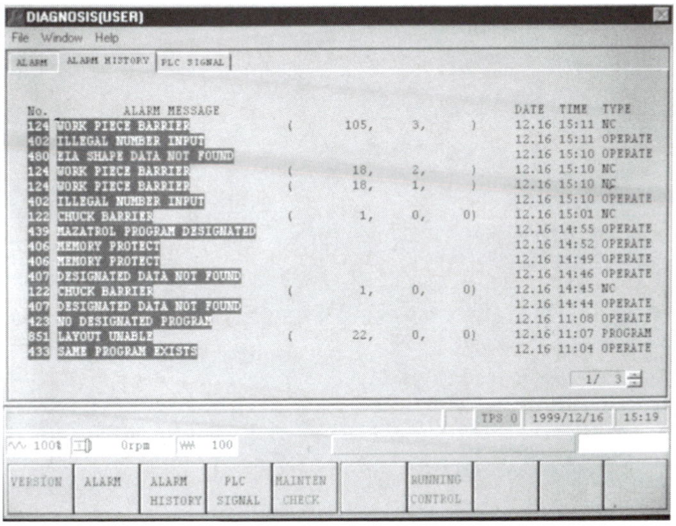

Figure 35 Machine Diagnostics Window
Courtesy MAZAK Corp.

417

scheduled maintenance is required and on-line support is available.

As technological advancements continue, at an unbelievable pace, the manufacturing industry will undoubtedly benefit. Innovative new methods for CAD/CAM and Conversational programming of CNC machines will continue to emerge. For the machinist/programmer, this means that the new methods and tools used for programming will require a life-long learning approach. This book has been intended to begin that learning process.

Part 7 Study Questions

1. A Turning Center program is made up of four basic parts:

   a. **BAR, CPY, CNR, EDG.**

   b. Program, Shape Check, Tool Path and Automatic Cycle.

   c. Common Data Process, Machining Process, Sequence Data and End Process.

   d. Common Data Unit, Coordinate System, Machining Units, Sequence Data and an End Unit.

2. A Machining Center program is made up of these basic parts:

   a. **BAR, CPY, CNR, EDG.**

   b. Program, Shape Check, Tool Path and Automatic Cycle.

   c. Common Data Process, Machining Process, Sequence Data and End Process.

   d. Common Data Unit, Coordinate System, Machining Units, Sequence Data and an End Unit.

3. If a 1/2—13 **THD** is to be turned, what is the lead?

4. What type of Machining Unit would be used to mill the contour of a part?

   a. Point Machining

   b. Line Machining

   c. Face Machining

   d. Manual Program

5. Mazatrol programming has a feature that will automatically develop tool sequences, based on the type of Machining Unit. T or F

6. In a Turning Center program, where do you set the Maximum Spindle RPM Limit?

   a. **END** Unit.

   b. Machining Process.

   c. Common Data Process

   d. Sequence Data

7. Graphic simulation capabilities are limited to tool path centerline on current controllers.

   T or F

Part 7 Mazatrol Conversational Programming

8. To set multiple Workpiece offsets to machine an array of parts, use the_____.

   a. **WPC**

   b. Multi 5x2

   c. Offset Type

   d. Manual Program

9. What does the acronym **SRV-Z** stand for?

10. When AUTO SET is pressed from the Menu keys for the **APRCH-X** and **APRCH- Y**, why are question marks inserted into the program for the X and Y coordinates?

11. When creating a Turning Center program, the AUTO SET Menu key is pressed for Surface Velocity and Feedrate. What criteria are these calculations based on?

   a. Work Offset and Tool Material

   b. Work Material and Tool Material

   c. Spindle Speed and Feedrate

   d. Stock Removal Amount and Depth of Cut.

Appendix

English Drill Sizes

DRILL/DECIMAL		Drill/ Decimal		Drill/Decimal		Drill/ Decimal	
80	.0135	40	.0980	2	.2210	33/64	.5156
79	.0145	39	.0995	1	.2280	17/32	.5312
1/64	.0156	38	.1015	A	.2340	35/64	.5469
78	.0160	37	.1040	15/64	.2344	**9/16**	**.5625**
77	.0180	36	.1065	B	.2380	37/64	.5781
76	.0200	7/64	.1094	C	.2420	19/32	.5938
75	.0210	35	.1100	D	.2460	39/64	.6094
74	.0225	34	.1110	E	.2500	**5/8**	**.6250**
73	.0240	33	.1130	**1/4**	**.2500**	41/64	.6406
72	.0250	32	.1160	F	.2570	21/32	.6562
71	.0260	31	.1200	G	.2610	43/64	.6719
70	.0280	**1/8**	**.1250**	17/64	.2656	**11/16**	**.6875**
69	.0292	30	.1285	H	.2660	45/64	.7031
68	.0310	29	.1360	I	.2720	23/32	.7188
1/32	.0312	28	.1405	J	.2770	47/64	.7344
67	.0320	9/64	.1406	K	.2810	**3/4**	**.7500**
66	.0330	27	.1440	9/32	.2812	49/64	.7656
65	.0350	26	.1470	L	.2900	25/32	.7812
64	.0360	25	.1495	M	.2950	51/64	.7969
63	.0370	24	.1520	19/64	.2969	**13/16**	**.8125**
62	.0380	23	.1540	N	.3020	53/64	.8281
61	.0390	5/32	.1562	**5/16**	**.3125**	27/32	.8438
60	.0400	22	.1570	O	.3160	55/64	.8594
59	.0410	21	.1590	P	.3230	**7/8**	**.8750**
58	.0420	20	.1610	21/64	.3281	57/64	.8906
57	.0430	19	.1660	Q	.3320	29/32	.9062
56	.0465	18	.1695	R	.3390	59/64	.9219
3/64	.0469	11/64	.1719	11/32	.3438	**15/16**	**.9375**
55	.0520	17	.1730	S	.3480	61/64	.9531
54	.0550	16	.1770	T	.3580	31/32	.9688
53	.0595	15	.1800	23/64	.3594	63/64	.9844
1/16	**.0625**	14	.1820	U	.3680	**1**	**1.0000**
52	.0635	13	.1850	**3/8**	**.3750**		
51	.0670	**3/16**	**.1875**	V	.3770	+64th	
50	.0700	12	.1890	W	.3860	increments	
49	.0730	11	.1910	25/64	.3906	up to 1-7/8"	
48	.0760	10	.1935	X	.3970		
5/64	.0780	19	.1960	Y	.4040		
47	.0785	8	.1990	13/32	.4062	+32nd	
46	.0810	7	.2010	Z	.4130	increments	
45	.0820	13/64	.2031	27/64	.4219	up to 2-1/4"	
44	.0860	6	.2040	**7/16**	**.4375**		
43	.0890	5	.2055	29/64	.4531		
42	.0935	4	.2090	15/32	.4688	+16th	
3/32	.0938	3	.2130	31/64	.4844	increments	
41	.0960	7/32	.2188	**1/2**	**.5000**	up to 4-1/4"	

METRIC DRILL SIZES

Drill / Decimal	Drill / Decimal	Drill / Decimal	Drill / Decimal
0.35mm .0138	2.50mm .0984	6.20mm .2441	**10.00mm** **.3937**
0.38mm .0150	2.55mm .1004	6.25mm .2461	10.20mm .4016
0.40mm .0157	2.60mm .1024	6.30mm .2480	10.50mm .4134
0.42mm .0165	2.65mm .1043	6.40mm .2520	10.80mm .4252
0.45mm .0177	2.70mm .1063	6.50mm .2559	**11.00mm** **.4331**
0.48mm .0189	2.75mm .1083	6.60mm .2598	11.20mm .4409
0.50mm .0197	2.80mm .1102	6.70mm .2638	11.50mm .4528
0.55mm .0217	2.90mm .1142	6.75mm .2657	11.80mm .4646
0.60mm .0236	**3.00mm** **.1181**	6.80mm .2677	**12.00mm** **.4724**
0.65mm .0256	3.10mm .1220	6.90mm .2717	12.20mm .4803
0.70mm .0276	3.20mm .1260	**7.00mm** **.2756**	12.50mm .4921
0.75mm .0295	3.25mm .1280	7.10mm .2795	**13.00mm** **.5118**
0.80mm .0315	3.30mm .1299	7.20mm .2835	13.50mm .5315
0.85mm .0335	3.40mm .1339	7.25mm .2854	**14.00mm** **.5512**
0.90mm .0354	3.50mm .1378	7.30mm .2874	14.50mm .5709
0.95mm .0374	3.60mm .1417	7.40mm .2913	**15.00mm** **.5906**
1.00mm **.0394**	3.70mm .1457	7.50mm .2953	15.50mm .6102
1.05mm .0413	3.75mm .1476	7.60mm .2992	**16.00mm** **.6299**
1.10mm .0433	3.80mm .1496	7.70mm .3031	16.50mm .6496
1.15mm .0453	3.90mm .1535	7.75mm .3051	**17.00mm** **.6693**
1.20mm .0472	**4.00mm** **.1575**	7.80mm .3071	17.50mm .6890
1.25mm .0492	4.10mm .1614	7.90mm .3110	**18.00mm** **.7087**
1.30mm .0512	4.20mm .1654	**8.00mm** **.3150**	18.50mm .7283
1.35mm .0531	4.25mm .1673	8.10mm .3189	**19.00mm** **.7480**
1.40mm .0551	4.30mm .1693	8.20mm .3228	19.50mm .7677
1.45mm .0571	4.40mm .1732	8.25mm .3248	**20.00mm** **.7874**
1.50mm .0591	4.50mm .1772	8.30mm .3268	20.50mm .8071
1.55mm .0610	4.60mm .1811	8.40mm .3307	**21.00mm** **.8268**
1.60mm .0630	4.70mm .1850	8.50mm .3346	21.50mm .8465
1.65mm .0650	4.75mm .1870	8.60mm .3386	**22.00mm** **.8661**
1.70mm .0669	4.80mm .1890	8.70mm .3425	22.50mm .8858
1.75mm .0689	4.90mm .1929	8.75mm .3445	**23.00mm** **.9055**
1.80mm .0700	**5.00mm** **.1969**	8.80mm .3465	23.50mm .9252
1.85mm .0728	5.10mm .2008	8.90mm .3504	**24.00mm** **.9449**
1.90mm .0748	5.20mm .2047	**9.00mm** **.3543**	24.50mm .9646
1.95mm .0768	5.25mm .2067	9.10mm .3583	**25.00mm** **.9843**
2.00mm **.0787**	5.30mm .2087	9.20mm .3622	
2.05mm .0807	5.40mm .2126	9.25mm .3642	+1.00mm increments up to 48mm
2.10mm .0827	5.50mm .2165	9.30mm .3661	
2.15mm .0846	5.60mm .2205	9.40mm .3701	
2.20mm .0866	5.70mm .2244	9.50mm .3740	
2.25mm .0886	5.75mm .2264	9.60mm .3780	+5.00mm increments from 50mm up to 105mm
2.30mm .0906	5.80mm .2283	9.70mm .3819	
2.35mm .0925	5.90mm .2323	9.75mm .3839	
2.40mm .0945	**6.00mm** **.2362**	9.80mm .3858	
2.45mm .0965	6.10mm .2402	9.90mm .3898	

Appendix Chart 2

ENGLISH THREADS

Thread Size*	Basic Major Diameter	Tap Drill Size
#0-80	.0600	3/64
#2-56	.0860	#50
#2-64	.0860	#50
#4-40	.1120	#43
#4-48	.1120	#42
#5-40	.1250	#38
#5-44	.1250	#37
#6-32	.1380	#36
#6-40	.1380	#33
#8-32	.1640	#29
#8-36	.1640	#29
#10-24	.1900	#25
#10-32	.1900	#21
1/4-20	.2500	#7
1/4-28	.2500	#3
5/16-18	.3125	F
5/16-24	.3125	I
3/8-16	.3750	5/16
3/8-24	.3750	O
7/16-14	.4375	U
7/16-20	.4375	25/64
1/2-13	.5000	27/64
1/2-20	.5000	29/64
9/16-12	.5625	31/64
9/16-18	.5625	33/64
5/8-11	.6250	17/32
5/8-18	.6250	37/64
3/4-10	.7500	21/32
3/4-16	.7500	11/16
7/8-9	.8750	13/16
7/8-14	.8750	49/64
1-8	1.0000	7/8
1-14	1.0000	15/16
1-1/8-7	1.1250	63/64
1-1/8-12	1.1250	1-3/64

Appendix Chart 3

* Coarse series shown in bold. Pitch callout not required on metric coarse series.
** Closest size for 75% theoretical thread.

METRIC THREADS

Basic Thread Size*	Drill Size**
M1.6x0.35	1.25mm or #55
M2x0.4	1 .60mm or #52
M2.5x0.45	2.05mm or #46
M3x0.5	2.50mmor#39
M3.5x0.6	2.90mm or #32
M4x0.7	3.30mm or #30
M5x0.8	4.20mm or #19
M6x1	5.00mm or #8
M8x1.25	6.80mm or H
M8x1	7.00mm or J
M10xl.5	8.50mm or R
M10xl.25	8.80mm or 11/32
Ml 2x1.75	10.20mm or 3/32
M12x1.25	10.80mm or 7/64
M14x2	12.00mm or 5/32
M14x1.5	12.50mm or 1/2
M16x2	14.00mm or 5/64
M16x1.5	14.50mm or 7/64
M18x2.5	15.50mm or 9/64
M18x1 .5	16.50mm or 1/32
M20x2.5	17.50mm or 1/16
M20x1 .5	18.50mm or 7/64
M22x2.5	19.50mm or 9/64
M22x1.5	20.50mm or13/16
M24x3	21.00mm or53/64
M24x2	22.00mm or 7/8
M27x3	24.00mm or15/16
M27x2	25.00mm or 1

Appendix Chart 4
Pitch callout not required on metric coarse series.
** Closest size for 75% theoretical thread.

COUNTERBORED HOLES

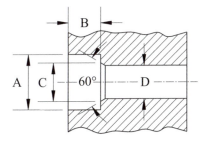

USA SOCKET-HEAD CAP SCREWS

SCREW DIA	A COUNTERBORE DIA	B COUNTERBORE DEPTH	C COUNTERSINK DIA	D CLEARANCE DIA NORMAL FIT	CLOSE FIT
#0	1/8	.060	.074	#49	#51
#2	3/16	.086	.102	#36	3/32
#4	7/32	.112	.130	#29	1/8
#5	1/4	.125	.145	#23	9/64
#6	9/32	.138	.158	#18	#23
#8	5/16	.164	.188	#10	#15
#10	3/8	.190	.218	#2	#5
1/4	7/16	.250	.278	9/32	17/64
5/16	17/32	.312	.346	11/32	21/64
3/8	5/8	.375	.415	13/32	25/64
7/16	23/32	.438	.483	15/32	29/64
1/2	13/16	.500	.552	17/32	33/64
5/8	1	.625	.689	21/32	41/64
3/4	1-3/16	.750	.828	25/32	49/64
7/8	1-3/8	.875	.963	29/32	57/64
1	1-5/8	1.000	1.100	1-1/32	1-1/64
1-1/4	2	1.250	1.370	1-5/16	1-9/32
1-1/2	2-3/8	1.500	1.640	1-9/16	1-17/32
1-3/4	2-3/4	1.750	1.910	1-13/16	1-25/32
2	3-1/8	2.000	2.180	2-1/16	2-1/32

Appendix Chart 5

COUNTERBORED HOLES

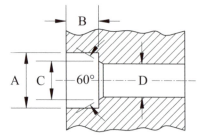

METRIC SOCKET-HEAD CAP SCREWS

SCREW DIA	A COUNTERBORE DIA	B COUNTERBORE DEPTH	C COUNTERSINK DIA	D CLEARANCE DIA	
				NORMAL FIT	CLOSE FIT
M1.6	3.50mm	1.6mm	2.0mm	1.95mm	1.80mm
M2	4.40mm	2mm	2.6mm	2.40mm	2.20mm
M2.5	5.40mm	2.5mm	3.1mm	3.00mm	2.70mm
M3	6.50mm	3mm	3.6mm	3.70mm	3.40mm
M4	8.25mm	4mm	4.7mm	4.80mm	4.40mm
M5	9.75mm	5mm	5.7mm	5.80mm	5.40mm
M6	11.20mm	6mm	6.8mm	6.80mm	6.40mm
M8	14.50mm	8mm	9.2mm	8.80mm	8.40mm
M10	17.50mm	10mm	11.2mm	10.80mm	10.50mm
M12	19.50mm	12mm	14.2mm	13.00mm	12.50mm
M14	22.50mm	14mm	16.2mm	15.00mm	14.50mm
M16	25.50mm	16mm	18.2mm	17.00mm	16.50mm
M20	31.50mm	20mm	22.4mm	21.00mm	20.50mm
M24	37.50mm	24mm	26.4mm	25.00mm	24.50mm
M30	47.50mm	30mm	33.4mm	31.50mm	31.00mm
M36	56.50mm	36mm	39.4mm	37.50mm	37.00mm
M42	66.00mm	42mm	45.6mm	44.00mm	43.00mm
M48	75.00mm	48mm	52.6mm	50.00mm	49.00mm

Appendix Chart 6

GEOMETRIC SYMBOLS AND DEFINITIONS

Feature-Control Frame

A specification box that shows a particular geometric characteristic (flatness, straightness, etc.) applied to a part feature and states the allowable tolerance. The feature's tolerance may be individual, or related to one or more datums. Any datum references and tolerance modifiers are also shown.

Datum Feature

A flag which designates a physical feature of the part to be used as a reference to measure geometric characteristics of other part features.

Datum Targets

Callouts occasionally needed to designate specific points, lines, or areas on an actual part to be used to establish a theoretical datum feature.

1.500

Basic Dimension

A box around any drawing dimension makes it a "basic" dimension, a theoretically exact value used as a reference for measuring geometric characteristics and tolerances of other part features.

Cylindrical Tolerance Zone

This symbol, commonly used to indicate a diameter dimension, also specifies a cylindrically shaped tolerance zone in a feature-control frame.

(M)

Maximum Material Condition

Abbreviation: MMC. A tolerance modifier that applies the stated tight tolerance zone only while the part theoretically contains the maximum amount of material permitted within its dimensional limits (e.g. minimum hole diameters and maximum shaft diameters), allowing more variation under normal conditions.

(L)

Least Material Condition

Abbreviation: LMC. A tolerance modifier that applies the stated tight tolerance zone only while the part theoretically contains the minimum amount of material permitted within its dimensional limits (e.g. maximum hole diameters and minimum shaft diameters), allowing more variation under normal conditions.

(S)

Regardless of Feature Size

Abbreviation: RFS. A tolerance modifier that applies the stated tight tolerance zone under all size conditions. RFS is generally assumed if neither MMC nor LMC are stated.

Projected Tolerance Zone

An additional specification box attached underneath a feature-control frame. It extends the feature's tolerance zone beyond the part's surface by the stated distance, ensuring perpendicularity for proper alignment of mating parts.

Appendix Chart 7

GEOMETRIC CHARACTERISTICS

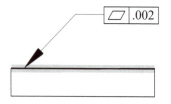

▱ Flatness

All points on the indicated surface must lie in a single plane, within the specified tolerance zone.

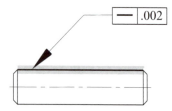

— Straightness

All points on the indicated surface or axis must lie in a straight line in the direction shown, within the specified tolerance zone

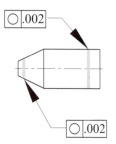

○ Circularity (Roundness)

If the indicated surface were sliced by any plane perpendicular to its axis, the resulting outline must be a perfect circle, within the specified tolerance zone.

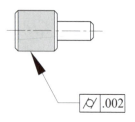

 **Cylindricity**

All points on the indicated surface must lie in a perfect cylinder around a center axis, within the specified tolerance zone.

Appendix Chart 8

GEOMETRIC CHARACTERISTICS

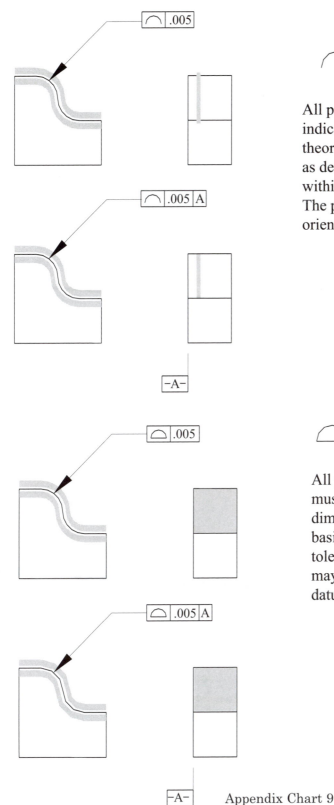

⌒ Linear Profile

All points on any full slice of the indicated surface must lie on its theoretical two-dimensional profile, as defined by basic dimensions, within the specified tolerance zone. The profile may or may not be oriented with respect to daatums.

◠ Surface Profile

All points on the indicated surface must lie on its theoretical three-dimensional profile, as defined by basic dimensions, within the specified tolerance zone. The profile may or may not be oriented with respect to datums.

Appendix Chart 9

GEOMETRIC CHARACTERISTICS

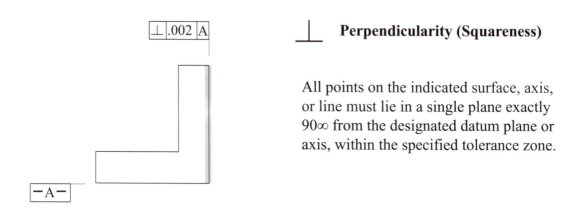

⊥ Perpendicularity (Squareness)

All points on the indicated surface, axis, or line must lie in a single plane exactly 90∞ from the designated datum plane or axis, within the specified tolerance zone.

∠ Angularity

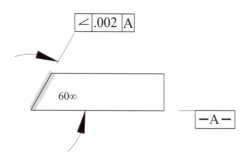

All points on the indicated surface or axis must lie in a single plane at exactly the specified angle from the designated datum plane or axis, within the specified tolerance zone.

// Parallelism

All points on the indicated surface or axis must lie in a single plane parallel to the designated datum plane or axis, within the specified tolerance zone.

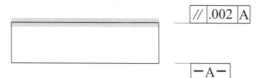

Appendix Chart 10

GEOMETRIC CHARACTERISTICS

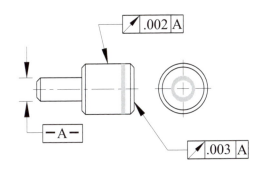

Circular Runout

Each circular element of the indicated surface is allowed to deviate only the specified amount from its theoretical form and orientation during 360∞ rotation about the designate datum axis.

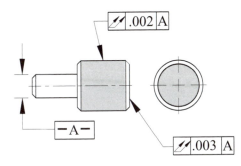

Total Runout

The entire indicated surface is allowed to deviate only the specified amount from its theoretical form and orientation during 360∞ rotation about the designated datum axis.

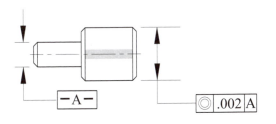

◎ **Concentricity**

If the indicated surface were sliced by any plane perpendicular to the designated datum axis, every slice's center of area must lie on the datum axis. within the specified cylindrical tolerance zone (controls rotational balance).

.500

−A−

.375

4 X ÿ .125±.002

⊕ .002 Ⓜ A

⊕ **Position (Replaces = Symetry)**

The indicated feature's axis must be located within the specified tolerance zone from its true theoretical position, correctly oriented relative to the designated datum plane or axis.

Appendix Chart 11

Appendix

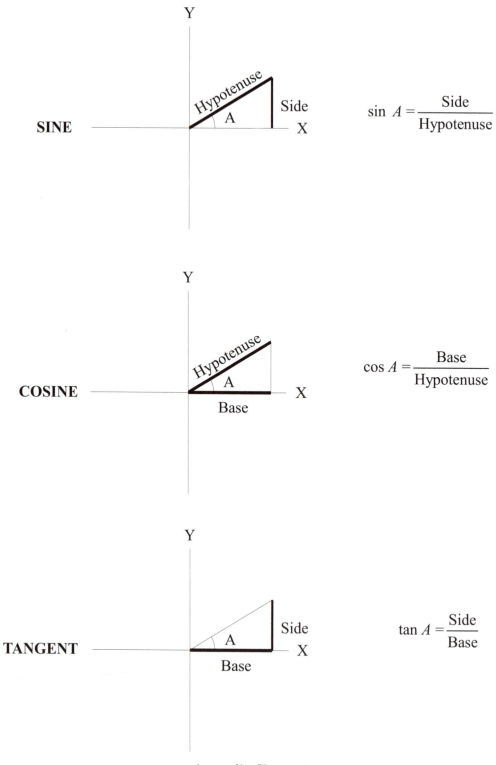

SINE

$$\sin A = \frac{\text{Side}}{\text{Hypotenuse}}$$

COSINE

$$\cos A = \frac{\text{Base}}{\text{Hypotenuse}}$$

TANGENT

$$\tan A = \frac{\text{Side}}{\text{Base}}$$

Appendix Chart 12

Appendix

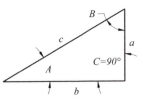

RIGHT TRIANGLES

Known Sides and	Unknown Sides and Angles			Area
a and b	$c = \sqrt{a^2 + b^2}$	$A = \arctan \dfrac{a}{b}$	$B = \arctan \dfrac{b}{a}$	$\dfrac{a \times b}{2}$
a and c	$b = \sqrt{c^2 - a^2}$	$A = \arcsin \dfrac{a}{c}$	$A = \arccos \dfrac{a}{c}$	$\dfrac{a \times \sqrt{c^2 - a^2}}{2}$
b and c	$a = \sqrt{c^2 - b^2}$	$A = \arccos \dfrac{b}{c}$	$B = \arcsin \dfrac{b}{c}$	$\dfrac{b \times \sqrt{c^2 - b^2}}{2}$
a and $\angle A$	$b = \dfrac{a}{\tan A}$	$c = \dfrac{a}{\sin A}$	$B = 90° - A$	$\dfrac{a^2}{2 \times \tan A}$
a and $\angle B$	$b = a \times \tan B$	$c = \dfrac{a}{\cos B}$	$A = 90° - B$	$\dfrac{a^2 \times \tan B}{2}$
b and $\angle A$	$a = b \times \tan A$	$c = \dfrac{b}{\cos A}$	$B = 90° - A$	$\dfrac{b^2 \times \tan A}{2}$
b and $\angle B$	$a = \dfrac{b}{\tan B}$	$c = \dfrac{b}{\sin B}$	$A = 90° - B$	$\dfrac{b^2}{2 \times \tan B}$
c and $\angle A$	$a = c \times \sin A$	$b = c \times \cos A$	$B = 90° - A$	$c^2 \times \sin A \times \cos$
c and $\angle B$	$a = c \times \cos B$	$b = c \times \sin B$	$A = 90° - B$	$c^2 \times \sin B \times \cos$

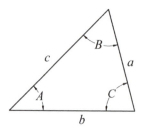

OBLIQUE TRIANGLES

Known Sides and Angles	Unknown Sides and Angles			Area
All three sides a, b, c	$A = \arccos \dfrac{b^2 + c^2 - a^2}{2bc}$	$B = \arcsin \dfrac{b \times \sin A}{a}$	$C = 180° - A - B$	$\dfrac{a \times b \times \sin C}{2}$
Two sides and the angle between them a, b, $\angle C$	$c = \sqrt{a^2 + b^2 - (2ab \times \cos C)}$	$A = \arctan \dfrac{a \times \sin C}{b - (a \times \cos C)}$	$B = 180° - A - C$	$\dfrac{a \times b \times \sin C}{2}$
Two sides and the angle opposite one of the sides a, b, $\angle A$ ($\angle B$ less than $90°$)	$B = \arcsin \dfrac{b \times \sin A}{a}$	$C = 180° - A - B$	$c = \dfrac{a \times \sin C}{\sin A}$	$\dfrac{a \times b \times \sin C}{2}$
Two sides and the angle opposite one of the sides a, b, $\angle A$ ($\angle B$ greater than $90°$)	$B = 180° - \arcsin \dfrac{b \times \sin A}{a}$	$C = 180° - A - B$	$c = \dfrac{a \times \sin C}{\sin A}$	$\dfrac{a \times b \times \sin C}{2}$
One side and two angles a, $\angle A$, $\angle B$	$b = \dfrac{a \times \sin B}{\sin A}$	$C = 180° - A - B$	$c = \dfrac{a \times \sin C}{\sin A}$	$\dfrac{a \times b \times \sin C}{2}$

Appendix Chart 14

Appendix

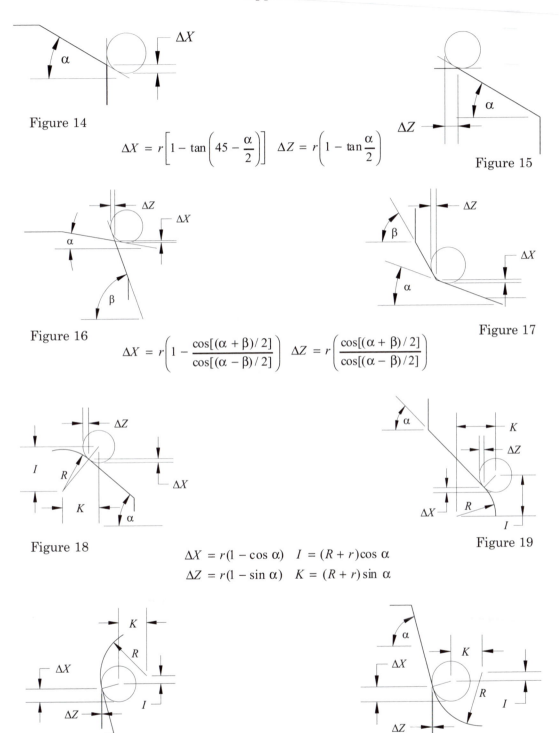

Figure 14

$$\Delta X = r\left[1 - \tan\left(45 - \frac{\alpha}{2}\right)\right] \quad \Delta Z = r\left(1 - \tan\frac{\alpha}{2}\right)$$

Figure 15

Figure 16

Figure 17

$$\Delta X = r\left(1 - \frac{\cos[(\alpha + \beta)/2]}{\cos[(\alpha - \beta)/2]}\right) \quad \Delta Z = r\left(\frac{\cos[(\alpha + \beta)/2]}{\cos[(\alpha - \beta)/2]}\right)$$

Figure 18

Figure 19

$$\Delta X = r(1 - \cos\alpha) \quad I = (R + r)\cos\alpha$$
$$\Delta Z = r(1 - \sin\alpha) \quad K = (R + r)\sin\alpha$$

Figure 20

Figure 21

$$\Delta X = r(1 - \cos\alpha) \quad I = (R - r)\cos\alpha$$
$$\Delta Z = r(1 - \sin\alpha) \quad K = (R - r)\sin\alpha$$

Appendix

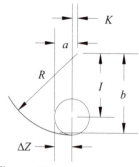

Figure 22

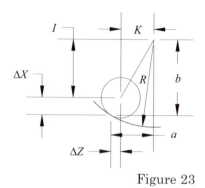

Figure 23

$$\Delta Z = r$$
$$\Delta X = r - [b - (R - r)^2 - (a - r)^2]$$
$$I = (R - r)^2 - (a - r)^2$$
$$K = a - r$$
$$\Delta X = r$$
$$\Delta Z = r - [a - (R - r)^2 - (b - r)^2]$$
$$I = b - r$$
$$K = (R - r)^2 - (b - r)^2$$

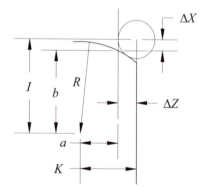

Figure 24

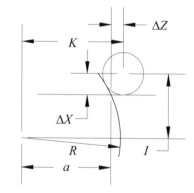

Figure 25

$$\Delta Z = r$$
$$\Delta X = r - [(R + r)^2 - (a + r)^2 - b]$$
$$I = (R + r)^2 - (a + r)^2$$
$$K = a + r$$
$$\Delta X = r$$
$$\Delta Z = r - [(R + r)^2 - (b + r)^2 - a]$$
$$I = b + r$$
$$K = (R + r)^2 - (b + r)^2$$

Appendix

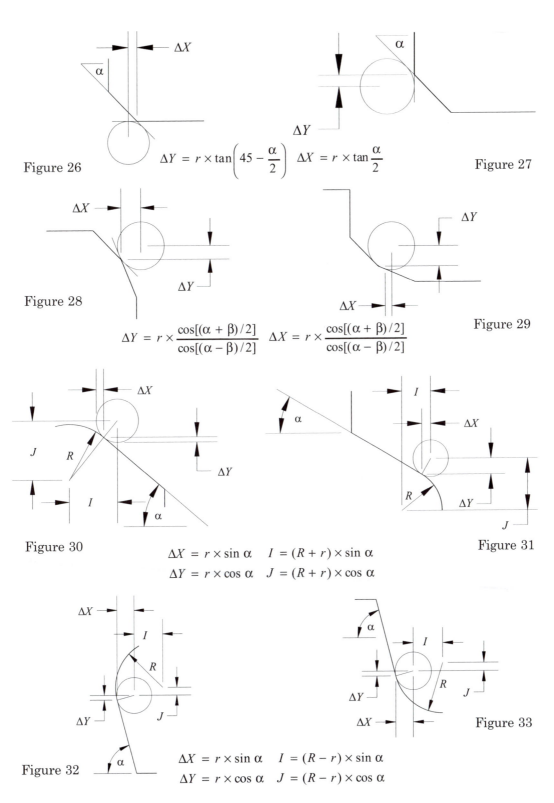

Figure 26

$$\Delta Y = r \times \tan\left(45 - \frac{\alpha}{2}\right) \quad \Delta X = r \times \tan\frac{\alpha}{2}$$

Figure 27

Figure 28

$$\Delta Y = r \times \frac{\cos[(\alpha + \beta)/2]}{\cos[(\alpha - \beta)/2]} \quad \Delta X = r \times \frac{\cos[(\alpha + \beta)/2]}{\cos[(\alpha - \beta)/2]}$$

Figure 29

Figure 30

$$\Delta X = r \times \sin\alpha \quad I = (R + r) \times \sin\alpha$$
$$\Delta Y = r \times \cos\alpha \quad J = (R + r) \times \cos\alpha$$

Figure 31

Figure 32

$$\Delta X = r \times \sin\alpha \quad I = (R - r) \times \sin\alpha$$
$$\Delta Y = r \times \cos\alpha \quad J = (R - r) \times \cos\alpha$$

Figure 33

Appendix

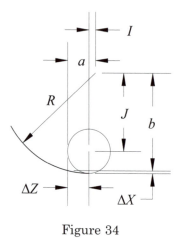

Figure 34

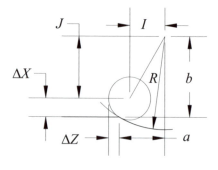

Figure 35

$$\Delta X = r$$
$$\Delta Y = b - (R - r)^2 - (a - r)^2$$
$$I = a - r$$
$$J = b - \Delta Y$$

$$\Delta X = a - (R - r)^2 - (b - r)^2$$
$$\Delta Y = r$$
$$I = a - \Delta X$$
$$J = b - r$$

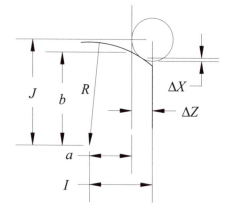

Figure 36

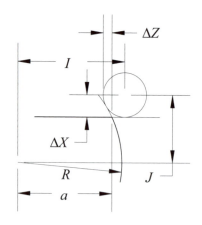

Figure 37

$$\Delta X = r$$
$$\Delta Y = (R + r)^2 - (a + r)^2 - b$$
$$I = a + r$$
$$J = (R + r)^2 - (a + r)^2$$

$$\Delta X = (R + r)^2 - (b + r)^2 - a$$
$$\Delta Y = r$$
$$I = (R + r)^2 - (b + r)^2$$
$$J = b + r$$

440

Glossary

Popular Acronyms

AFR	Automatic Feature Recognition
AOS	Algebraic Order System
APT	Automatically Programmed Tools
ATC	Automatic Tool Changer
ANSI	American National Standards Institute
ASCII	American Standard Code for Information Interchange
ASME	American Society of Mechanical Engineers
AVI,	Audio Video Interleave
BCD	Binary Coded Decimal
CAD	Computer Aided Design
CAM	Computer Aided Manufacturing
CAD/CAM	Computer Aided Design & Computer Aided Manufacturing
CBN	Cubic Boron Nitride
CIM	Computer-Integrated Manufacturing
CW	Clockwise
CCW	Counterclockwise
CD-ROM	Compact Disc-Read Only Memory
CD-RW	Compact Disc-Re-Writable
CDC	Cutter Diameter Compensation
CRC	Cutter Radius Compensation
CNC	Computer Numerical Control
CMM	Coordinate Measuring Machine
CPL	Coordinate Position Locator
CPU	Central Processing Unit
CRT	Cathode Ray Tube
CSS	Constant Surface Speed
DNC	Direct Numerical Control

Glossary

DVD	Digital Video Disc
DXF	Drawing Exchange Format
EDM	Electronic Discharge Machine
EIA	Electronics Industries Association
EOB	End of Block
FMS	Flexible Manufacturing System
GB	Gigabit
G-Code	Preparatory Functions (commands)
GD&T	Geometric Dimensioning & Tolerancing
GHz	Gigahertz
GUI	Graphical User Interface
HP	Horsepower
HMI	Human Machine Interface
HSS	High Speed Steel
ID	Inside Diameter
IGES	Initial Graphics Exchange Specification
IPM	Inches per Minute (also in/min)
IPR	Inches per Revolution (also in/rev)
ISO	International Standards Organization
KW	Kilowatt
LAN	Local Area Network
LED	Light Emitting Diode
LCD	Liquid Crystal Display
MB	Megabit
M-Codes	Miscellaneous Functions
MCU	Machine Control Unit
MDI	Manual Data Input
MHz	Megahertz
MPG	Manual Pulse Generator
Mm	Millimeters

Glossary

NC	Numerical Control
OD	Outside Diameter
PC	Personal Computer
PCMCI	Portable Computer Memory Card Interface
PCD	Polycrystalline Diamond
PLC	Programmable Logic Controller
PSI	Pounds per Square Inch
PVD	Physical Vapor Deposition
PWE	Parameter Write Enable
RAM	Random Access Memory
RPM	Revolutions per Minute (also rev/min or r/min)
RISC	Reduced Instruction Set Computer
ROM	Read Only Memory
SFPM	Surface Feet per Minute
SME	Society of Manufacturing Engineering
STL	Sterolithography
R-8	Taper Designation
RS232	Industry Standard Cabling Interface
TiN	Titanium Nitride
TiCN	Titanium Carbon Nitride
TLO	Tool Length Offset
TNRC	Tool Nose Radius Compensation
USB	Uniform Serial Bus

Please also see Part 7 of this text for acronyms specific to Mazatrol Conversational Programming.

Definitions

A-Axis

The A-axis is an auxiliary rotary axis that rotates about the X axis. Angular movements are specified in decimal degrees. Positive angular values refer to counterclockwise rotation and negative angular values indicate clockwise rotation.

Absolute Dimension

All numerical values (dimensional measurements) are derived from a fixed origin or datum in the coordinate system.

Address

Commonly referred to as letter address, because in programming, each program word is preceded by a letter, in order to identify what function is to be executed. Examples of letter address are: S for spindle speed designation, T for tool identification, M for miscellaneous functions, G for preparatory functions, etc.

Auxiliary axis

An auxiliary axis is any axis that is in addition to the primary axes of X, Y, and Z. These axes can be rotary (A,B or C) or linear (U, V or W). They are also called secondary axes.

Axis

The axis is the primary identifier of the cutting tool direction of movement in relationship with the machine type and orientation. The three linear axes for a machining center are X, Y and Z, and they are perpendicular to each other. The rotary axes are; A,B and C.

B-Axis

The B-axis is an auxiliary rotary axis that rotates about the Y axis. Angular movements are specified in decimal degrees. Positive angular values refer to counterclockwise rotation and negative angular values indicate clockwise rotation.

Block

A single line of CNC code consisting of program words that identify what activities the machine is to execute. Generally, each block is preceded by a block or sequence number (N) and is followed by an "End of Block" (EOB) character, represented by the semicolon (;).

Block Skip

Block Skip is sometimes called Block Delete or Optional Block Skip. When the Block Delete, or Optional Block Skip button or switch is on, the controller skips execution of the program blocks that are preceded with "/"and end with the end of block (;) character. If the button or switch is off, the machine will execute the programmed blocks and disregards the "/" symbol.

Burr

A sharp edge created as a result of machining.

C-Axis

The C-axis is an auxiliary rotary axis that rotates about the Z axis. Angular movements are specified in decimal degrees. Positive angular values refer to counterclockwise rotation and negative angular values clockwise.

Canned Cycle

The function of a given cycle is defined as a set of operations assigned to one block and performed automatically without any possibility of interruption. Examples of canned cycles are: Cycle G81, which will perform a simple drill cycle and G84, which will perform tapping. Canned Cycles require additional information such as coordinate locations, reference plane values, peck amounts, etc. Canned Cycles simplify the part program by decreasing programming time. Another name for Canned Cycles is Fixed Cycles.

Cartesian Coordinates

A coordinate system that consists of three axes (X, Y and Z) that are perpendicular to each other. A grid is formed consisting of numerical graduations, representing the distances from the intersection of the three axes called origin.

Chamfer

A chamfer is a beveled edge, normally machined at 45° to the adjacent surfaces. This operation is commonly used to remove burrs and eliminate sharp corners around holes or part edges.

Circular Interpolation

A programming feature that enables programming of two axes simultaneously in order to create arcs and circles. Information generally needed is the location of the arc center, the arc radius, the starting and ending points of the arc and the direction of cutting motion.

Coordinates

Numerical values that define the positional location of points from a predetermined zero point or origin from within the Cartesian Coordinate System.

Definition

Datum

A datum is an exact point, axis, or plane. A datum is the origin from which either the dimensional location and/or the characteristics of features of a part are established.

Deburr

To remove a burr or sharp edge from machined surfaces.

Dry Run

Sometimes the CNC part program is executed with no part mounted, to verify the programmed path of the tool under automatic operation. The typical form of Dry run is set by activating the DRY RUN function on the control during automatic cycle, where all of the rapid and work feeds are changed to the rapid traverse feed set in the parameters instead of the programmed feed. DRY RUN is also used to check a new program on the machine without any work actually being performed by the tool. This is particularly useful on programs with long cycle times so the operator can progress through the program more quickly.

Dwell

Dwell is determined by the preparatory function G04 and by using the letter address P or X, which corresponds to the time duration of dwell (also U for lathes). When used, it causes a pause in the machining operation for the length of time indicated in seconds (X or U) milliseconds (P) or revolutions (depending on parameter setting).

End of Block Character

A special character represented by the semicolon (;) that identifies the end of a program block. Known by the abbreviation or acronym of: EOB or E-O-B.

End of program

A miscellaneous function (M30) is placed in the last line of a program to indicate the end of the part program. At this command, the spindle, coolant, and feed are stopped and the program is returned to its start.

F-Word

The F-Word is utilized to determine the work feed rates (cutting feedrates). This program word is used to establish feed rate values and precedes a numeric input for the feed amount in Inches per Minute (IPM) or, Inches per Revolution (IPR) and Meters per Minute (m/min) or Millimeters per Revolution (mm/rev) for metric programs. The value that is set by this command stays effective until changed by reentering a new value.

Feedrate Override

Feedrate Override allows control of the traverse feedrate by adjustment of a rotary dial. This function allows the control of the cutting feedrates defined by the F-word in the program by increasing or decreasing the percentage of the value entered in the program. It can also be used to control feedrates during jog mode function.

Fillet

A fillet is a convex machined radius where two surfaces meet at an angle.

Fixed-Cycle

See Canned Cycle

G-Codes

Preparatory functions (G-Codes) are programmed with an address G, typically followed by two digits, to establish the mode of operation in which the tool moves.

Incremental Dimension

An Incremental Dimension is a position within the coordinate system in which each numerical value is taken from the previous point.

Jog

Activating the JOG feed mode allows the selection of manual feeds along a single axis X, Y, or Z (rotary axes may be jogged, as well). With the mode activated, use the Axis/Direction buttons and the Speed/Multiply buttons to move the desired axis at the chosen feed rate (in/min or mm/min) and amount.

Linear Interpolation

This function allows programming of one, two or three axes simultaneously, that allows movement along a straight-line path, or at an angle in plane or space.

Machine Home

A reference position located within the machine tool working envelope that is determined by the manufacturer, in order to establish a measurement system for the machine.

Machine Lock

An operation panel control, usually a button or a toggle switch that allows the operator to lock all of the axes movements in order to check long programs for errors. If an error is encountered during this process, an alarm will be displayed on the control.

Manual Data Input (MDI)

The MDI mode enables the automatic control of the machine, using information entered in the form of blocks through the control panel without interfering with the basic program.

Miscellaneous Function

Miscellaneous functions are used to command various auxiliary operations such as activating coolant flow (M08) or starting clockwise spindle rotation (M03). The code consists of the letter M, typically followed by two digits. The M-Code is normally the last entry in a block.

Definition

Modal Commands

Modal commands remain in effect until they are replaced by another command from the same group. The F-Word is modal as are many G-Codes.

Origin

A starting point for the coordinate system used to machine parts; a fixed point on a blueprint from which dimensions are taken.

Polar Coordinate System

A rotational coordinate system that locates points within a plane with respect to their distance from a fixed point of origin or pole by angular and radial values.

Preparatory Function (G-Word)

See G-Codes

Quadrant

A quadrant is one fourth (1/4) of a 2 dimensional grid in a plane of the rectangular coordinate system for measurement. It also represents an arc of 90∞ that is one fourth of a circle. Quadrant 1 is located in the upper right corner and 2 through 4 proceed in the counterclockwise direction.

Rectangular Coordinate System

See Cartesian Coordinates

Right Hand Rule

Using the right hand with palm up, the thumb will be pointing in the positive linear axis direction for X, then the little and ring fingers of the hand are folded over to touch the palm, the middle finger is allowed to point upwards (positive Z axis) and the index finger is pointing in the Y axis positive direction (vertical machine orientation).

Sequence Number (N-Word)

Sequence numbers (also called the block numbers) are identified by the letter N and are followed by one to five digits. A block number provides easier access to information contained in the program. The arrangement of block numbers in a given program can be random, but typically is sequenced in increments of one, two, five or ten. The most common step increments are five or ten. Block or sequence numbers can be omitted from a block (except in special cases, such as in some lathe cycles). The logical location then, for sequence numbers is at tool changes enabling the restart of that tool.

Single Block

The execution of a single block of information in the program is initiated by activating a switch or button on the control panel. While in this mode, each time the cycle start button is pressed, only one block of information will be executed.

Definition

Sub-Program

The subprogram is a subordinate program to the main program. It is registered in the controller memory with the letter O, followed by a four or five-digit number, same as the main program. In the main program, a subprogram is called by using the M98 function with the P-address to identify the sub-program number and then M99 (called for in the subprogram), is the function that ends a subprogram. Subprograms greatly simplify programming and decrease the amount of data that must be placed into the controller memory.

Tool Changer

A mechanical apparatus used to automatically change cutting tools by program control on CNC machines. The tool changer may be a magazine-type, with random access or, a carousel. The magazine-type uses a device similar to a robot arm that transfers tools from the magazine to the spindle and vice versa.

Tool Length Offset (TLO)

Tool Length Offsets "TLO", are called in the milling program by the H word. The measured values representing the difference between the spindle face gage line and the tools tip are input into corresponding offset registers and are needed for proper positioning of the tool along the Z axis.

U-Axis

An additional linear axis parallel to X axis.

V-Axis

An additional linear axis parallel to Y axis.

W-Axis

An additional linear axis parallel to Z axis.

Work envelope

The working envelope is the maximum area of machining travel for each two axes in all four directions.

Workpiece Zero

A starting point, work piece zero is the point from which dimensions on the work piece are established, sometimes referred to as part zero.

X-Axis

The axis of motion that is always horizontal and parallel to the machine tool table. For a vertical milling machine this axis moves left or right.

Y-Axis

The axis of motion that is perpendicular to both X and Z axes in relation to the machine tool table. For a vertical milling machine this axis moves forward and backward.

449

Definition

Z-Axis

The machine tool axis of motion that is always parallel to the primary spindle. For a vertical milling machine this axis moves up and down.

Programming of CNC Machines Third Edition
Answers to Study Questions

Part 1, 1. T, **2.** c, **3.** b, **4.** c, **5.** d, **6.** a, **7.** F, **8.** 3, **9.** T, **10.** T, **11.** X0Y0, X0Y5.0, X2.5Y5.0, X4.0Y3.5, X4.0Y0, X0Y0 **12.** X0Y0, Y5.0, X2.5, X1.5Y-1.5, Y-3.5, X-4.0 **13.** Daily

Part 2, 1. T, **2.** Position (All) or Program (Check), **3.** Active commands, **4.** Pressing the Input key enters a whole number where +Input enters an incremental amount, **5.** c, **6.** d, **7.** T, **8.** b, **9.** a, **10.** Manual Data Input, **11.** Offset, **12.** T, **13.** Jog

Part 3, 1. c, **2.** T, **3.** The first two digits refer to the geometry offset and the second two are to the wear offset. **4.** b, **5.** d, **6.** T, **7.** b, **8.** T, **9.** a, **10.** b, **11.** c, **12.** b, **13.** b, **14.** a negative sign, **15.** T, **16.** d, **17.** b, **18.** d, **19.** b

Part 4, 1. b, **2.** d, **3.** c, **4.** d, **5.** b, **6.** d, **7.** c, **8.** T, **9.** d, **10.** c, **11.** d, **12.** a, **13.** T, **14.** F, **15.** b, **16.** b, **17.** a, **18.** c, **19.** d, **20.** a, **21.** a. M00, b. M01, c. M30, d. M03, e. M04, f. M05, g. M08, h. M09, i. M19, j. M98, k. M99, l. M06, **22.** a. G01, b. G02, c. G00, d. G04, e. G90, f. G91, g. G80, h. G83, i. G81, j. G41, k. G42, l. G40, m. G28, n. G20, o. G21, p. G84, q. G92, r. G54, s.G43, **23.** a. O, b. N, c. G, d. M, e. X, f. Y, g. Z, h. R, i. F, j. S, k. L, l. T, m. H, n. D o. A, p. B, q. C, r. P, s. Q, t. I, u. J, v. K, **24.** c, **25.** T, **26.** T

Part 5, 1. T, **2.** T, **3.** d, **4.** A fillet is a convex machined radius where two surfaces meet at an angle. **5.** via the Menu Bar by pressing Create and then Fillet or the fillet button on the Sketcher Toolbar. **6.** A chamfer is a beveled edge, normally machined at 45° to the adjacent surfaces. This operation is commonly used to remove burrs and eliminate sharp corners around holes or part edges. **7.** via the Menu Bar by pressing Create and then Chamfer or the Chamfer button on the Sketcher Toolbar. **8.** c, **9.** c, **10.** a, **11.** T, **12.** d, **13.** To aid in the selection of specific characteristics of entities by snapping to them when activated and picked with the mouse. **14.** The software compensates the tool path by the radial amount based on the tool selected. **15.** a or d, **16.** T, **17.** Angle, **18.** T

Part 6, 1. T, **2.** T, **3.** b, **4.** d, **5.** b, **6.** b, **7.** F, **8.** T, **9.** b, **10.** a.

Part 7, 1. c, **2.** d, **3.** .0769, **4.** b, **5.** T, **6.** c, **7.** F, **8.** b, **9.** the amount of stock removal in Z, **10.** at this point the tool to be used has not been selected. The values will be input when the tool path is checked after the unit is completed. **11.** b

Index

Index

Index

Index

Index